MW01630161

SYRACUSE CHINA

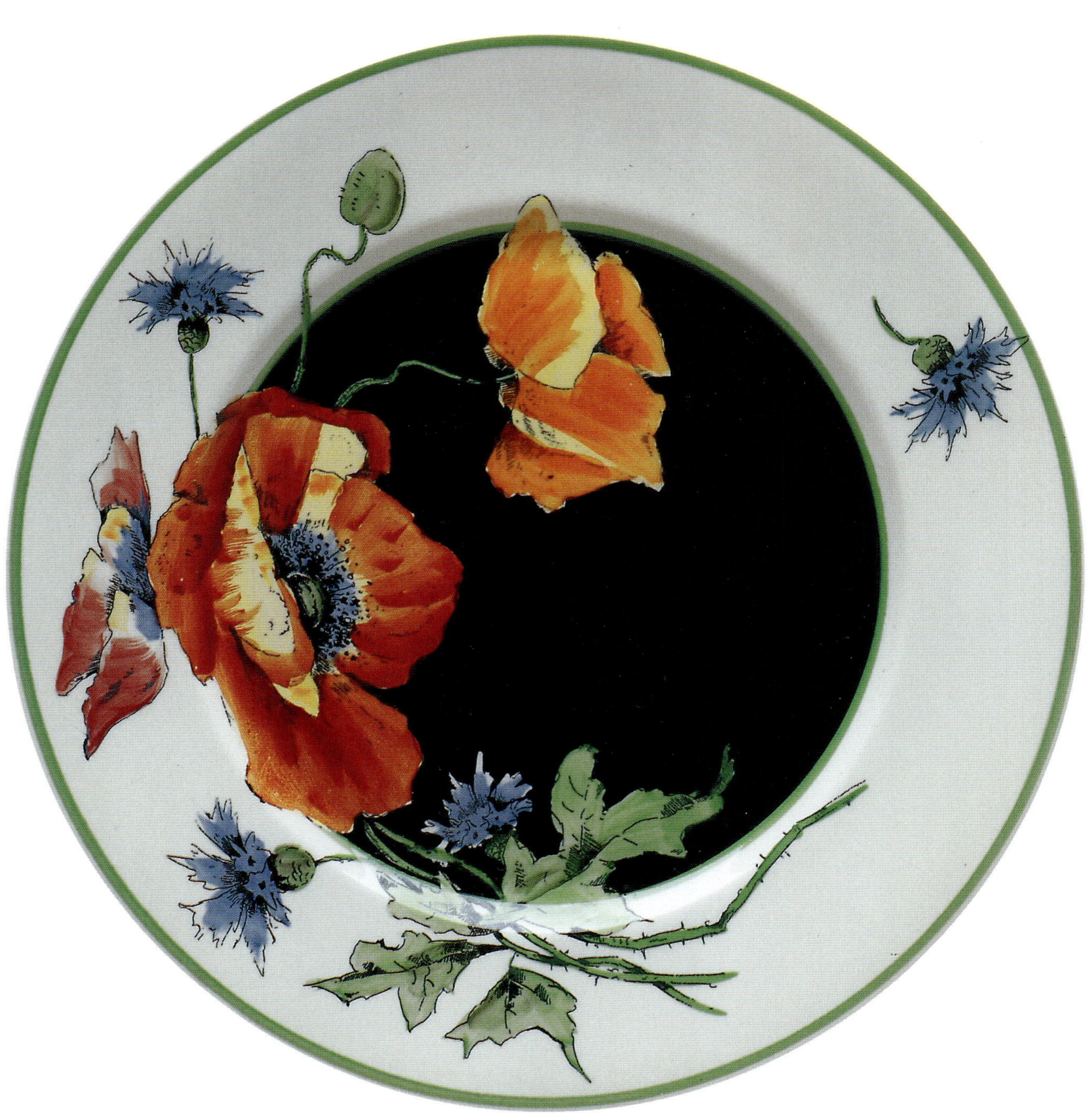

SYRACUSE CHINA

Cleota Reed and Stan Skoczen

With an Afterword by Ruth Pass Hancock

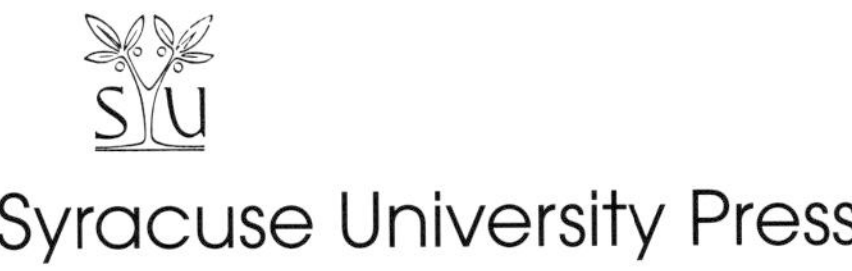

Syracuse University Press

Copyright © 1997 by Syracuse University Press, Syracuse, New York 13244-5160

All Rights Reserved

First Edition 1997

97 98 99 00 01 02   6 5 4 3 2 1

Photographs from the Syracuse China archives, unless otherwise noted.
Courtesy of Syracuse China Company, a unit of Libbey Inc.

The paper used in this publication meets the minimum requirements
of American National Standard for Information Sciences—Permanence of
Paper for Printed Library Materials, ANSI Z39.48-1984. ⊗

Library of Congress Cataloging-in-Publication Data

Reed, Cleota.
Syracuse China / Cleota Reed and Stan Skoczen; with an afterword
by Ruth Pass Hancock. — 1st ed.
       p.   cm.
Includes bibliographical references and index.
    ISBN 0-8156-0474-2 (cloth : alk. paper)
    1. Syracuse China Company.   2. Pottery, American—New York
(State)—Syracuse.   I. Skoczen, Stan.   II. Title.
NK4210.S97R44 1997
738'.09797'66—dc21   97-14548

Printed in Canada

To the workers of Onondaga Pottery and

Syracuse China, past and present, who

have made possible the telling of this story.

CLEOTA REED interrupted her career as a studio potter to become
a historian of American ceramics and of the Arts and Crafts Movement
in America. Among her publications are *The Arts and Crafts Ideal: The
Ward House—An Architect and His Craftsmen* (1978), *Henry Keck Stained
Glass Studio* (1985), and *Henry Chapman Mercer and the Moravian Pottery
and Tile Works* (1987). She is a Scholar Affiliate of Syracuse University's
Department of Fine Arts and consultant archivist of the Syracuse China
Company historical collection.

STAN SKOCZEN began working for the Onondaga Pottery/Syracuse
China Company in 1937 and never left. For many years he headed quality
control and product standards. He has long been the preeminent authority
on the history of the company's products. A resident of Syracuse, he
lectures widely on Syracuse China and remains a consultant to the
company's archives.

The building of the Onondaga Pottery Company has been the work
of many people during the three generations since the company was
founded in 1871. The success of this organization has been due primarily
to the character of the men and women who comprise the Onondaga
pottery today, to those who have made up this organization in the past,
and to the spirit of goodwill, fairness, and mutual helpfulness in which
we have worked together.

—Richard Pass, 1956

Contents

Illustrations

Preface

THIS book's primary concern is the ceramic dinnerware produced for more than a century and a quarter by the Onondaga Pottery and Syracuse China Company, but its secondary subjects are nearly as important. They include the people who made and sold the ware and they also include in broader scope the place of ceramic dinnerware in American life since the middle of the nineteenth century. The authors hope that they have begun to illuminate a long neglected but highly interesting aspect of American material culture: ceramic dinnerware as an intersection of art, industry, and social history.

In the 1890s, Syracuse China had the distinction of being the first vitreous china dinnerware produced successfully in America on an industrial scale. Over the decades the company remained in the forefront of the American ceramics industry in design, decoration, manufacturing, marketing, and employee relations. The fine china it produced for home use from the 1890s through the 1960s, highly prized in its time and since, remains a prime example of how an industry in a prospering democratic society both served and shaped public taste. The somewhat heavier, more durable china pioneered by Onondaga Pottery for use in hotel dining rooms and restaurants, which was also widely admired and emulated, remains in production to this day.

Compared to other kinds of ceramic art, high quality dinnerware of the kind made in the United States by Onondaga Pottery/Syracuse China and its competitors has attracted relatively little scholarly attention over the years. The work of many studio potters and the art pottery of the Arts and Crafts Movement have both received far more extensive study. Recent monographs on ceramic manufacturing in New Jersey and Ohio have contributed much to an understanding of the industry in general, and of its labor and management history in particular, but relatively little of the ware produced in these centers has been studied in depth.[1]

Perhaps the ubiquity of ceramic dinnerware in everyday life has argued against its study. The plates, cups, and saucers from which we dine daily seem too familiar to need close attention. Perhaps, too, the fact that production ceramics are a product of industry as much as art has presented a barrier of sorts to those inclined to study ceramics only as art. But a more important reason for the relative neglect of the history of American production ceramics as a field for serious study has undoubtedly been the lack of extensive documentary evidence. For most potteries, few (if any) primary sources have survived to explain their history. The Syracuse

China archives and its historical collection of ware have gone far to erase this lack of information. They have provided the authors with an abundance of documentation that sheds light not only on the history of one firm's production but also on the practices and issues of the American ceramics industry in general. The archives contain business records, correspondence, photographs, drawings, diaries, patents, factory orders, tools, moulds, advertisements, production orders, customer orders, house publications, catalogues, motion pictures, memorabilia, and much else. So far as is known, the Syracuse China archives are unique in size, scope, and significance within the American ceramics industry.[2]

Since 1987, the authors, aided by Ruth Pass Hancock and several Syracuse University graduate students, have sorted, identified, and catalogued the company's historical collection of thirteen thousand ceramic artifacts, as well as its extensive archives. The book's narrative has come largely from sources in the archives (with specific documents cited only when they are quoted). To have nearly all of the company's published catalogues available as a tool for identifying the ware has been invaluable.

Though the collection and archives are not open to the public, the major Syracuse China exhibition installed at the Onondaga Historical Association in 1998 draws on the archives for its historical and interpretive materials. One purpose of this book has been to complement the exhibition.

Another of the book's purposes has been to correct a historical tendency to overlook the firm and its ware in popular accounts of the American ceramic dinnerware industry. Within a generation of its founding in 1871, Onondaga Pottery had become a national leader in its field. Its fine china had rapidly earned a prestigious reputation, and its hotel ware was one of the industry's great successes. Despite this reputation, publications about American dinnerware that appeared throughout the first half of the twentieth century only rarely gave more than brief mention to the company, and sometimes none at all. To some extent this may have been the result of the pottery's location in central New York State, well apart from most of the rest of the production ceramics industry and off the beaten track of writers of the time about ceramic art. But it probably also reflected the company's belief in these years that it was different from the norm in the industry. Its management was disinclined to encourage writers who were apt to view the company as a typical industrial-scale American pottery.

The distinctiveness of the pottery and its products came in good part from the leadership of five men, four of them related by blood or marriage, who for nearly a century led the firm as president. Each encouraged the firm's spirit of individuality within its industry and gave a long-lasting emphasis to the company's character. The first of these, and the most phenomenal, was James Pass. As the Syracuse China archives make clear, Pass was a seminal influence in his field in so many ways that he needs to be seen as an important historical figure in the larger world of American ceramics. At Onondaga Pottery, he set a pattern of personal leadership—micromanagement—that persisted for decades. For that reason, the chapters pertaining to the history of the company prior to 1971—the "old" company—have in large part been organized around the company's presidents and other key figures.

The arrival of the "new" company in 1971 moved Syracuse China into a different, rapidly changing world in leadership, technology, merchandising, labor relations, and design. It is too soon to set this recent period in historical perspective, but one observation seems sound: the challenges of making and selling ceramic dinnerware have, to a remarkable degree, remained the same over more than a century. In both the 1890s and the 1990s the company's major concerns were competition from foreign imports, the imperative need to be in the forefront of manufacturing technology, and the search for new markets and new responses to changes in taste. Of course, major differences also exist between the two eras. What was once a highly labor-intensive industry has steadily become first a mechanized one and then a computerized one, dramatically so since the 1970s.

Because processes of production have changed so much over the past half-century, it seems useful to include as a chapter an edited version of the company's own descriptive account of how it made its ware in the period between 1920 and 1960. In addition to its historical interest, this account, "The Story of a Piece of China," is a reminder that clay has been the great constant throughout the firm's history, indeed throughout the industry's history.

Clay has a life of its own. There are many things that it will not do and cannot be made to do. It insists on respect. It records not only the impression we make on it by hand or machine but also something of the saga of our life on this earth. Many of the earliest impressions of human life can be found in the clay objects that this life left behind.

We spend a significant and often pleasant part of our lives dining from fired clay. We tend to take it for granted. It rarely occurs to us that the manufacturing processes and technologies used to make ceramic dinnerware have always needed to adapt to clay's potentials. Formed, fired, decorated, and sent out into the world, fine ceramic dinnerware can rank among the most appealing and useful of the decorative arts. Even the more ordinary ceramic dinnerware has a long-established central role in the civilizing rituals of dining. Fine or ordinary, ceramic dinnerware's history can be found in microcosm in the history of Syracuse China. This book keeps that ware in the forefront while considering many of the people, events, and circumstances that contributed to its success.

Patrons and Donors

Patrons

In memory of William Root Salisbury, from his family:

> Ethel G. Salisbury, Marilyn and Tom Heebink, Judith Salisbury,
> William L. and Karen Salisbury, Todd Heebink, and Paige Heebink

The Robert Mills Salisbury Family

Syracuse China, a subsidiary of Libbey Inc.

Donors

Lee and Chet Amond

Bond, Schoeneck & King, LLP

Mr. and Mrs. William G. Curran, Jr., in memory of Ruth P. and Richard H. Pass

Richard Pass Curran, John Henry Curran, and Charles Edward Curran

Charles and Karen Goodman

Robinson B. Gorley and Darrell and Eunice Pass Carpenter, in memory of Ann Pass Gorley

Hon. and Mrs. Stewart F. Hancock, Jr.

Martin B. and Esther G. Shellenberger

Eleanor and Jere Williams and Family, in honor of Ruth Pennock Pass and Richard Henry Pass

Dorothy Crouse Witherill

Acknowledgments

So many people have contributed in so many ways to this book that it is impossible to thank them all by name, but special mention must go to a few for recalling forgotten processes and nearly forgotten incidents, and for contributing crucially in other ways. Among them are former Syracuse China employees Pete Baldwin, Sally Brown, Betty DeCerce, Sandy Hayes, William and Dorothy Jamelske, Matt Rizzo, George Schultz's family, and John Wilkinson. Among those who went out of their way to help us were Robert Grannis, Warren Hoffmann, Gordon Huggins, Wayne Long, John Morris, Leon Rocco, and Ralph Work. Robert Hoffman shared his memories in a long interview, which has been preserved in the archives. Others who in their daily contacts with us helped our work to preserve and document the company's archives included Avis Adams, Debby Black, Phil Harvard, Barbara Jacobs, Joe Walto, Linda Williams, Janice Michalak, Bob Kern, and Terry Weslowki. Many others remained keenly alert to our mission to rescue tools, ware, and documents from the past for preservation in the archives. We will not mention their names for fear of excluding someone, and we appreciate their interest and care.

Special thanks go to Mark Bassett and Victoria Naumann for sharing their Guy Cowan research. Dennis Connor, director of the Onondaga Historical Association, and his staff have encouraged our project and ably helped to coordinate it with the association's Syracuse China exhibition. The association's archivist, Edward Lyon, and research coordinator, Judy Haven, were constant sources of information, encouragement, and good cheer. At the Syracuse University Photo Center, David Broda made superb color plates of historic ware, and Rich Pitzeruse, Jr., supervised the fine black-and-white production. Syracuse China Company of Libbey Inc. and the William Fleming Educational Fund generously contributed toward the expense of photographic work.

Sophia Papapanu, Mary Kasper, Irene Donovan, and Richard and Rosemarie Lewis have been splendidly helpful sources of information over the years, and Mrs. Papapanu generously contributed ware to be photographed. Among others, James Blackaby, Regina Blaszczyk, Bert Denker, Amy Doherty, Bruce Manwaring, Jeanette Mattson, Susan Myers, and Polly Stetler have lent their expertise and encouragement along the way.

Our thanks go especially to the historians of American ceramics and decorative arts, Ellen

Denker, William Gates, and Coy Ludwig, who read the manuscript, offered helpful insights, and wished our project well. Dick Case, a pioneer investigator of early Onondaga County pottery, devoted time to reading a draft of the manuscript. The counsel of all four authorities has made this a better book. The authors alone are responsible for any inacuracies or omissions that remain.

Syracuse China's designers, Lucie Wellner and Joanne Capella, have been constantly resourceful consultants at every turn. We could not have completed the project without their help.

Charles Goodman, president of Syracuse China, and Martin Shellenberger, then its senior vice president of Marketing, initiated the Syracuse China archives project in the 1980s along with Barbara Perry, then Curator of Ceramics at the Everson Museum of Art. This book grew from that project. We are especially grateful for their constant encouragement and support over the years and for their helpful suggestions for the brief history of the "new company" that appears as chapter 10. Others who provided important information on the subject are Robert Theis, Richard Besse, William Calnan, and Steve Unger. All have our thanks.

Ruth Pass Hancock, whose forebears played such a vital part in the company's history for four generations, came to help us in the archives in 1987. The three of us have worked ever since as a team, searching records, cataloguing ware, and in every way working to complete the historical record. We have answered hundreds of inquiries by people wanting to know about their Syracuse China. No one has cared more about the company and its history than she has. The thousands of hours she has contributed to the archives as a labor of love has enriched us all.

Finally, to Stan's sister, Ann Burdick, and to Cleota's husband, David Tatham, and her daughters, Heila Martin and Ragen Tiliakos, we are profoundly indebted for their many contributions to this project and for their patience and understanding in seeing us through to a successful end.

SYRACUSE CHINA

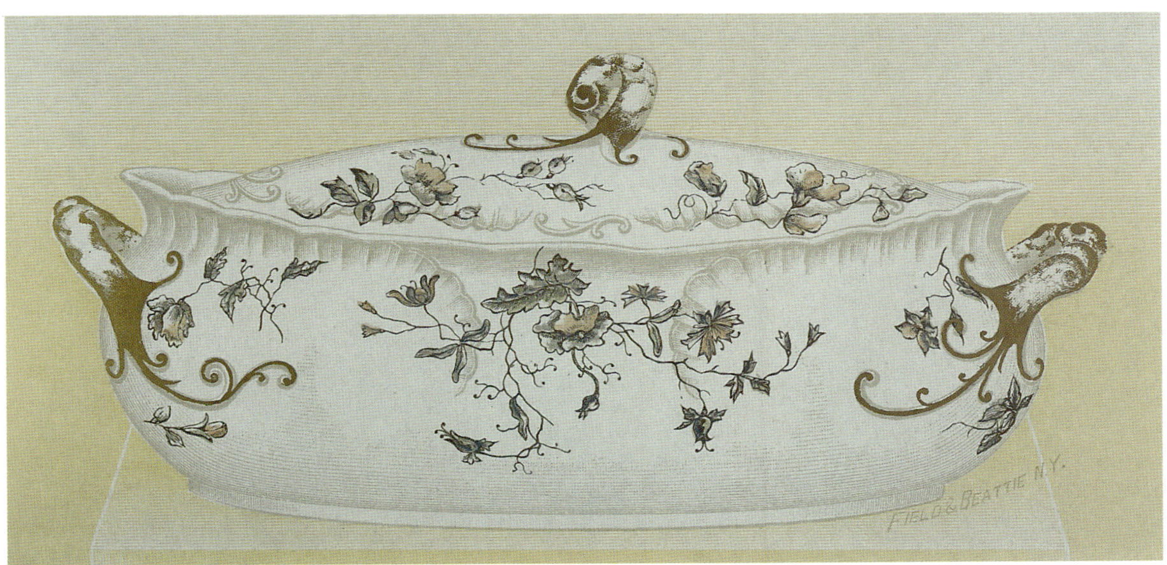

1. *Marmora* casserole with V-609 decoration. Color lithograph for 1893 catalogue.

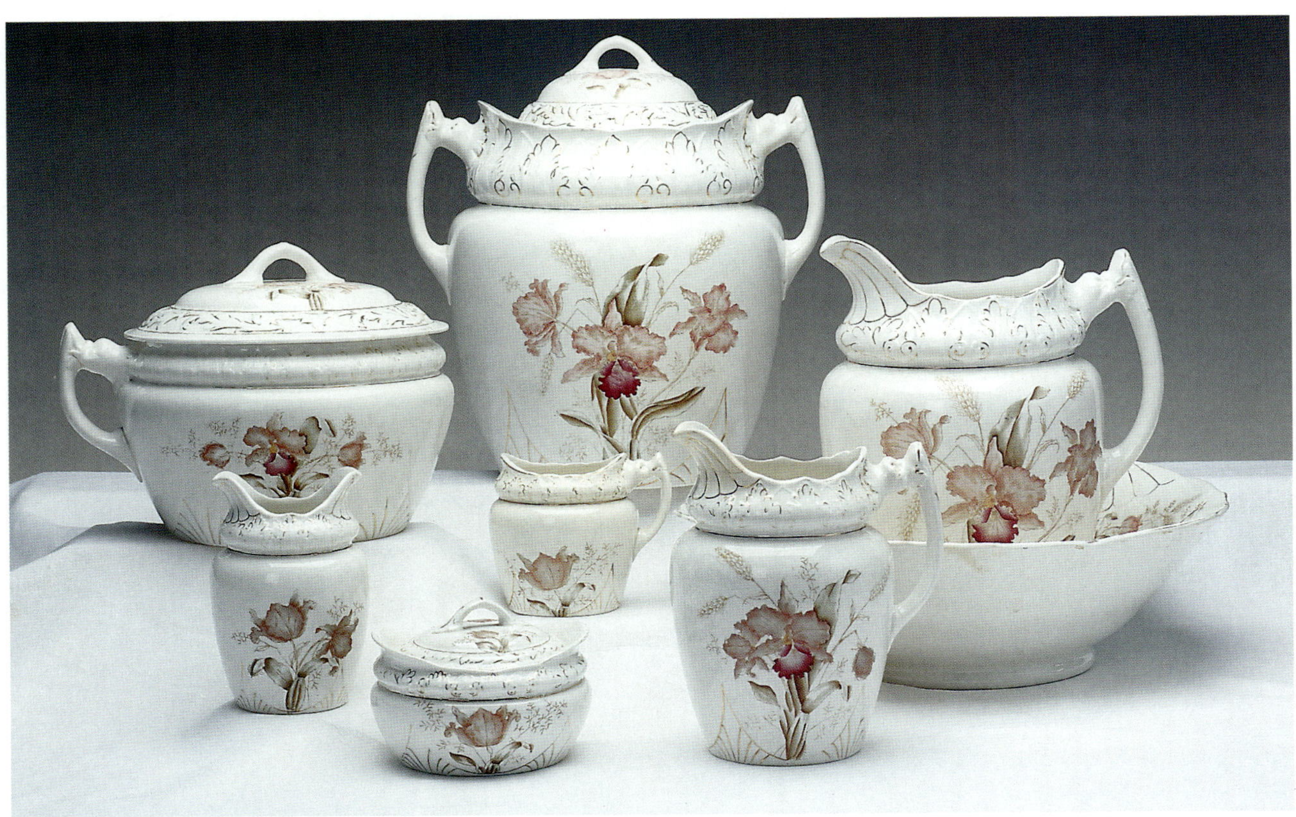

2. *Juno* toilet ware in *O.P.China* with transfer printed, hand-filled, gilded "Orchid" pattern.

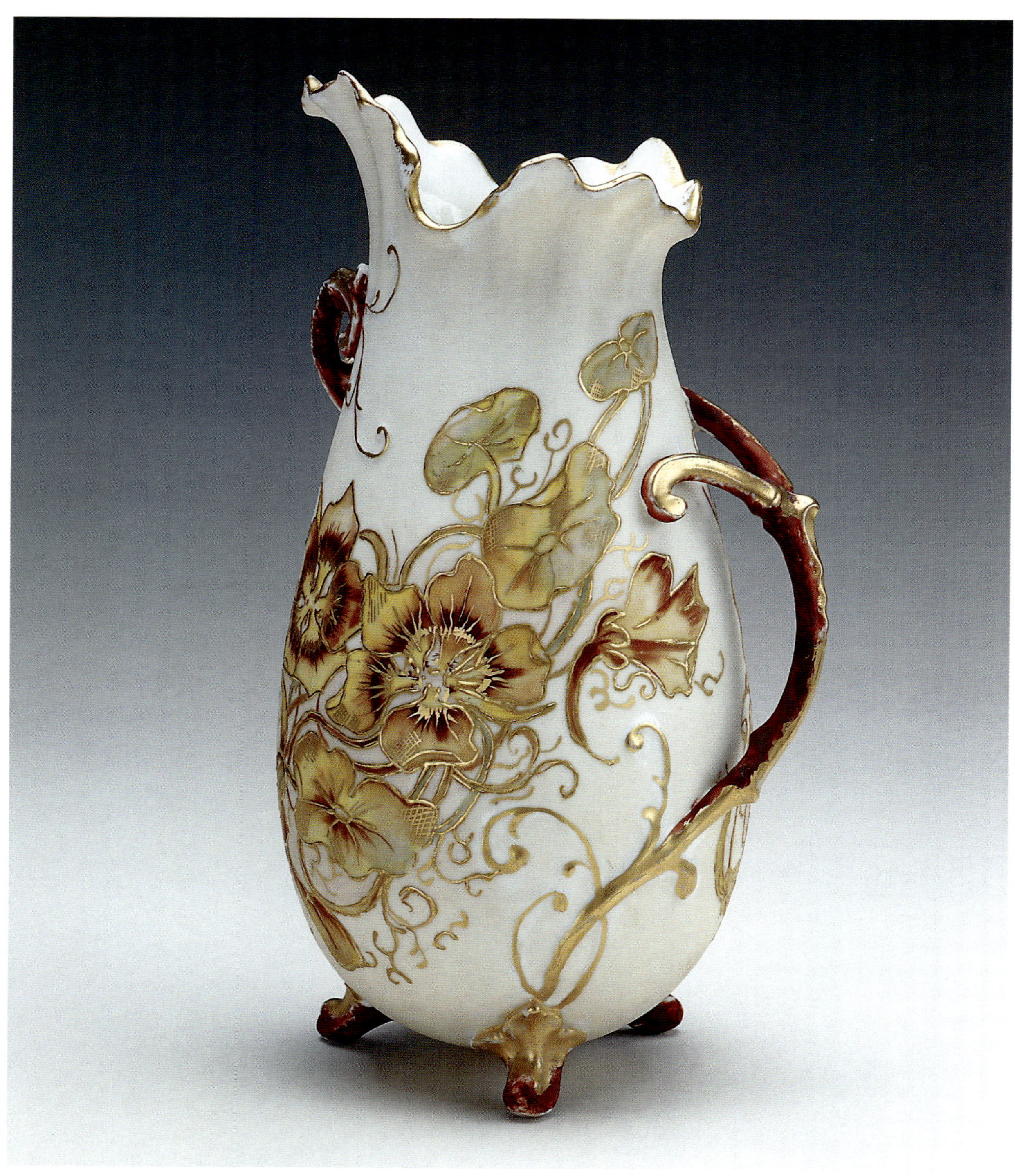

3. *Imperial Geddo* vase decorated with Worcester finish, gilded raised paste decoration, and hand coloring. Elmer Walter possibly decorated it for display at World Columbian Exposition, Chicago, 1893.

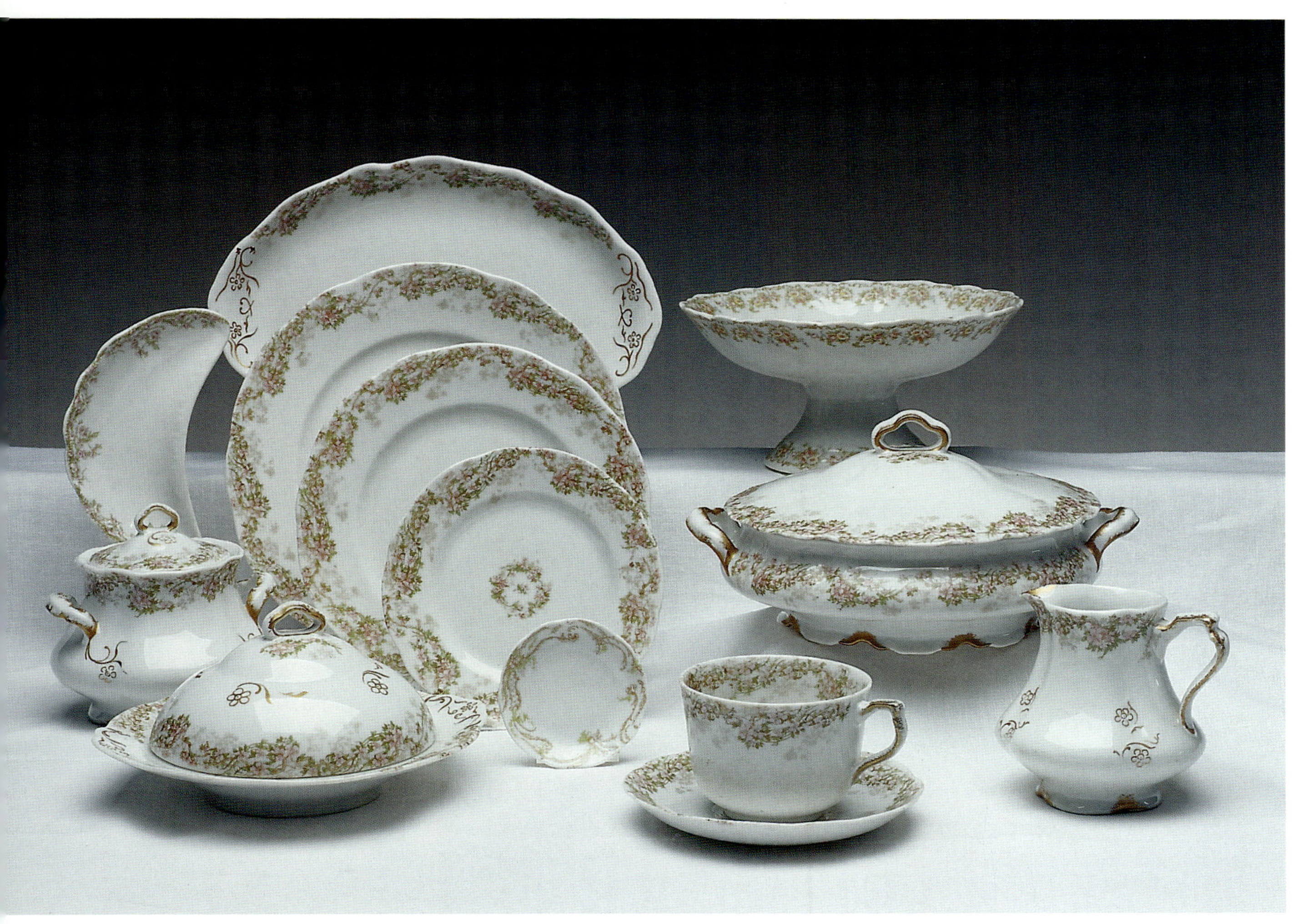

4. A selection of **Plymouth** ware in Syracuse China with early decalcomania decoration.

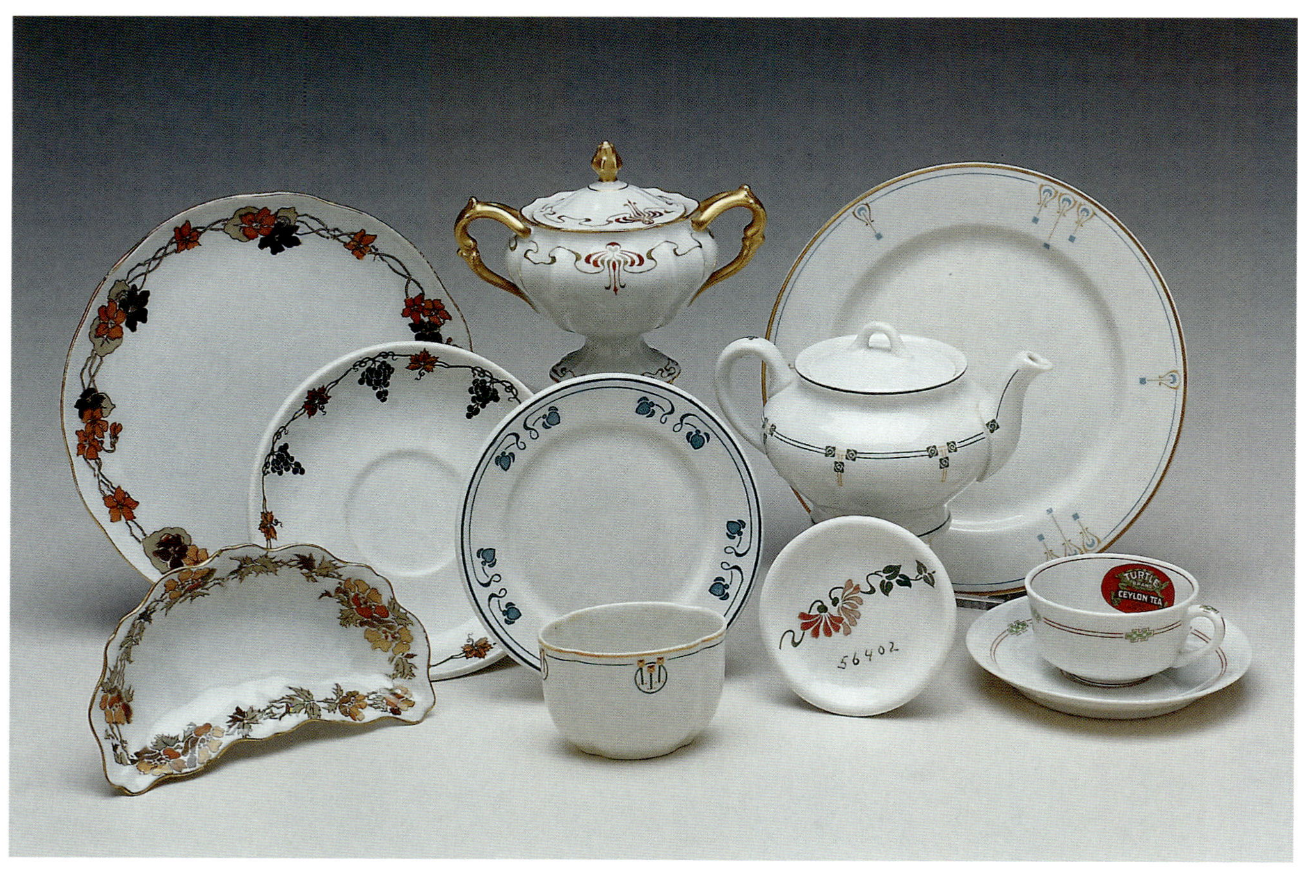

5. Examples of Art Nouveau and Arts and Crafts Movement patterns on Syracuse China. *From top left:* Adelaide Alsop Robineau's "Violet Border" on **Plymouth** plate; "Grape Border" on **Round Edge** saucer; "Wild Rose Border" on **Puritan** bone dish; *from top center:* unique Art Nouveau pattern on **Puritan** tête-à-tête sugar; "Modern Art" print on **Round Edge** plate; "Marathon" on **Hospital** sugar; *from top right:* "Art Pendant" on **Mayflower** plate; "Tudor Rose" on **Mayflower** teapot; "Chelsea" on litho shop sample **Round Edge** individual butter pat; "Squares" and crest on **Mayflower** five o'clock teacup and saucer for Turtle Ceylon Tea.

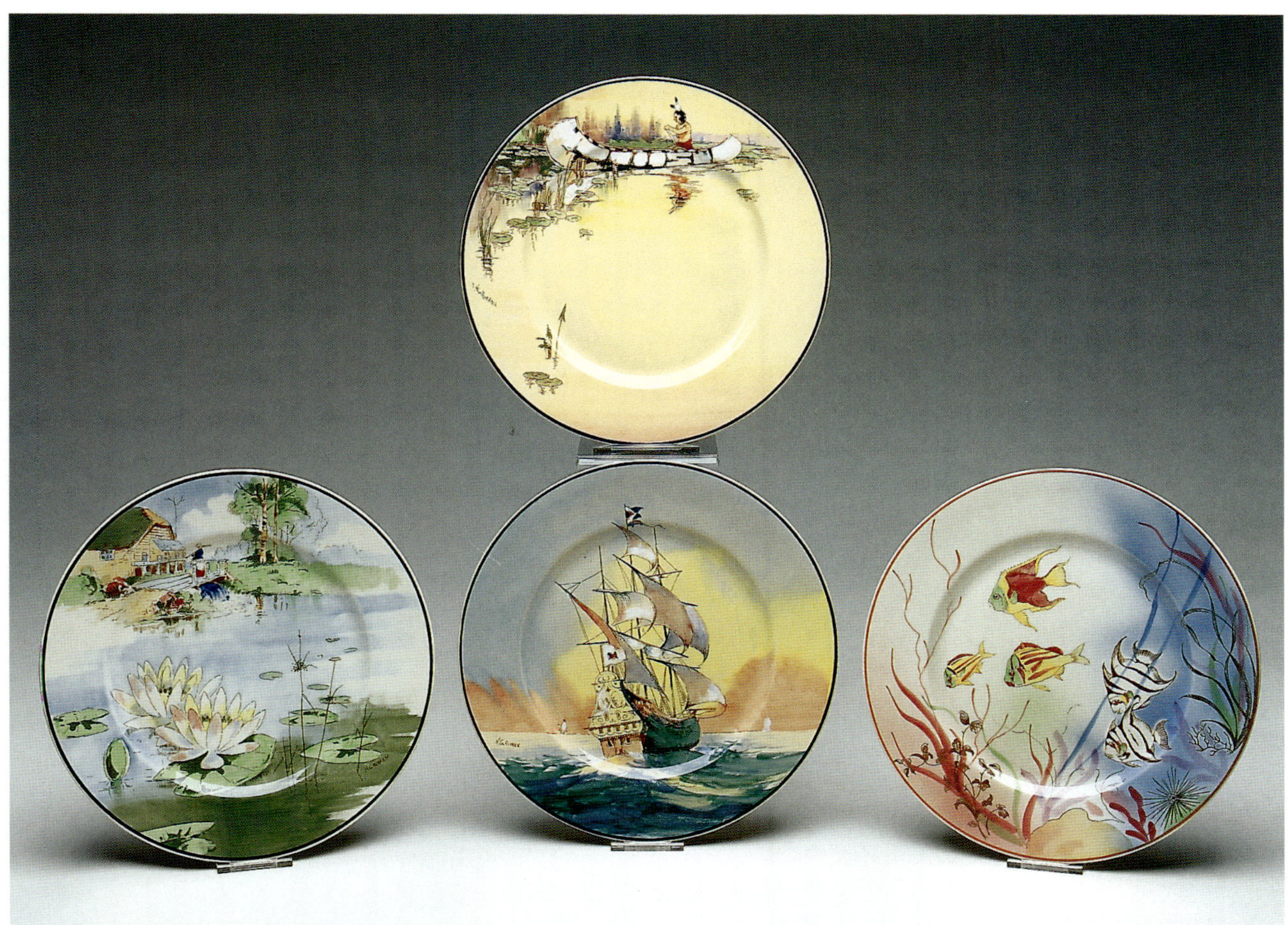

6. Harry Aitken transfer-printed and hand-painted water themes on service plates.
Top: "Indian in Canoe" (unique), *bottom, left to right:* "Water Lily" (one of six designs),
"Spanish Galleon" (one of four), "Deep Sea Gardens" (one of four). Private collection.

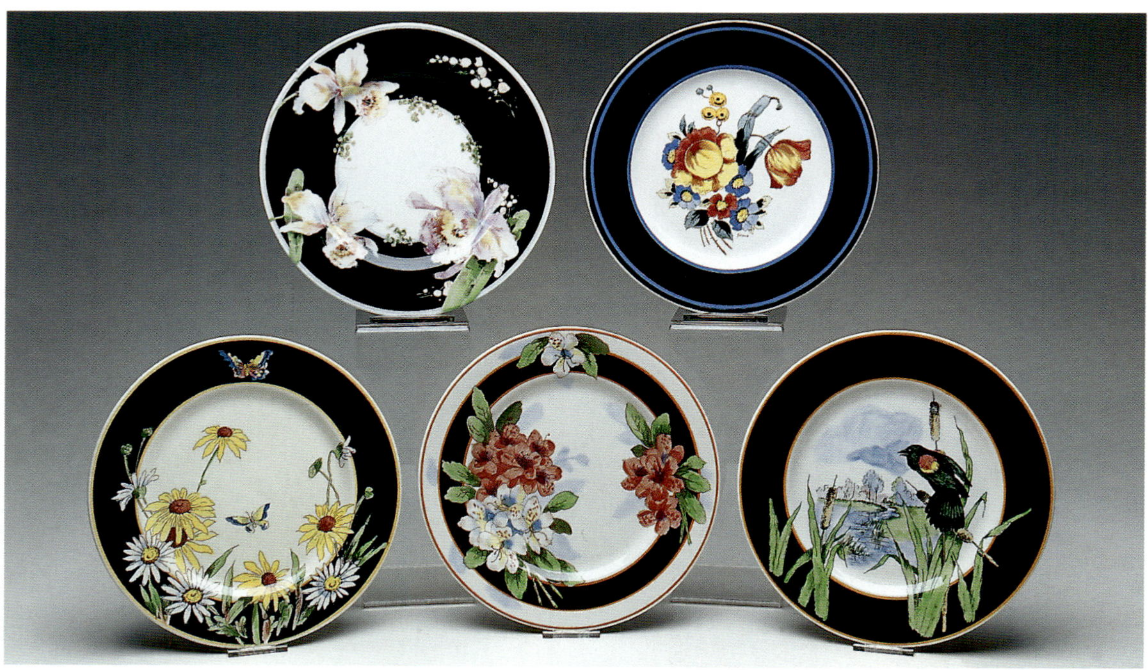

7. Harry Aitken hand-painted designs on dinner and service plates. *From top left:* "Orchids" (one of four) and "Garden Flowers" (one of six). *From bottom left: Nature Studies* (of forty): "Field Daisies and Black-Eyed Susans," "Rhododendron," and "Red-winged Blackbird." Private collection.

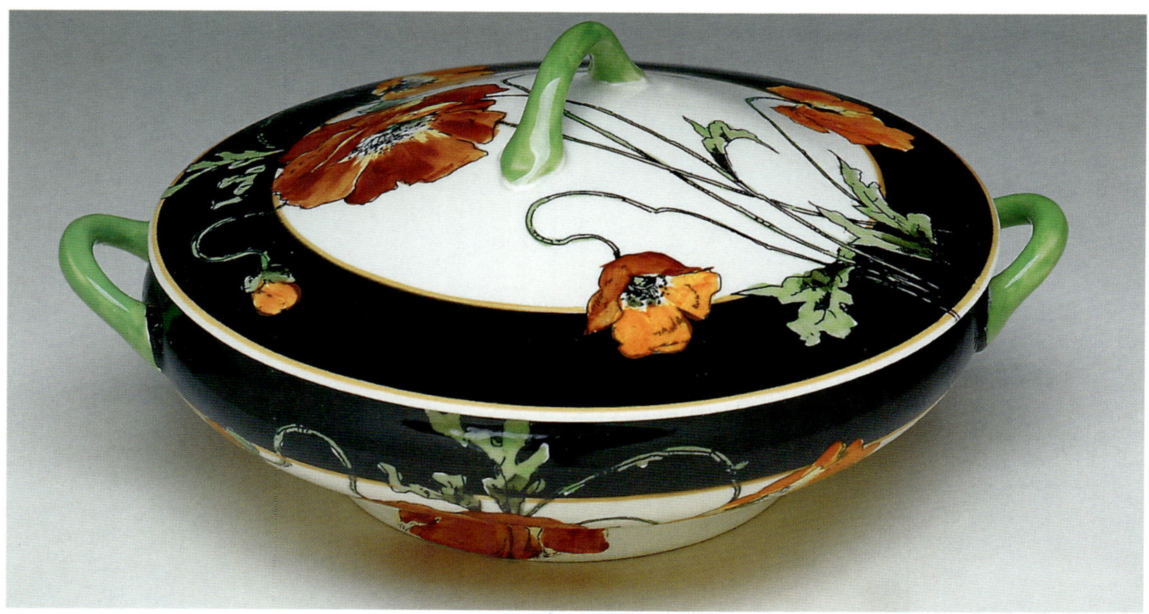

8. Harry Aitken. Unique ***Mayflower*** casserole hand-decorated with *Nature Studies* "Poppy" motif. Private collection.

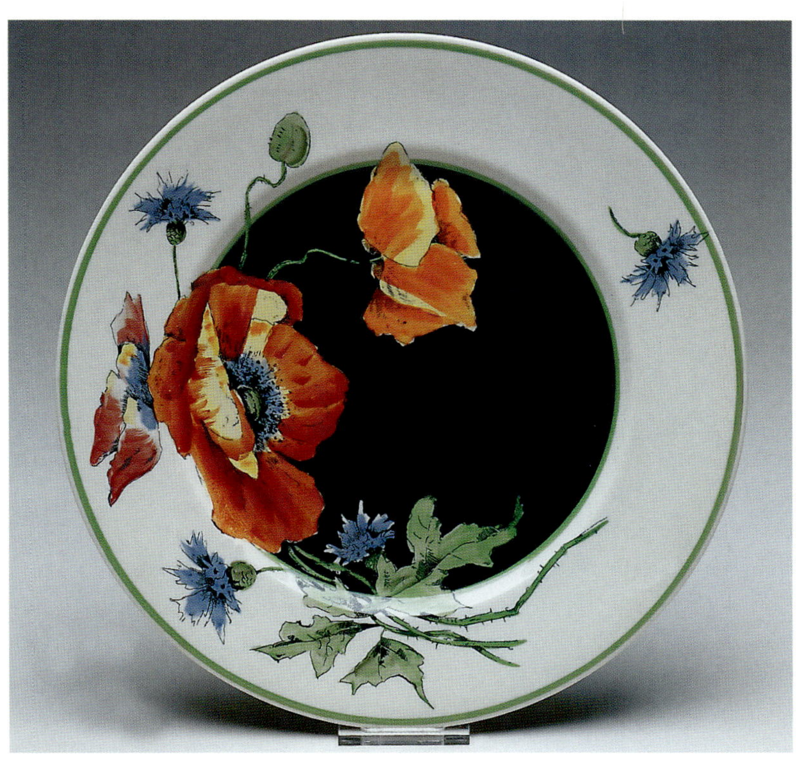

9. Harry Aitken. *Nature Studies* "Poppy."

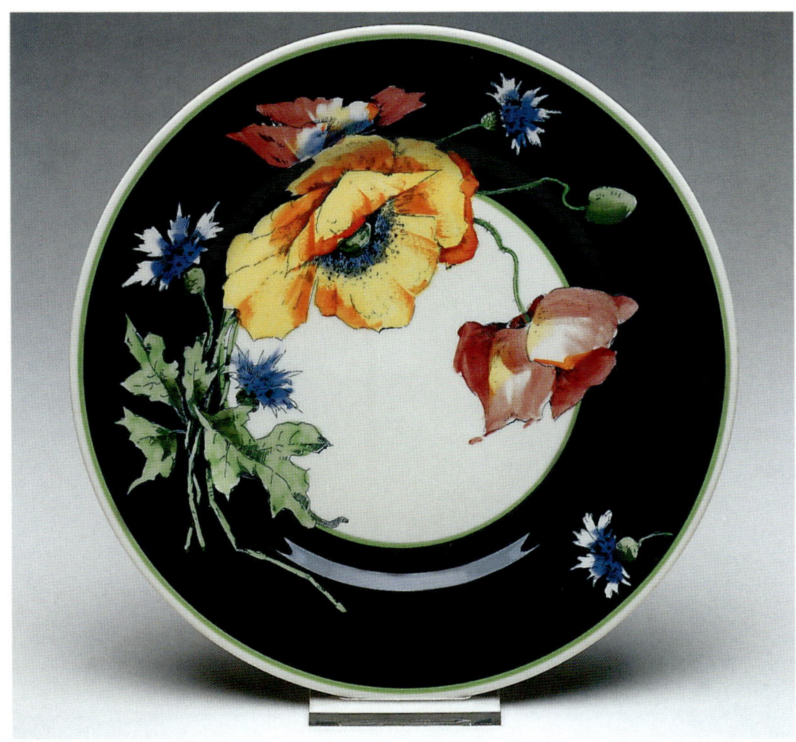

10. Harry Aitken. *Nature Studies* "Poppy."

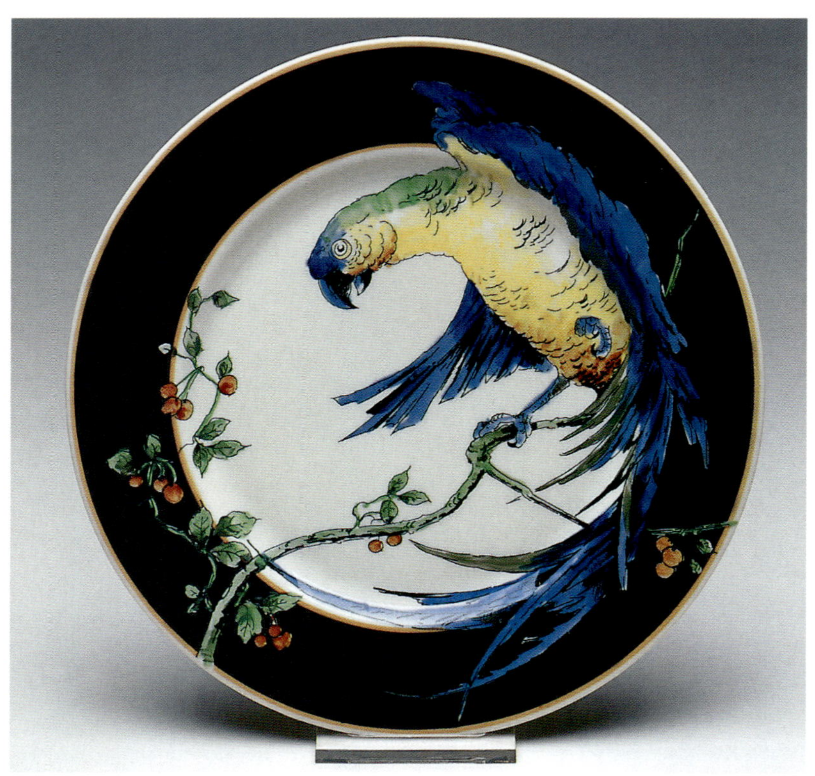

11. Harry Aitken. *Bird Series:* "Parrot" (one of six). Private collection.

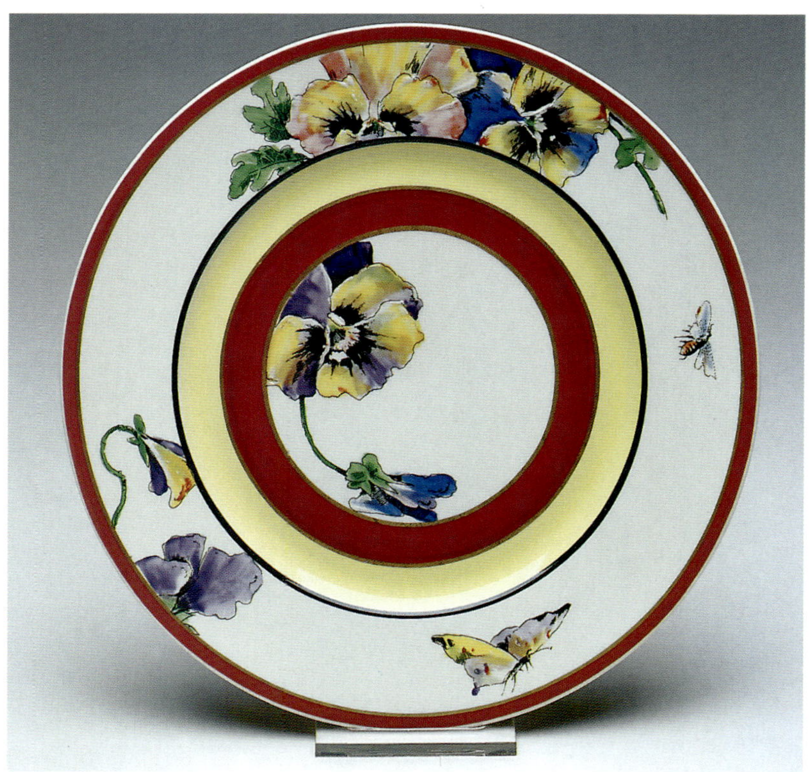

12. Harry Aitken. "Pansy" plate. Unique hand-painted decoration. Private collection.

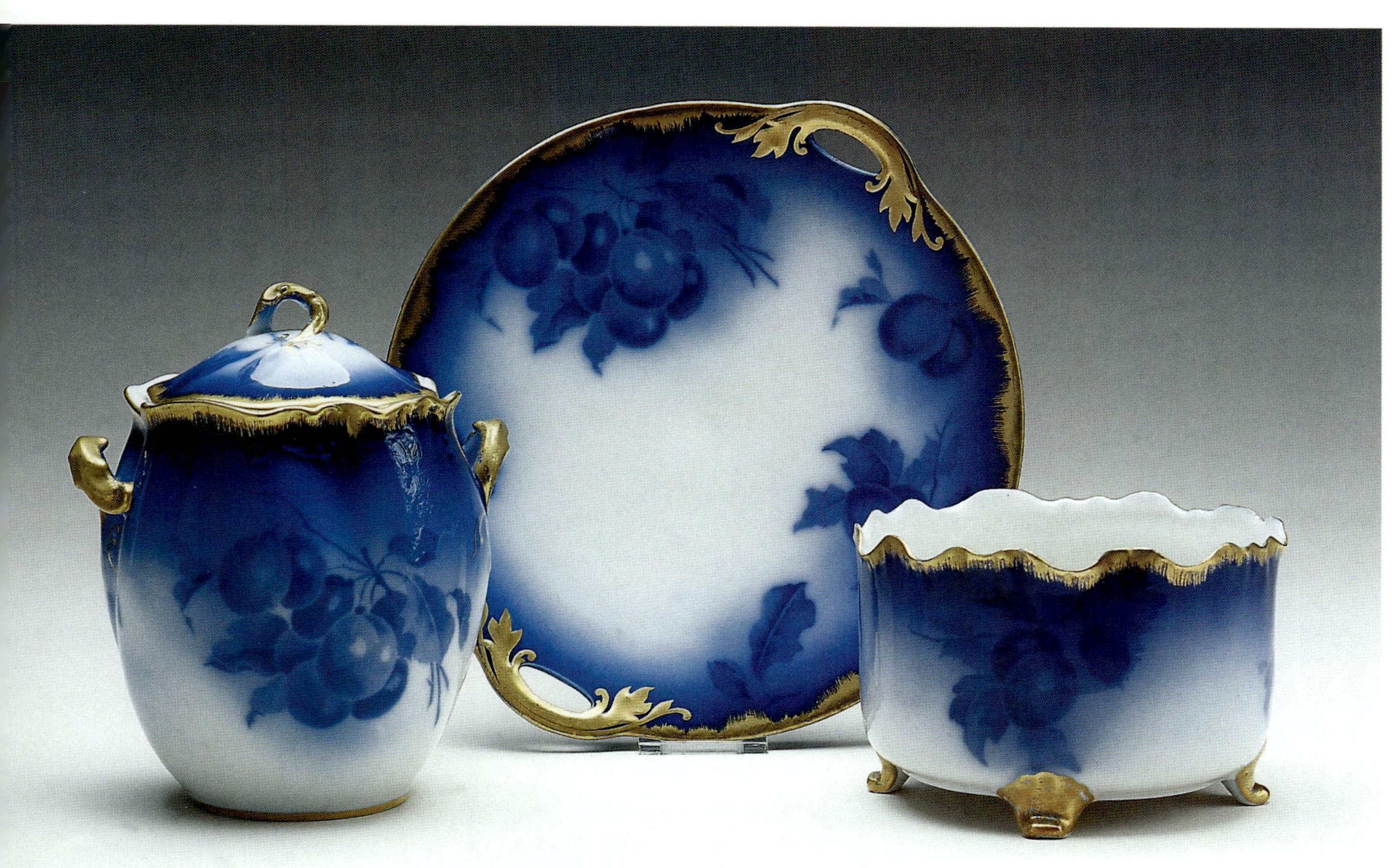

13. Harlow Pierce and Elmer Walter. "Blue Plum" ware. **Imperial Moro** biscuit jar, **Mayflower** cake plate, and **Imperial** unhandled nut bowl.

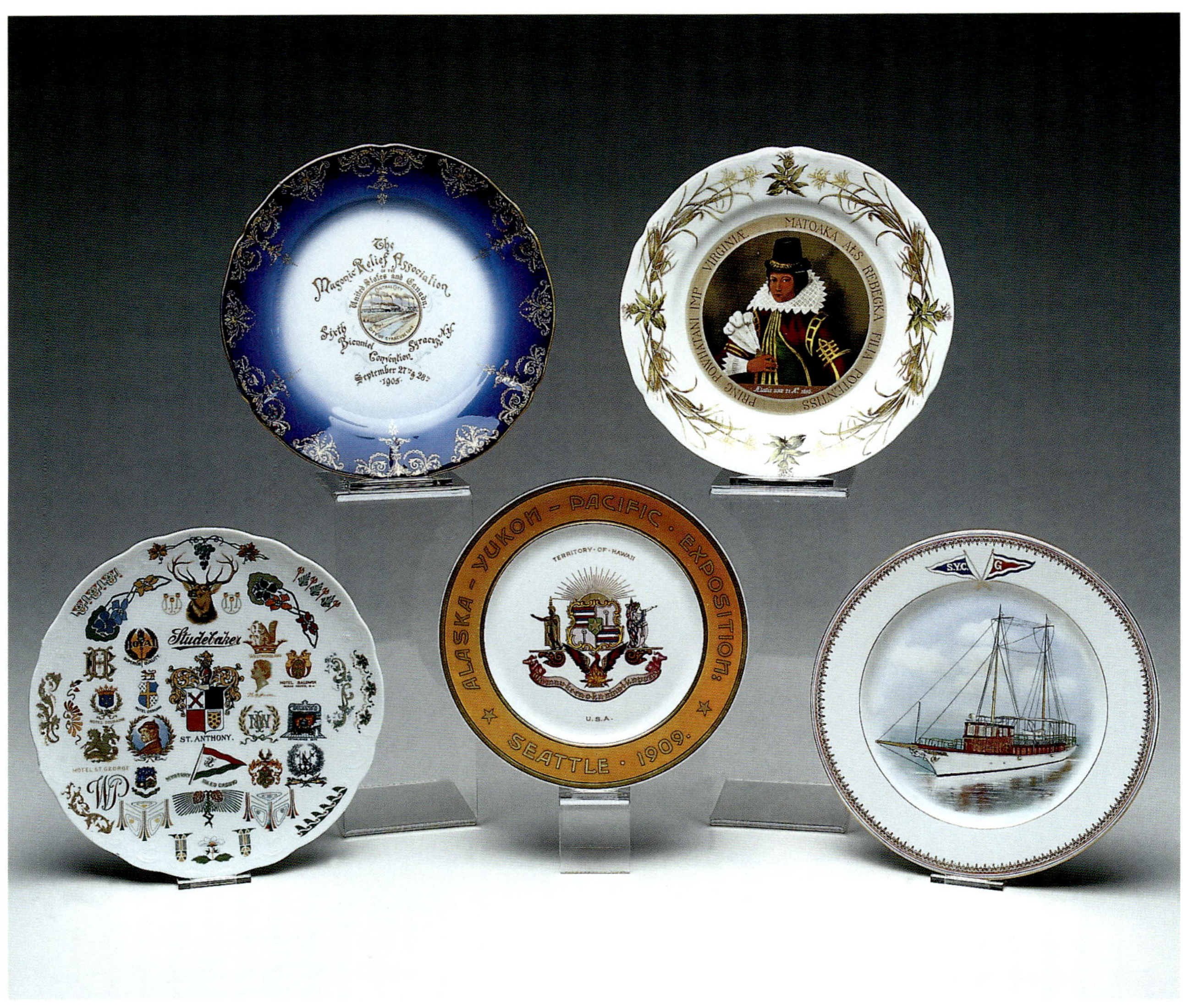

14. Custom and commemorative plates decorated with decalcomania crests. *From top left:* souvenir plate to commemorate the Sixth Biennial Convention of the Masonic Relief Association of the U.S. and Canada, with "City of Syracuse" center, 1905, overglaze decoration: spray blue, gold print and lines, and decalcomania center crest; "Pocahontas," overglaze decalcomania on **Puritan** plate for unknown client, 1905; *from bottom left:* sales-man's sample plate with examples of decal crests and pattern strikes on **Plymouth** cake plate, 1909; souvenir plate to commemorate Alaska-Yukon Pacific Exposition in Seattle in 1909, lustre glaze, gold line, and overglaze crest; "Josephine," custom plate carrying flags of yacht club and private yacht, overglaze decalcomania with enamel highlights and gold print border, 1911.

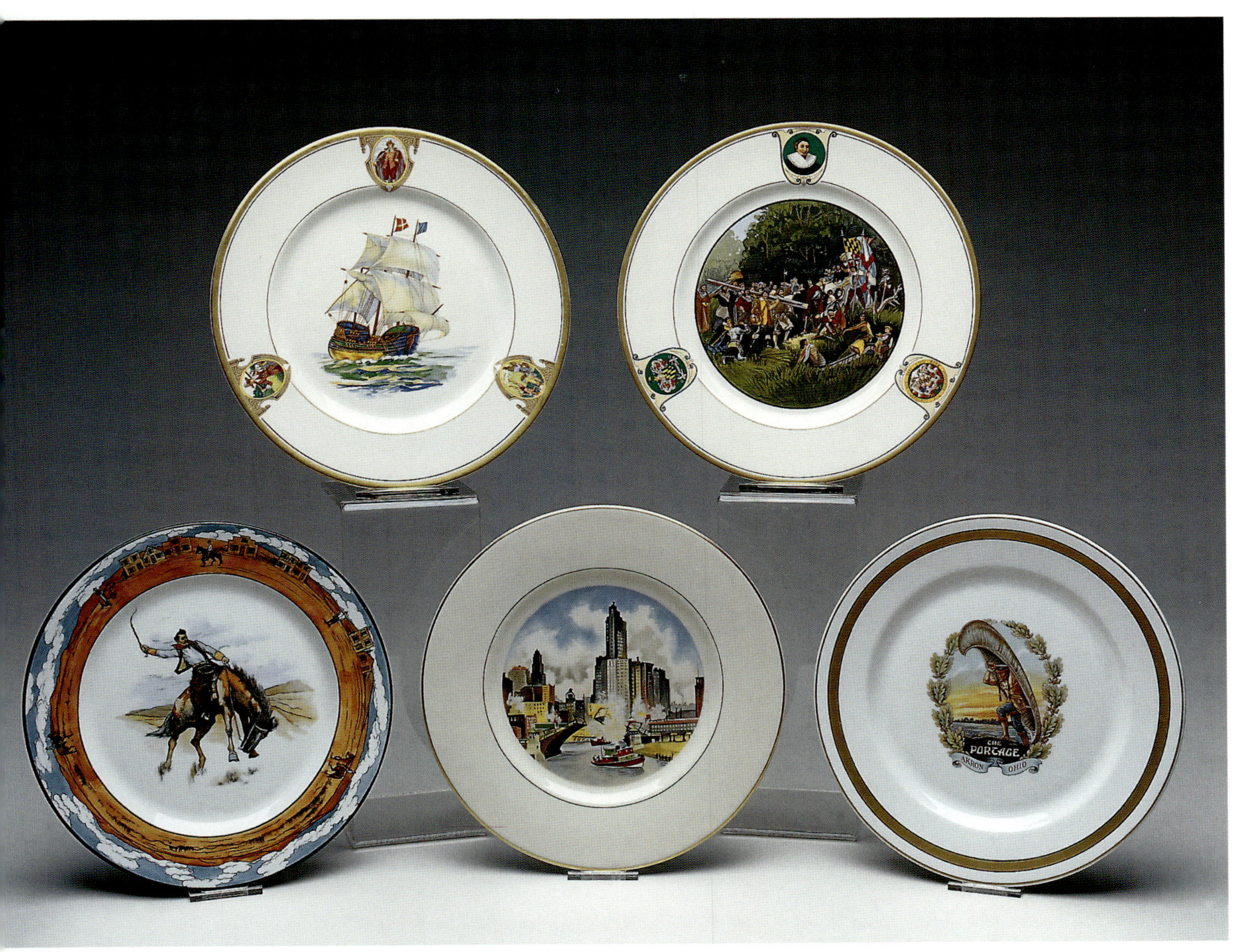

15. Custom service plates depicting American themes. *From top left:* "Mayflower," overglaze decalcomania with gold band, service plate for Mayflower Hotel, Akron, Ohio, 1930; "The Planting of the Colony of Maryland, March 25, 1634," after a painting by F. B. Mayer, overglaze decalcomania with gold band, service plate for Lord Baltimore Hotel, 1928; *from bottom left:* "Cowboy," underglaze decalcomania, service plate for unknown client; Adolph Dehn, *American Scene:* "Chicago," one of twelve plates for B. Altman & Co., New York, 1952; custom service plate for Portage Hotel, Akron, Ohio, 1926.

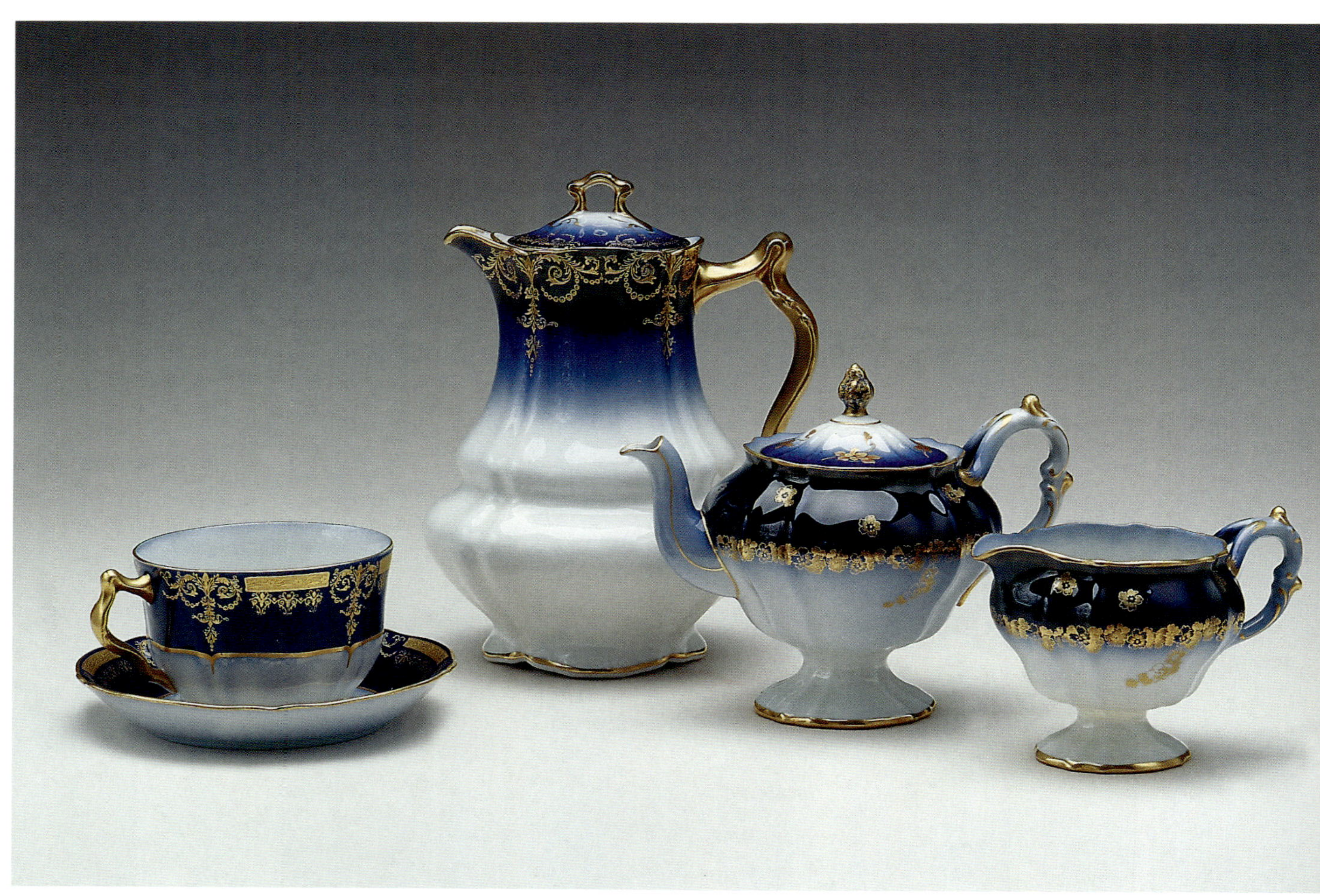

16. **Puritan** ware: Coffee cup, chocolate pot, tête-à-tête tea pot and creamer. The chocolate pot is decorated with spray blue, coin gold transfer print, and hand gilding. The cup and saucer are decorated with blue ground-lay, coin gold transfer print, and acid etch techniques. The tête-à-tête pieces are decorated with ground lay and sprayed cobalt blue and gold transfer print.

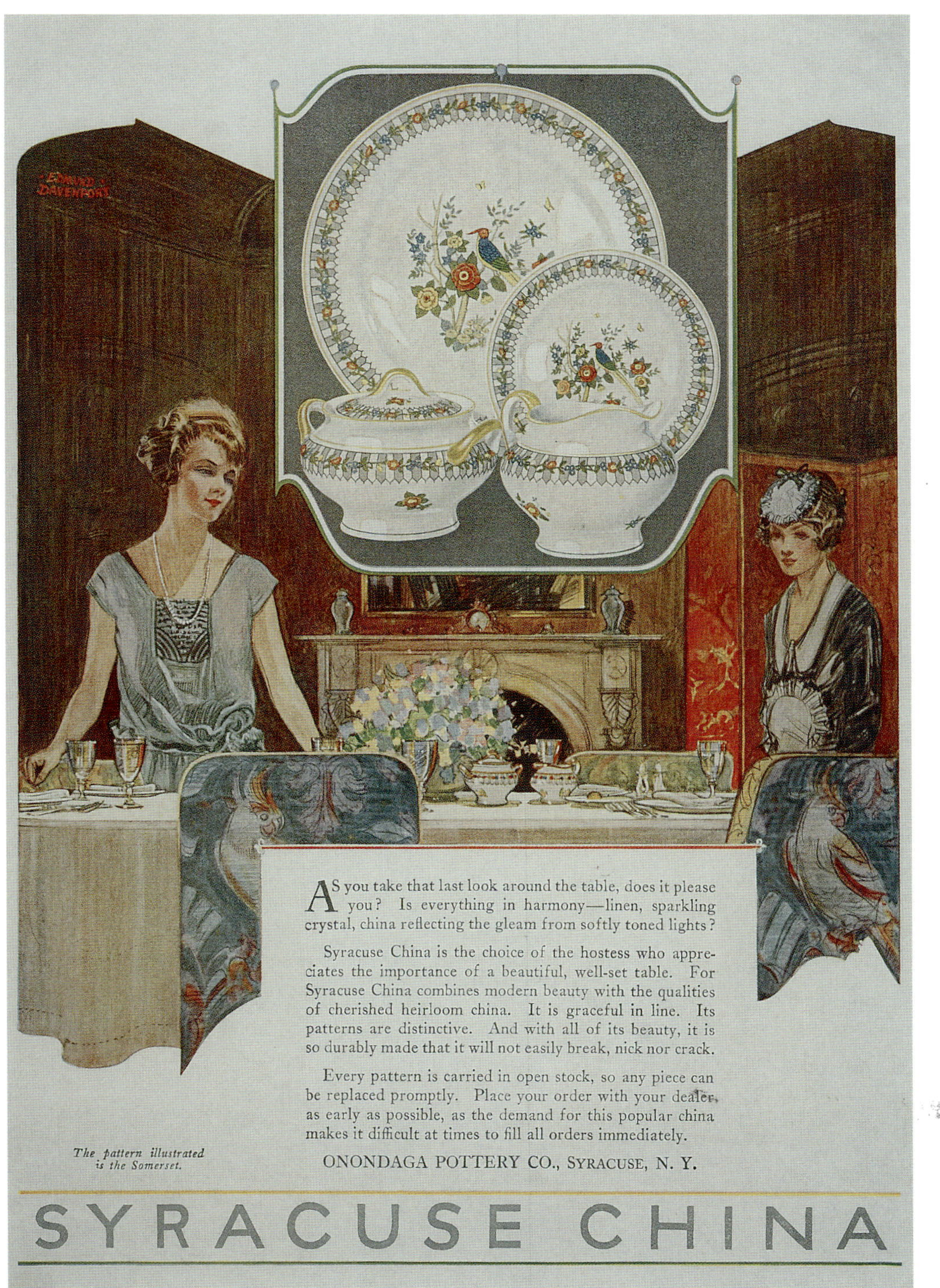

17. Advertisement in popular woman's magazine of the 1920s for *Mayflower* featuring "Somerset" pattern.

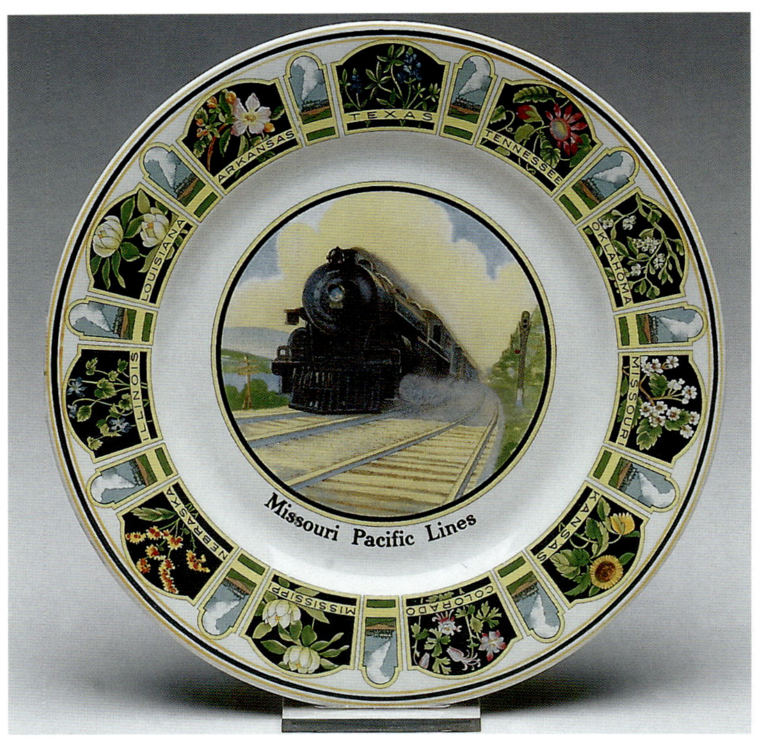

18. "The Sunshine Special." After a painting by William Harnden Foster. Service plate for Missouri-Pacific steam train showing official flowers for eleven states through which the train passed, 1930–49.

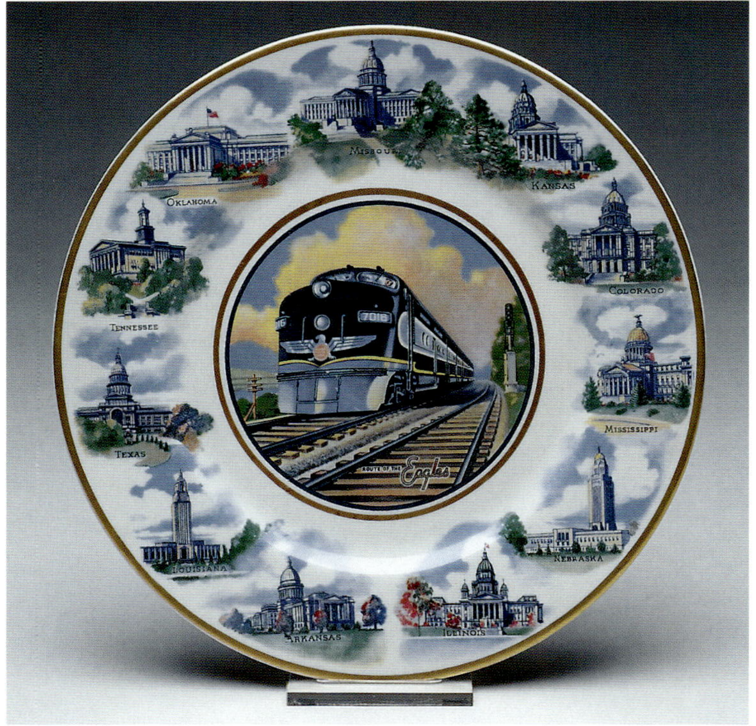

19. "Flight of the Eagle." Service plate for Missouri-Pacific diesel service, taken in the same location as earlier plate depicting steam engine. Rim designs depict Capitol buildings of eleven states through which the train passed, 1949–61.

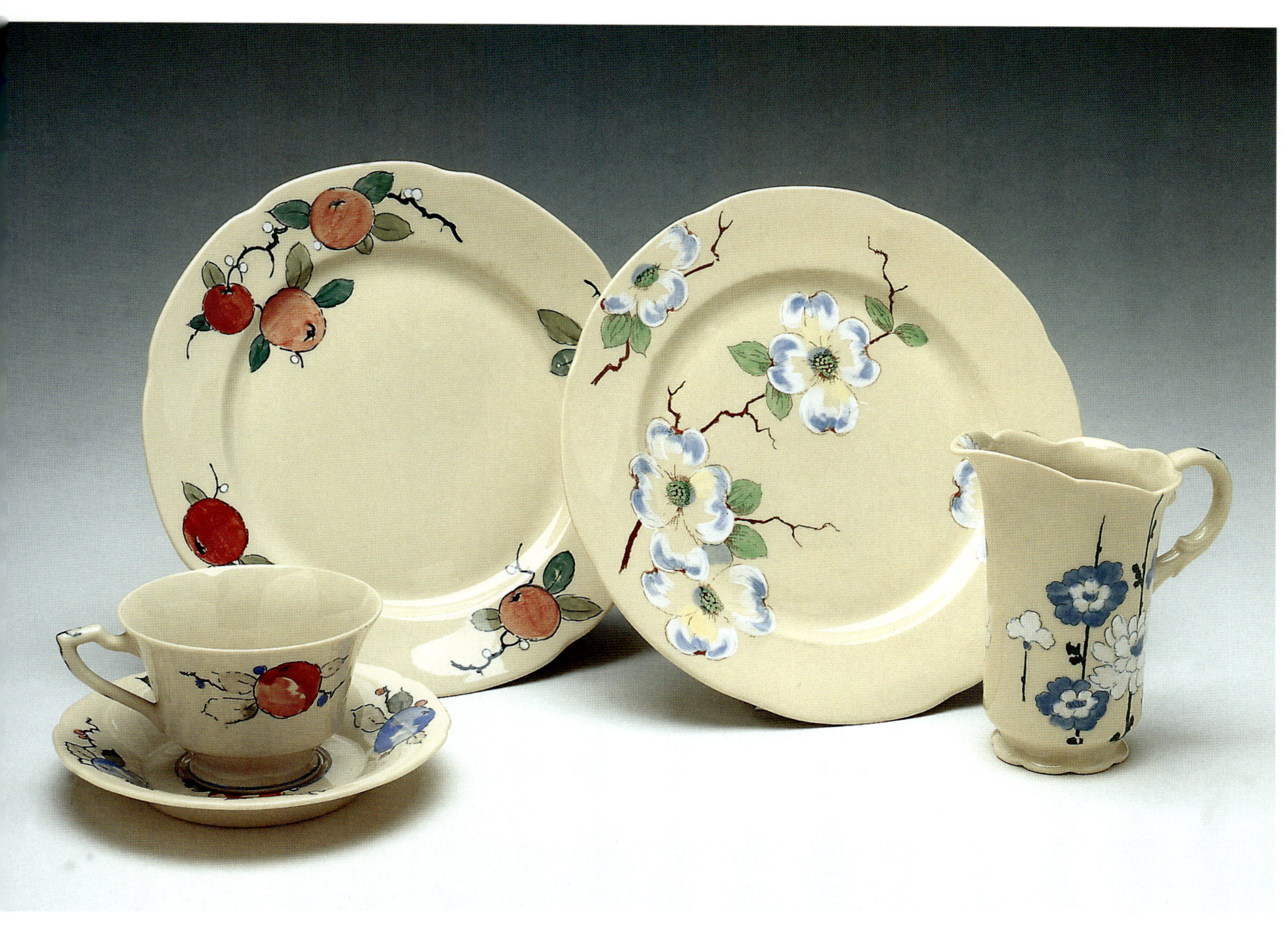

20. ***Governor Winthrop*** and ***Clinton*** shapes with hand-paint (Peasantry P-series) decorations on *Adobe* body. Shapes shared flatware but had different hollowware.

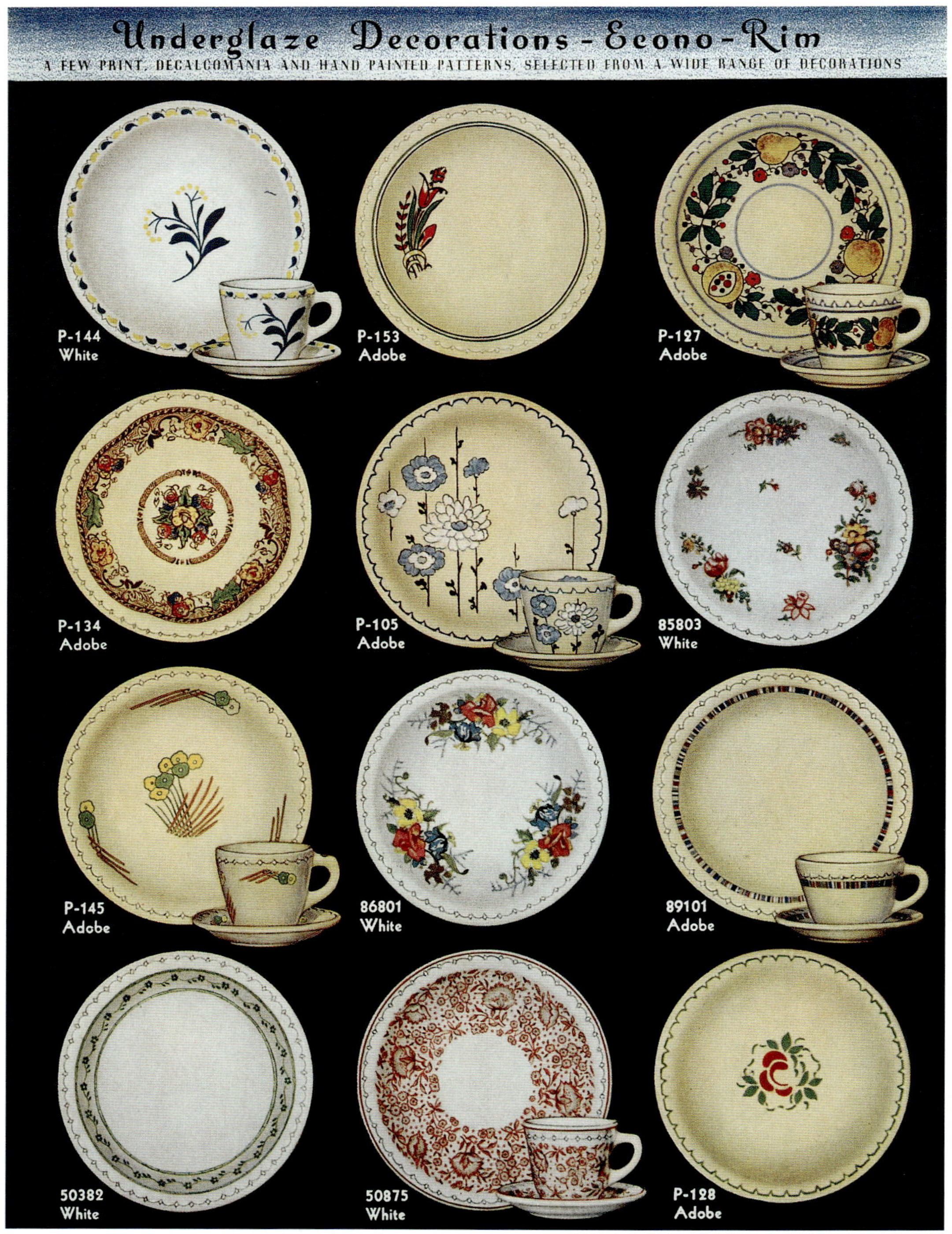

21. "Underglaze Decorations on Econo-rim." Page from Onondaga Pottery 1937 catalogue showing hand-paint (P-series), decalcomania, hand lining, and print decorations on commercial hotel ware.

16

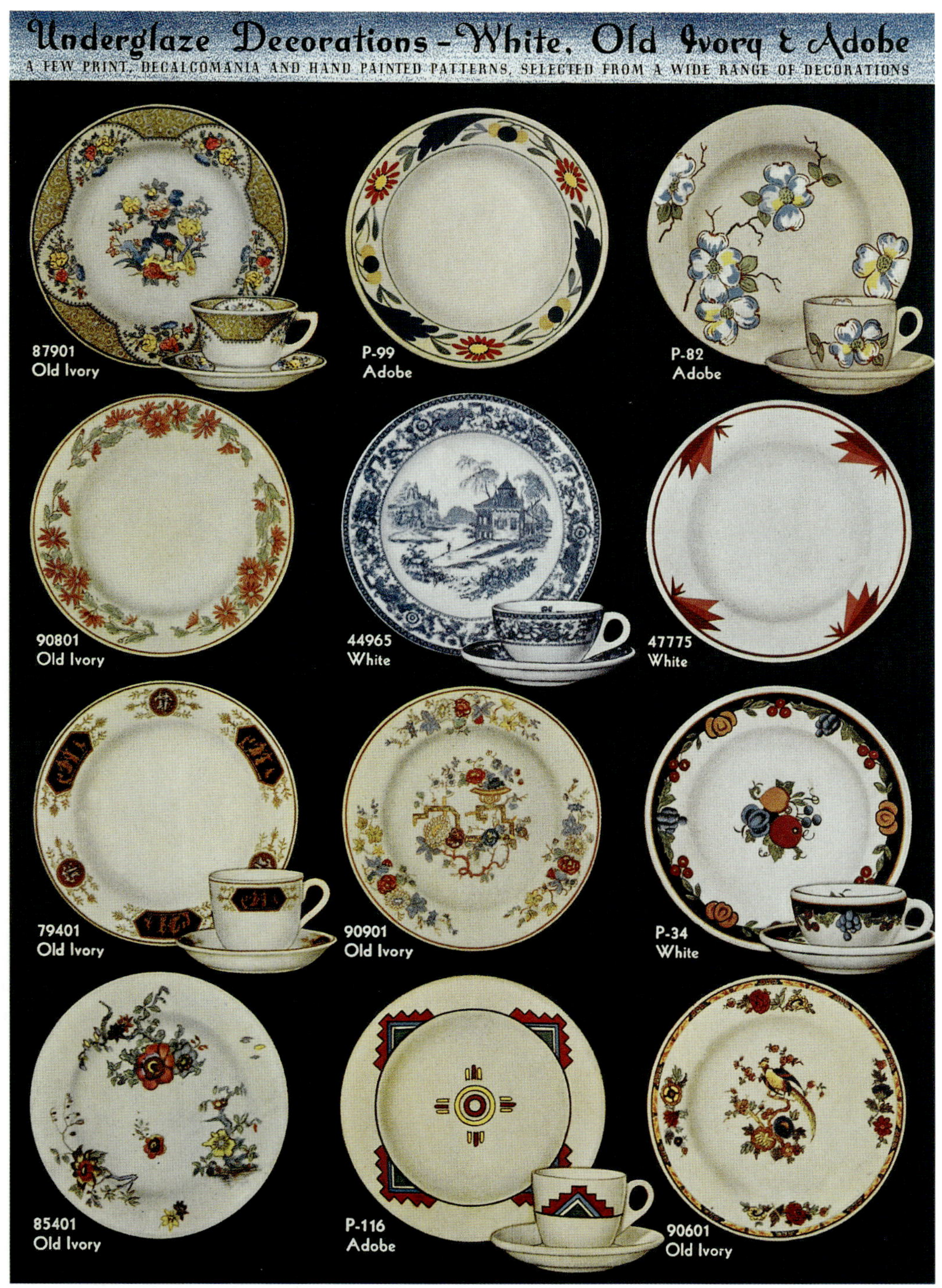

22. "Underglaze Decorations—White, Old Ivory, and Adobe." Page from Onondaga Pottery's 1937 catalogue showing suggestions for hand-paint, decalcomania, hand lining, and print decorations on commercial hotel ware.

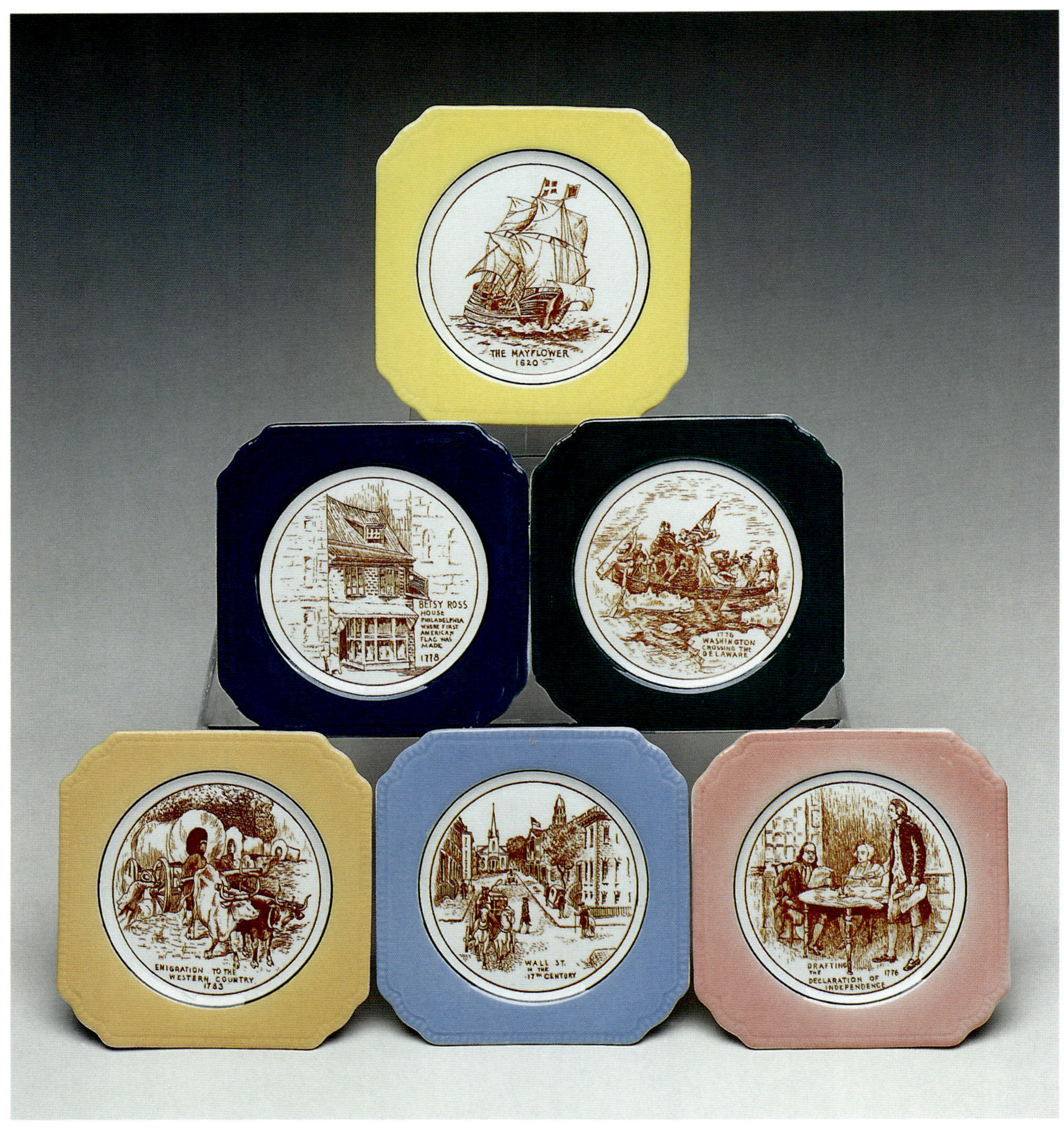

23. "American History" series, based on original prints. Series of six four-inch **Field** plates with *Vitritone* borders. *From top:* "The Mayflower 1620," "Betsy Ross House 1778," "Washington Crossing the Delaware 1776," "Emigration to the Western States 1783," "Wall Street in the 17th Century," and "Drafting the Declaration of Independence 1776."

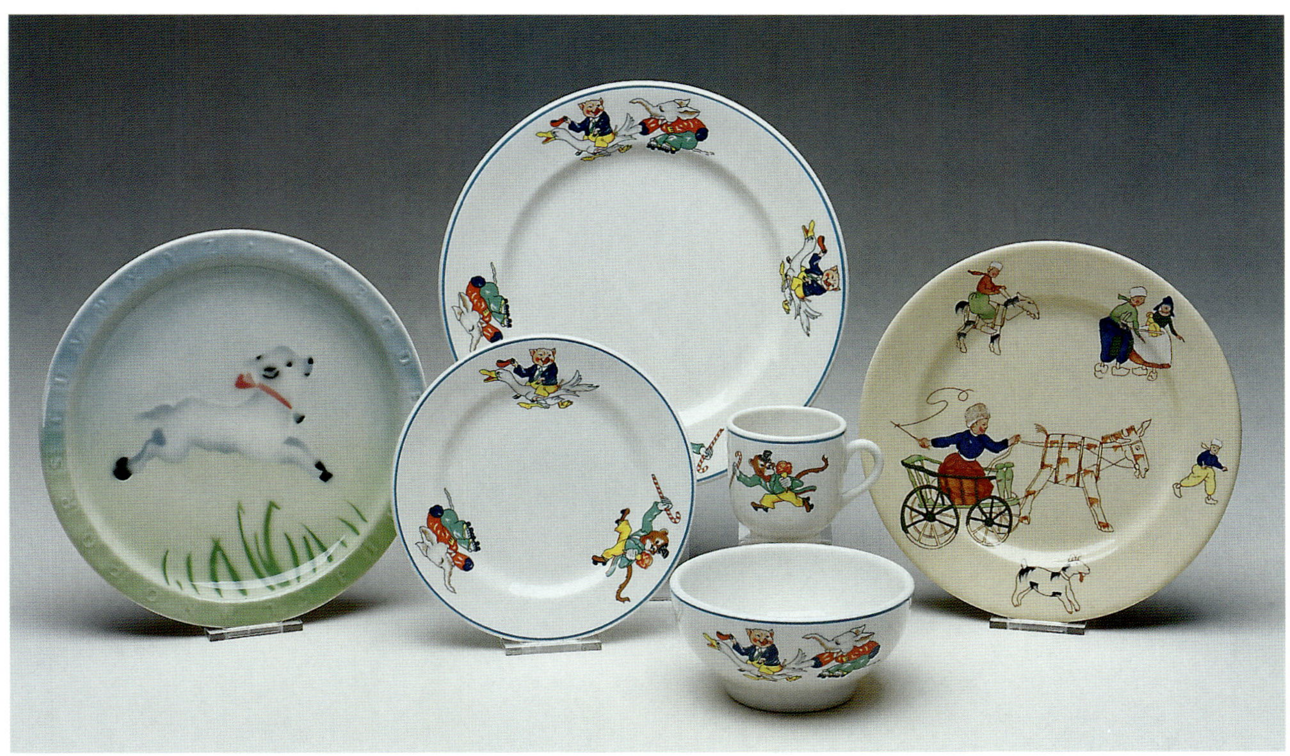

24. Examples of "Juvenile" ware. *Left to right:* "Lamb," Shadowtone pattern No. 114 on special ***Juvenile*** shape; 1994 reproduction of stock pattern used by the Great Northern Railroad in 1941; J-series pattern, 1930s.

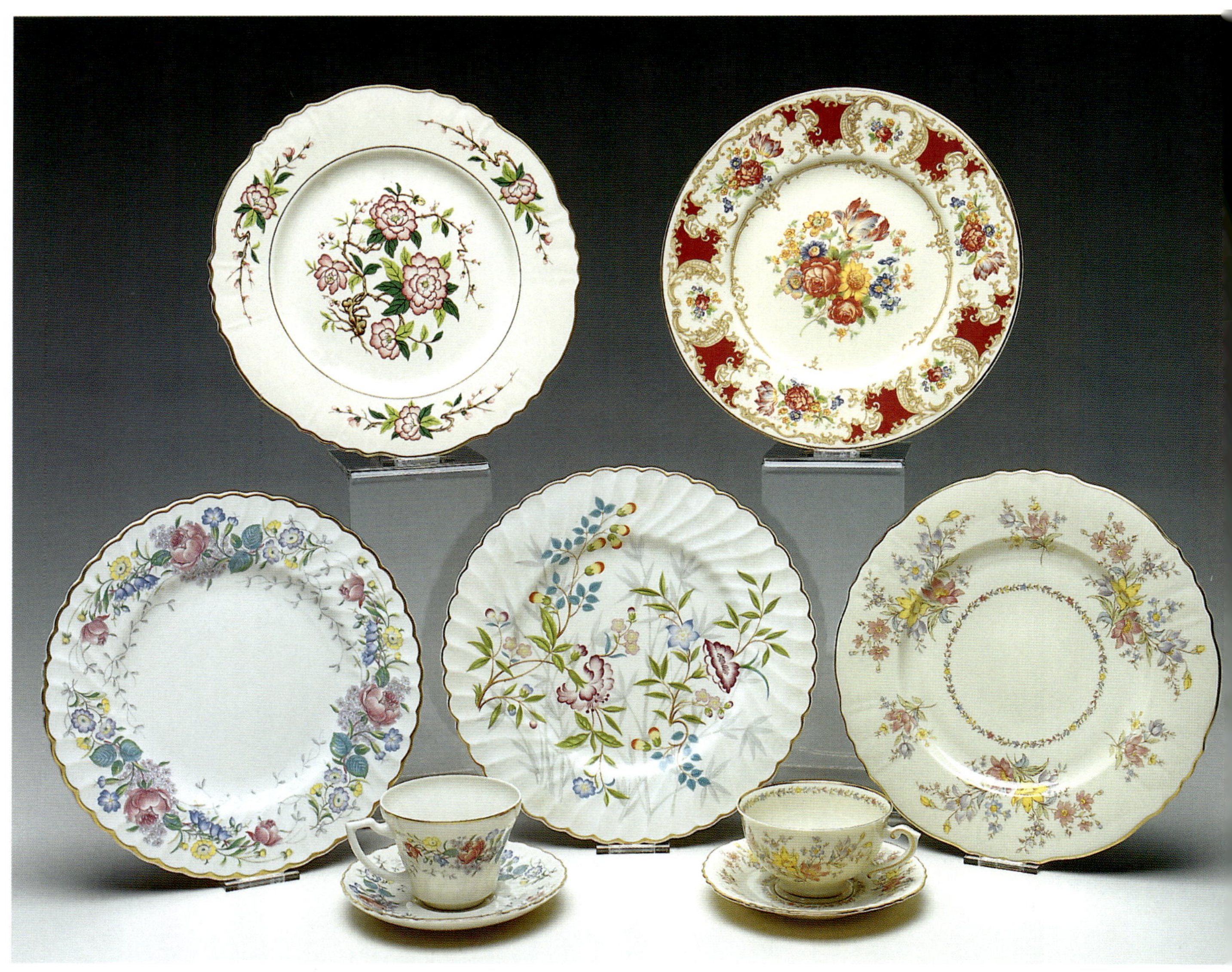

25. Examples of dinnerware shapes and patterns. *From top left:* "Arlington" on **Paul Revere;** "Maroon Romance" on **Virginia;** *from bottom left:* "Lilac Rose" and "Jewel Tree" on **Berkeley;** "Briarcliff" on **Federal.**

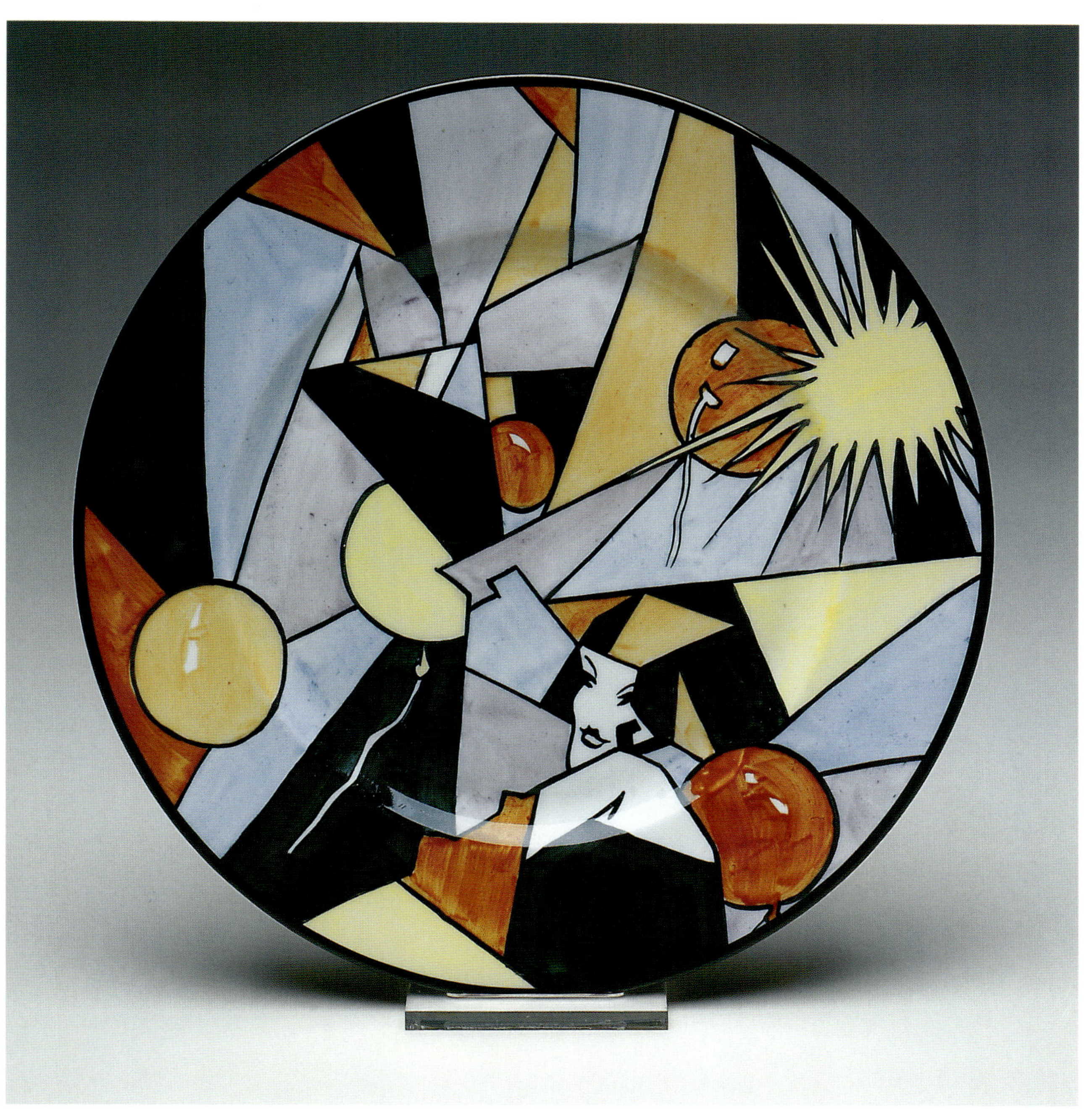

26. "Night Club." Unique jazz-era plate with underglaze decalcomania decoration.

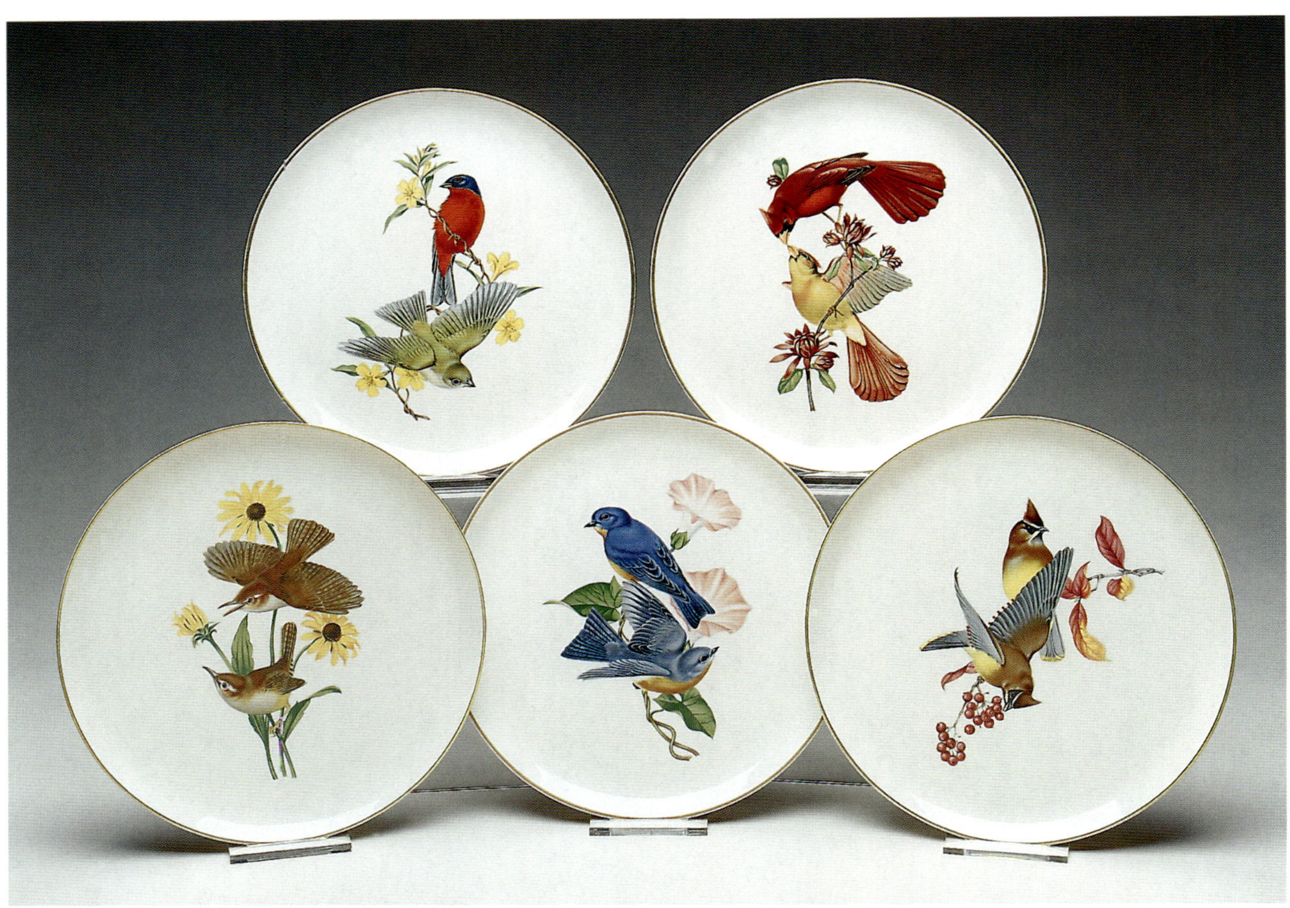

27. "American Song Birds," after original paintings by Athos Menaboni. Five of series of eight limited-edition plates for Cerebral Palsy School in Atlanta, Georgia. Overglaze decals on *California* dinner plates. "Painted Bunting with Carolina Jessamine," "Cardinal with Sweet Shrub," "Carolina Wren with Black-eyed Susan," "Eastern Bluebird with Hedge Bindweed," and "Cedar Waxwing with Red Chokeberry."

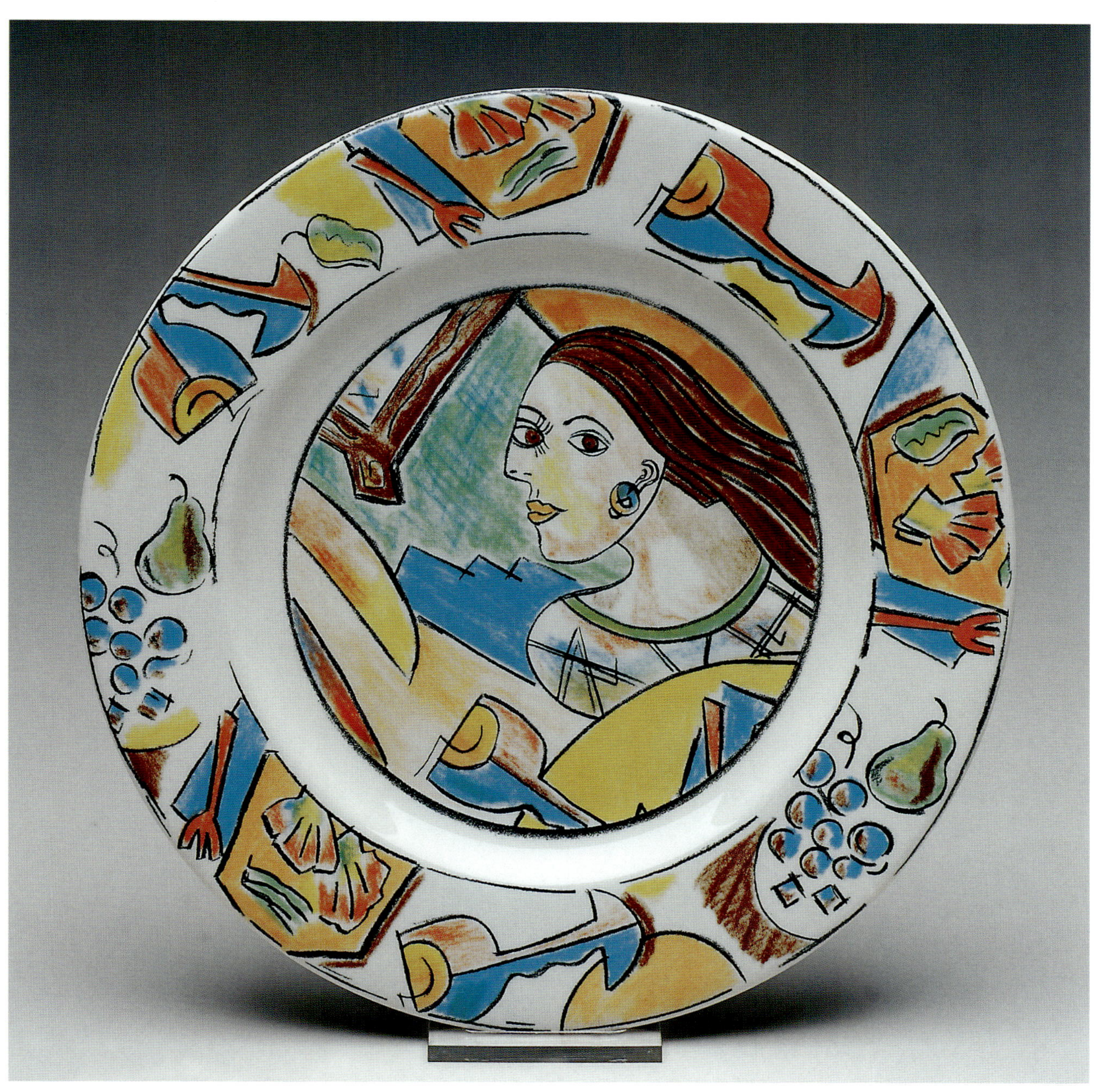

28. "Picasso." Prototype service plate based on a menu cover design by Associates Design in 1995.
Awarded Excellence in Design for Institutional Ceramic Tableware, 1996.

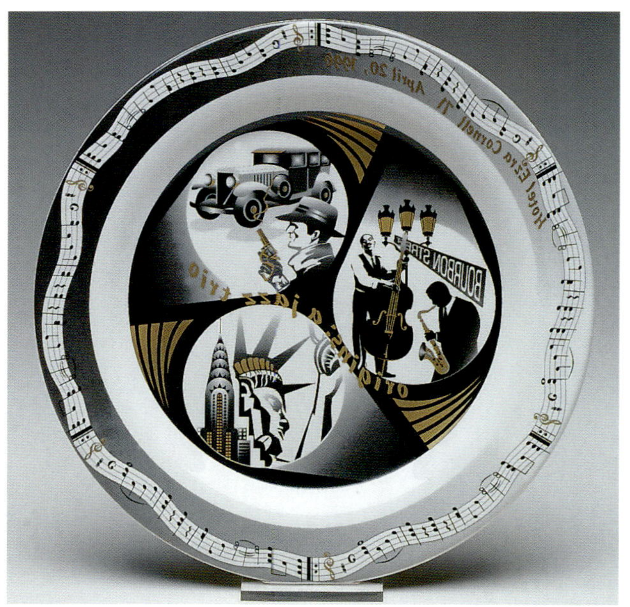

29. Service plate to commemorate seventy-fifth anniversary dinner at "Hotel Ezra Cornell," Cornell University School of Hotel Management, October 1996. The plate received the Award in Design Excellence for Institutional Ceramic Tableware from Society of Glass and Ceramic Decorators, 1996.

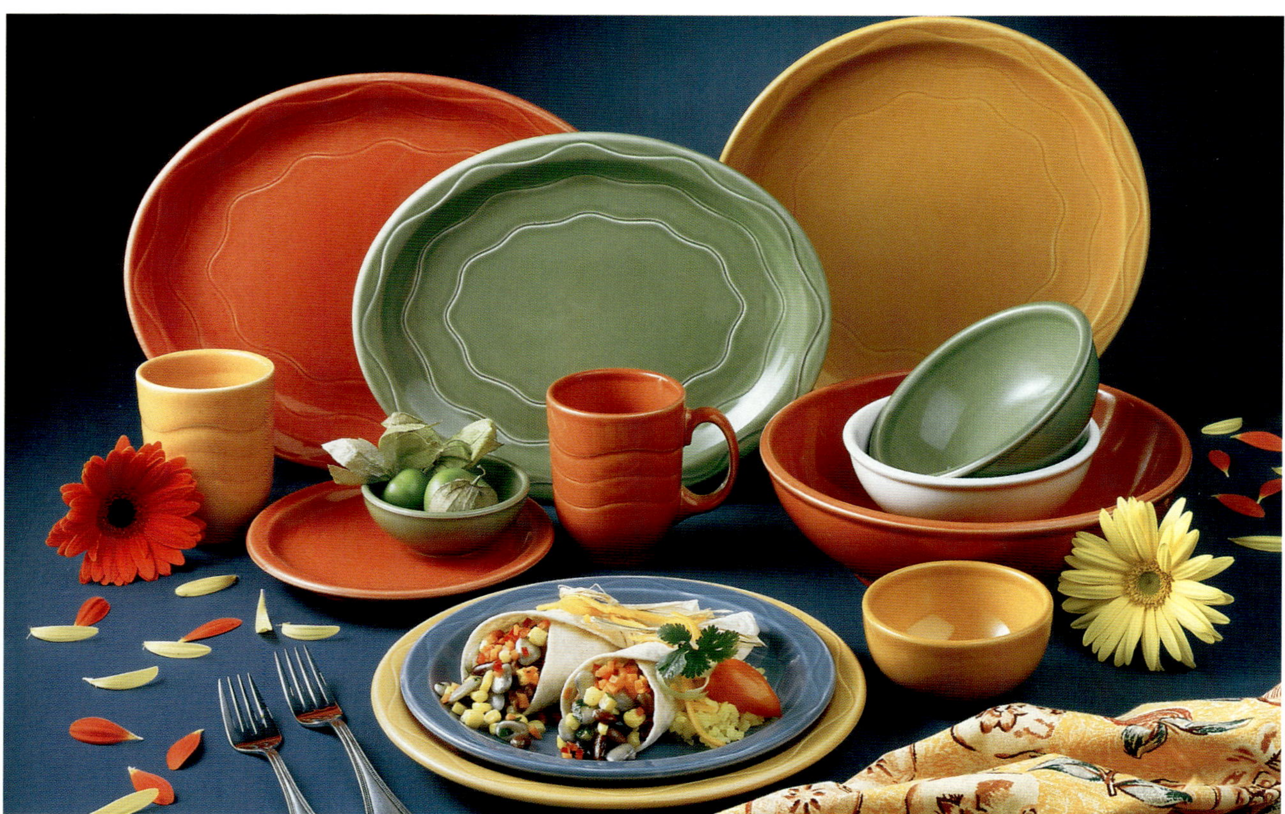

30. Lucie Wellner, **Cantina.** Casual service in four colors plus white, 1996. Photo by Hal Silverman Studio, Syracuse.

One

Onondaga Pottery Comes into Being
1871–1884

Turn, turn, my wheel!
  Turn round and round
Without a pause, without a sound:
    So spins the flying world away!
This clay, well mixed with marl and sand,
  Follows the motion of my hand;
for some must follow, and some command,
    Though all are made of clay!

—Henry Wadsworth Longfellow

IN 1857, in the canal village of Geddes on the western edge of Syracuse, New York, the potter Williams H. Farrar climbed to the roof of his pottery warehouse and secured to its peak a large painted copper jug. He meant this to be his new trade sign. Seventeen years earlier he had begun working as an artisan stoneware potter nearby on the Genesee Turnpike (later West Genesee Street), not far from the Erie Canal. Now he had moved his pottery to a new location on Furnace (later West Fayette) Street even closer to the canal. The jug that perched above his business announced that Farrar intended to produce a different type of ware from that which he had made before (fig. 1.1).[1]

The jug remained at that place for 113 years. It survived a disastrous fire in 1883 and then rose higher to surmount the brick factory buildings that the Onondaga Pottery Company erected on Farrar's site and the land around it. Where Farrar in his small facility had made ceramic ware chiefly for Syracuse and its environs, Onondaga Pottery made it for the nation.

In 1970, just before a wrecker's ball began to reduce Onondaga Pottery's West Fayette Street factory buildings to rubble, the jug was rescued and carried to Lyncourt on Syracuse's northern edge. There stood the seventeen-acre plant of the Syracuse China Corporation, as the Onondaga Pottery Company had been re-named. Video cameras whirred while the long arm of a fire department "snorkel" truck lifted William Salisbury, the corporation's president, high above the roof,

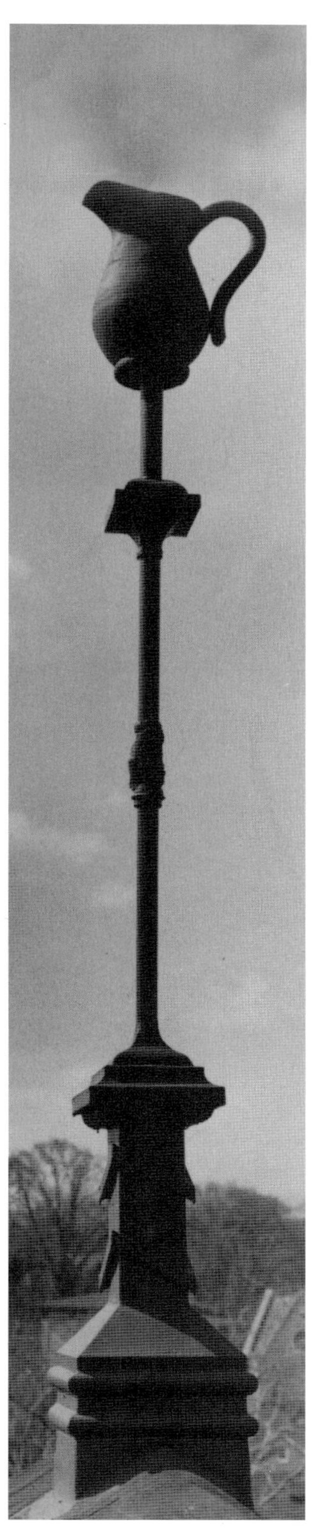

1.1. W. H. Farrar's jug atop the Syracuse China factory.

where he placed the jug atop a new pole. It has remained there ever since, a treasured symbol of pottery-making in Syracuse and of the history of one of the nation's leading manufacturers of ceramic tableware. That history embraces ceramics research, manufacturing technology, the arts of design and decoration, strategies of marketing and sales, labor and management relations, and the crucial contributions of a number of remarkable individuals, all influenced by ever-changing patterns in American dining habits.

Farrar had come to Geddes from Vermont in 1839. He bought property and established his pottery on it in 1841. He had learned his craft as a member of an extended family of stoneware potters active in northern New England and Canada. The rapid development of towns and villages along the Erie Canal drew him to Central New York. Even before the completion of the waterway across the state in 1825, fast-growing Syracuse had become a town full of opportunities for artisans and entrepreneurs. For a potter, the canal offered good transport for both raw materials and finished ware.

Like nearly all the many artisan potters of his time throughout the country, Farrar made pottery for a regional market. In Geddes, he produced wheel-thrown, salt-glazed, heavy, utilitarian stoneware: urns, crocks, mugs, bowls, pie plates, and jugs of the kind that since colonial times had been a staple of American households. A small work force helped him produce his ware. At his busiest he seems to have employed only six men, at an average monthly wage of thirty dollars. Still, as the industrial census for 1855 made clear, for a self-employed artisan, Farrar operated a substantial stoneware business. In that year, he used 225 tons of clay (from New Jersey) costing $1,350 and five hundred cords of wood ($1,500) in order to fire forty kiln loads of stoneware, which sold for $9,360.

By the late 1850s, however, popular taste increasingly demanded a finer ceramic tableware than this heavy and old-fashioned pottery. In 1857, when Farrar sold his Genesee Turnpike pottery to an employee (Joseph Shepard) and began again on Furnace Street, one of his motives seems to have been to leave the fading market for ordinary stoneware. He began to make various types of glazed red earthenware in fancier items such as molded figures of animals and spittoons with mottled yellow and brown "Rockingham" glazes. As indicated by his jug sign, he probably also made some inexpensive common tableware of yellowish earthenware.

In the 1850s, the rising demand for thinner, lighter-weight tableware for American homes was satisfied largely by imports, chiefly from England. This imported ware came not from artisan potters like Farrar who made ware from start to finish, but from factories whose workforces were divided into several skilled specialties, with each worker responsible for only one or a few stages in the production of a piece of ware. The wages of even the best-skilled English workers were well below those of American potters, and this kept the prices of the English imports low, despite a substantial protective tariff. England's industrialized potteries presented a challenge greater than Farrar or other individual American potters could meet. By the late 1850s there was little doubt that the future of ceramic tableware production in America lay in factory operations.

Farrar had spent his career as a potter remote from the place that in the 1850s became the nation's first center of factory ceramic production on the English model: Trenton, New Jersey.[2] Situated on the Delaware River, Trenton had good access to eastern Pennsylvania coal and New Jersey, Delaware, and Pennsylvania clay deposits. By the late 1860s, fourteen potteries operated in Trenton, some employing over a hundred workers. East Liverpool, Ohio, which had access to clay and coal via the Ohio River, soon became a second, multiple-pottery industrial center. Within a few years, the potteries of Trenton and East Liverpool accounted for nearly 50 percent of the tableware produced in America. Trenton mainly served the east. East Liverpool served the growing regions west of the Appalachians. Although a few potteries in both places made tableware of good quality by American production standards of the day, more turned out quantities of inexpensive pieces for ordinary dining. Some of it was cheap in appearance as well as price. Much Trenton ware was prone to "crazing." After a period of use, its glaze developed minute cracks, exposing the porous clay body underneath to discoloring food stains. Despite the presence of a few manufacturers of high quality ware, by the 1880s both cities had a reputation for cheap goods.[3]

Syracuse seemed a much less promising location for large-scale ceramic production. Coal and good clays needed to be brought in from great distances. Beginning in the 1850s, the arrival of railroads promised to ameliorate that problem somewhat, but even so, the thought that a pottery in Syracuse could compete with those in Trenton seemed a little presumptuous. Nevertheless, the idea took hold.

In 1868, Farrar sold his Furnace Street business to Peter Coykendall, a native of New Jersey who had come to Syracuse in the early 1850s. He has been described as a potter and may have been associated with Farrar. He owned the property on which Farrar's pottery stood. It seems clear that when he purchased Farrar's business he envisioned a larger operation of his own, one that would move the pottery closer to the kind of mass-production factory operation that prospered in Trenton.

Coykendall and a small group of backers set out to raise capital. On 4 August 1868, they announced in the *Syracuse Journal* that books would soon open for subscriptions to the amount of $75,000 in the capital stock of the Empire Crockery Manufacturing Company. Four days later, the *Journal* called attention to a publication titled *The Crisis,* perhaps a prospectus, which contained a stock subscription form and promotional text. The *Journal* reported that the text included "much matter of interest, tending to prove this to be the most lucrative branch of business in the country—the safest and best in all respects."

At this point, the projected Empire company was in actuality C. W. Coykendall and Company.[4] Lyman W. Clark, an able potter who managed the company's production, was one of the partners. He would also oversee the early production of Empire's successor, Onondaga Pottery. The capital raised probably amounted to much less than the desired $75,000, for Empire expanded Farrar's facilities only modestly. A view of his pottery drawn for a map of Geddes published around 1858 shows what amounts to a two-and-a-half-story building, converted from a dwelling, with a single kiln shed attached, and a small barn (the warehouse) nearby (fig. 1.2).

Despite its limited manufacturing facilities, Empire set itself in direct competition with the

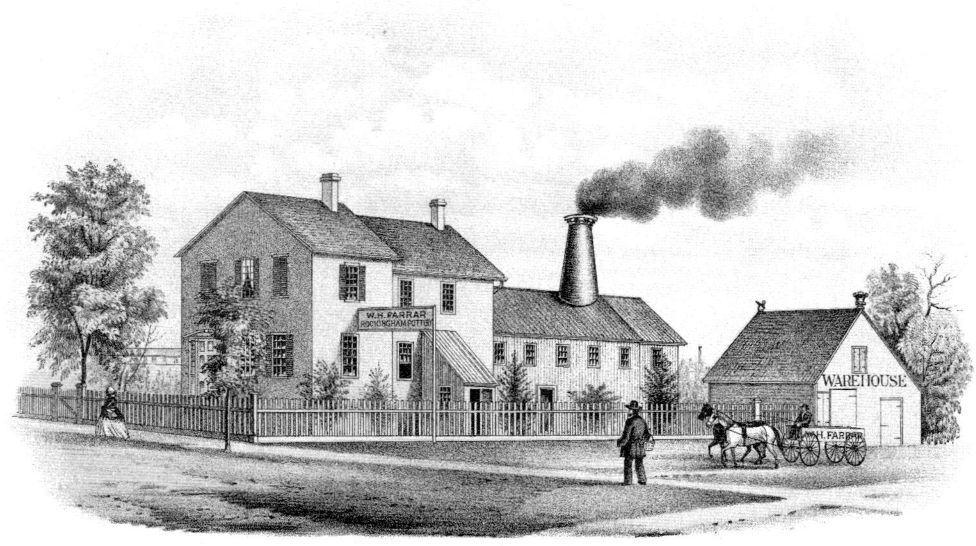

1.2. W. H. Farrar's pottery
in 1858, adapted from
a map of Geddes.

Trenton potteries. Under Clark's management, the company began to produce common "white-ware," a glazed, undecorated (with few exceptions), porous white earthenware of the kind whose popularity had risen steadily in American households in the years following the Civil War. By the late 1860s, several Trenton potteries specialized in relatively inexpensive whiteware that had begun to compete successfully with English imports. They had accomplished this not only by adopting the English division of labor into specialized skills but also by attracting from the English pottery towns skilled workers and managers well-versed in the industry's most advanced manufacturing techniques. These workers left their homeland, often with their families, to gain better wages, improved working conditions, greater opportunities, and a higher standard of living in America. To a large extent, the success of the American ceramic industry in the second half of the nineteenth century depended on this emigration of expertise from England. Empire's successor, the Onondaga Pottery Company, would also exploit this source. It would do so even as late as the turn of the century, when restrictions on immigration were tighter. But in the late 1860s, Clark, Coykendall, and their associates began to compete with Trenton using, so far as is known, local American workers.

In its short history, the Empire company met with mixed success. *Childs' Onondaga County Directory* assessed the new firm during the first months of its existence, reflecting Empire's aims more than its accomplishments. The report indicated that the company was "at present making $1600 worth of ware per month, but expect[s] soon to enlarge their works to make this amount per week. They make C.C. [cream-colored] ware, and various kinds of granite ware, pearl white, corn colored and decorated, all of superior quality, and far surpassing the same grades of foreign manufacture. Rockingham ware is also made by the same company."[5]

Empire apparently never backstamped a trademark on its ware. Omitting an identifying backstamp was a common practice in the industry in the 1860s, probably stemming from the belief that the public would resist buying American-made ware. An alternative practice was to imitate an English trademark. No identifiable products of Empire's kilns are known. In spite of its claim that it manufactured a variety of products, it apparently issued neither a catalogue nor a price list.

The company probably limited its output largely to common whiteware. The report of the 1870 census of Geddes says as much, describing the company's inventory as consisting chiefly of "white earthenware," valued at $25,000. The census report also states that the company's inventory included one hundred tons of New Jersey clay and fifty tons of North Carolina clay, as well as seventy-five tons of flint and fifty of feldspar. These are the major components of clay bodies and it is significant that, as was true also for the East Liverpool potteries, none came from local sources. The census reported that the Empire company had paid fourteen men, two women, and nine children a total of $7500 in annual wages. Without knowing how much each had earned, and for how much work, it is not possible to compare these wages with those of the Trenton potteries, but there was probably little difference.[6]

Tradition holds that the Empire company could not keep up with the demand for its products in the rapidly growing towns along the routes of the Erie and Oswego Canals. It needed more space, another kiln, and better machinery. But Coykendall and his partners found themselves without the capital needed to grow. They decided to sell.

On 1 July 1871, a group of sixteen men of Geddes with money to invest formed a new pottery from the old one. On 11 July, the *Syracuse Journal* boosted this development.

> Under the able management of Lyman Clark, a skillfully bred potter, the [Empire company] undertaking proved a success, and the manufactory turned out an article for everyday use second to none. . . . The reputation of the "Empire Ware" in the market is such that for the past year the demand for the ware has been far beyond the capacity of the works. Realizing the necessity for more extended facilities for manufacturing, and laboring under the disadvantage of the lack of capital, Messrs. Coykendall & Co. proposed to transfer the works and business to other hands possessed of capital sufficient to meet the requirements of a rapidly growing trade. A few of the young businessmen of Geddes took the matter in hand, and proposed a stock company with a capital of $50,000.

The prospects suggested by Empire's reported inability to satisfy the market for tableware "for everyday use" in upstate New York were apparently so attractive that some would-be investors had to be turned away. The sixteen subscribers, all related by blood, marriage, or business, and most of them from established Geddes families, met a week after the stock issue on 8 July. The minutes of that meeting recorded their purpose: "to organize for the purchase of the "Empire Pottery, . . . with its business and appurtenances, and to continue the same as a manufactory of white earthen ware." They named their new organization the Onondaga Pottery Company, commemorating both the local Iroquois tribe and the county in which Geddes and Syracuse were located.[7]

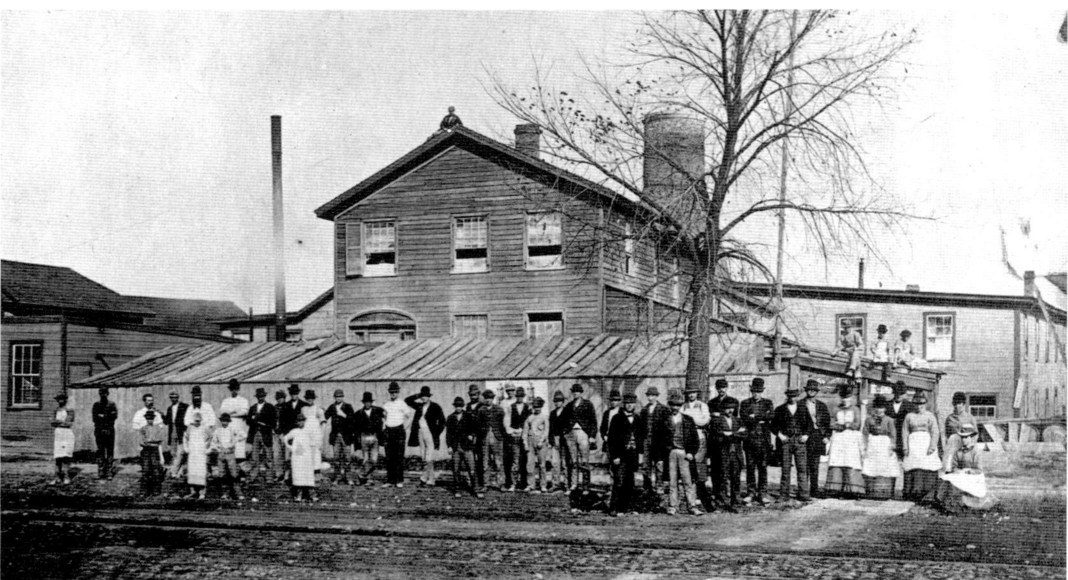

From their ranks the shareholders elected nine directors.[8] At least a few of the subscribers seem to have been withdrawing from the local salt industry, which had begun to decline at the end of the Civil War. The directors elected R. Nelson Gere as their president. He added this responsibility to his others as president of the Merchants National Bank, the Syracuse Iron Works, the Syracuse Gas and Light Company, and the Geddes and Syracuse Street Railway.

The directors set aside twenty shares of capital stock for the benefit of Lyman Clark, who became superintendent of manufacturing. He was to purchase it with acquired dividends, plus interest. The directors inventoried Empire Pottery's property and effects before deciding whether to purchase its finished or unfinished ware, or both. On 20 July 1871 they filed Articles of Incorporation with the State of New York and on that day, the Onondaga Pottery Company officially came into existence. It purchased Empire Pottery, lock, pots, and barrels, including all the inventory.[9]

A photograph taken about this time shows the employees of the newly organized Onondaga Pottery posed in front of Farrar's old pottery building. It was still intact and little-changed by Empire's three-year occupancy, but was now virtually surrounded by lean-tos and other extensions. A single kiln rises on the site (fig. 1.3). This was hardly a facility capable of large-scale production. Onondaga Pottery's cash ledger shows that in its very first week, the company began construction of a steel-and-brick shed to house a second kiln. The new kiln was operative by December.

Even by Trenton standards, the pottery was still a relatively primitive operation. Its crude facilities assured rough working conditions. An early employee, Joe Weiss, who began wedging clay as a youth, recalled that in the converted dwelling that Farrar and Empire had used, there was so little room that finished goods and materials had to be stored in a Methodist church nearby. On winter mornings, after lighting the heating stoves, the workmen's first job was to thaw the clay sufficiently so that they could mould it. The machinery consisted of one plate jigger and one cup and saucer jigger; other pieces of ware were hand-pressed in plaster moulds. Half a century later, in 1921, Weiss remembered of those earliest days and Syracuse's long winters that

> we worked with tallow candles after dark. . . . Everything was drawn by one horse. In those days there was no elevator, but there was a pulley attached to the rafters and there was a board . . . on which the clay was hoisted to the different floors. . . . In the winter when we put the pitchers under the bench before putting the handles on, we would find them frozen. We had to put them around the stove to thaw.[10]

As Onondaga Pottery prepared to manufacture its own ware, the Empire inventory kept old customers, mostly local merchants, supplied. To make new lines of a higher quality, Clark hired several potters, some of them English-trained. Five months into the operation, the pottery was making white granite ware. The *Syracuse Journal* reported on 19 December 1871 that "the goods manufactured are of two kinds, one known as the C.C., or common table and toilet ware, and the other as white granite, second only to china ware in strength and appearance." White granite and "CC" (cream-colored) were both forms of earthenware. They were the basic stock-in-trade of the tableware potteries in Trenton, East Liverpool, and elsewhere. It was relatively inexpensive, usually undecorated, and a far cry from "true" vitreous fine china.

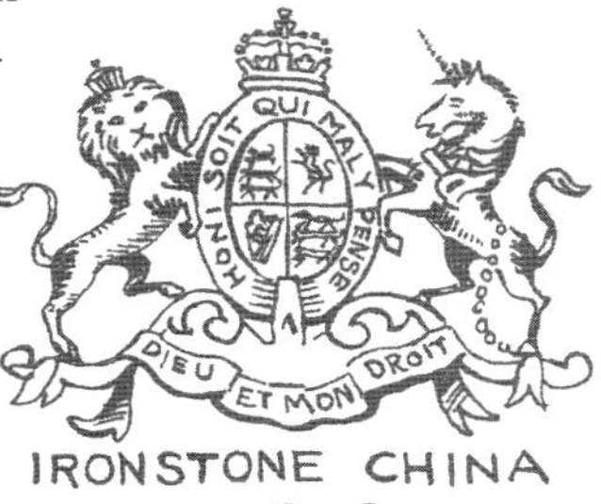

IRONSTONE CHINA
O. P. Co.

Lion and Unicorn Arms
of England (1871–73)

Although Empire Pottery had not marked its ware, Onondaga Pottery decided to do so, beginning late in 1871. Like most other ceramic manufacturers in America and Europe, it marked the underside of each piece of ware with a backstamp. In the trade these stamps typically consisted of a pictorial symbol that identified the manufacturer of the ware, applied to the bottom of most items with a rubber stamp after the biscuit firing and before the glaze was applied.

The pottery's new white granite ware got a nod of approval late in 1871 by receiving the company's first trademark backstamp. (CC ware went unmarked.) The backstamp consisted of the *Lion and Unicorn Arms of England* over the words "*Ironstone China/O.P.Co.*" (abbreviating the pottery's name to initials). Many American potteries used a variation of these arms. The arms encouraged the public to think that the ware came from England, or at least that it was "English" in appearance. This was not wholly a matter of deception, because the ware was made in the English manner by potters who were often English by birth and training. The ware was of an English type, and different from the heavier redware and yellowware many Americans recalled. Only a few examples of Onondaga Pottery's ware stamped with this trademark survive.[11]

One of the company's directors, Henry Case, served as the pottery's first general manager. In 1873, George W. Oliver took his place. A native of New Hampshire, he had worked in New York in the rubber and leather trade before moving to Syracuse in his midthirties to join the pottery.

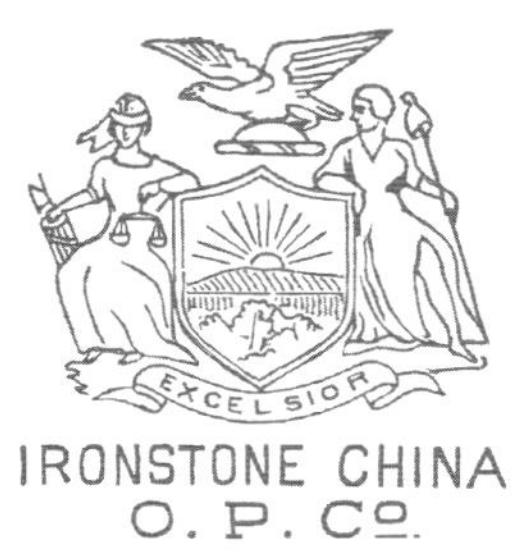

Great Seal of State of
New York (1873–97)

He ran the company's business affairs with aplomb for many years while Clark oversaw its manufacturing operations. Clark, whose tenure would be briefer, rapidly improved the ware, so much so that by 1873 the company felt compelled to adopt a new backstamp.

The pottery used the *Great Seal of the State of New York* as a backstamp to mark its new white granite ware. This mark also read "*Ironstone China/O.P.Co.*" The company already felt confident enough with this white granite to claim it as "a new whiter, stronger body." From 1873 on, the New York State seal marked all of its white granite ware.

In dropping reference to England in favor of a bold statement of New York State as its home, the company took an important step in promoting acceptance of American-made ceramic products. It was among the first, if not the very first, to do so. In addition, with this mark's clear reference to New York, the pottery turned its back not only on England, but also on Trenton. Five years later, while addressing the third annual meeting of the American Ceramic Society in 1878, Oliver spoke about the assertion of national identity through backstamp designs. He said:

> It is a well known fact that Americans distrust their own workmanship, especially in art, and American crockery is found bearing a foreign stamp . . . . Until within a few years, nearly all this crockery . . . bore the English trademark of the lion and unicorn. The trade demanded it, and the people would have it. The Geddes [Onondaga] pottery early adopted a distinctive American trademark, determined to live or die sailing under its own colors, and the result has justified the wisdom of this course. Now every manufacturer . . . uses an American stamp or trademark, one enthusiastic maker going so far as to use a stamp representing the British lion on his back, and the American eagle perched upon him picking his eyes out.[12]

At Onondaga Pottery (as in a few other potteries) the backstamp took on a function beyond asserting national and company identity. In steady succession over the years the company's new and altered backstamp designs marked newly introduced technological developments, not only in its ware's clay body composition, but also in other aspects of the product.

In marketing as well as backstamps, Onondaga Pottery differed from the industry mainstream. The Trenton whiteware potteries tended to sell their product largely through wholesale jobbers in New York and Philadelphia. Onondaga Pottery, because of its distance from the major East Coast jobbers, took a more active interest in marketing its own ware. In March 1873, Oliver hired a sales agent to go on the road with the product. This "commercial traveler," carrying suitcases heavy with samples, used the network of railroads that had recently been built throughout New York State and well beyond to reach his markets. In these early years, the pottery sold its ware directly to both individual merchants (usually proprietors of hardware and dry goods stores) and distributors as far west as Peoria, Illinois. For many years, the pottery had received its raw materials on the Erie Canal (see fig. 1.7). Now it shipped much of its ware by canal boat. Though this waterway was a slow means of transport, it was free of the shocks and shakings of rail travel. After 1905, when the Erie Canal ceased to operate through the center of Syracuse, railroads carried ware to more far-flung customers, and to closer ones as well.[13]

In 1876, the pottery celebrated the nation's centennial with a float displaying its best ware, some of it decorated (fig. 1.4). Although it did not show its products at the United States Centennial International Exposition of that year in Philadelphia, the pottery decorated some ware for local observances with images of the exposition's buildings and portraits of George and Martha Washington for the occasion (fig. 1.5).

The pride shown in its products was timely, for the pottery was now earning a profit. A year earlier, in 1875, the board had set Oliver's annual salary as general manager at a very considerable $2500 plus expenses. The directors then stipulated that he move his place of residence closer to the pottery. He demurred and resigned. The company urged him to stay and rescinded its residency requirement, and he

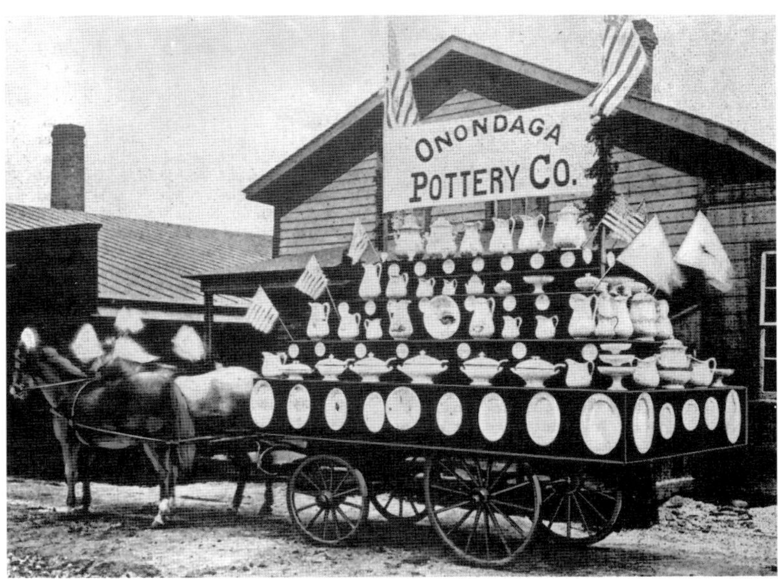

1.4. Onondaga Pottery's Centennial float for parade in Syracuse, 1876.

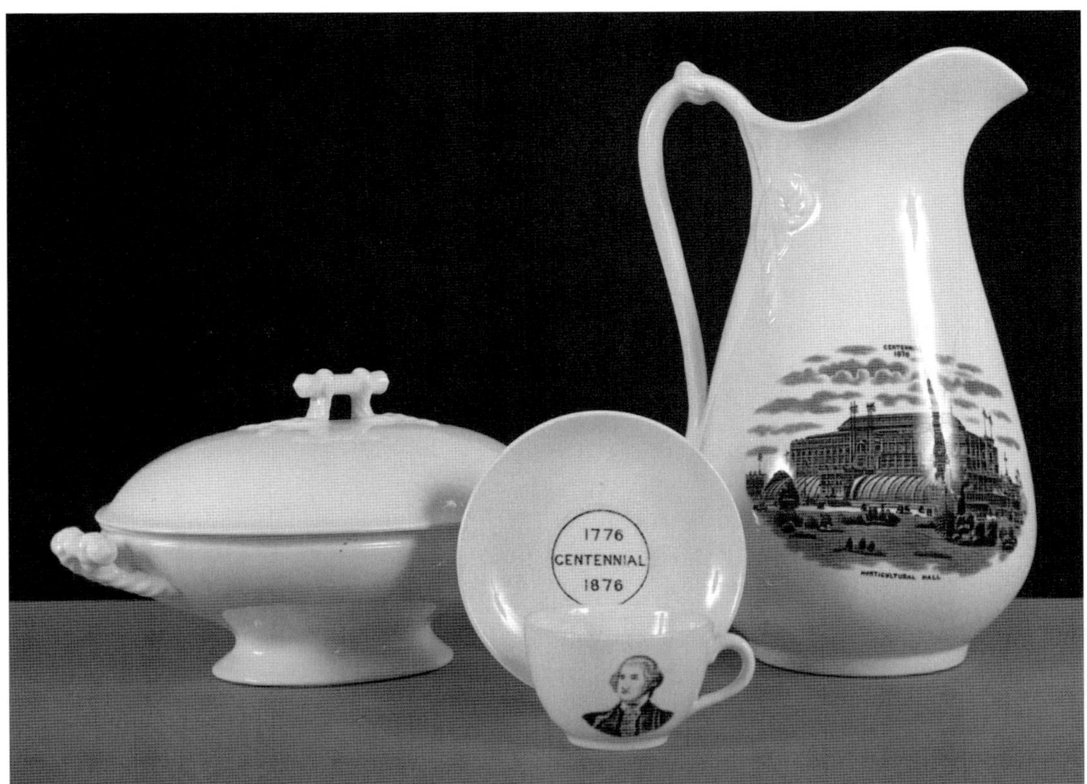

1.5. Centennial ware on *Cable* shape, 1876.

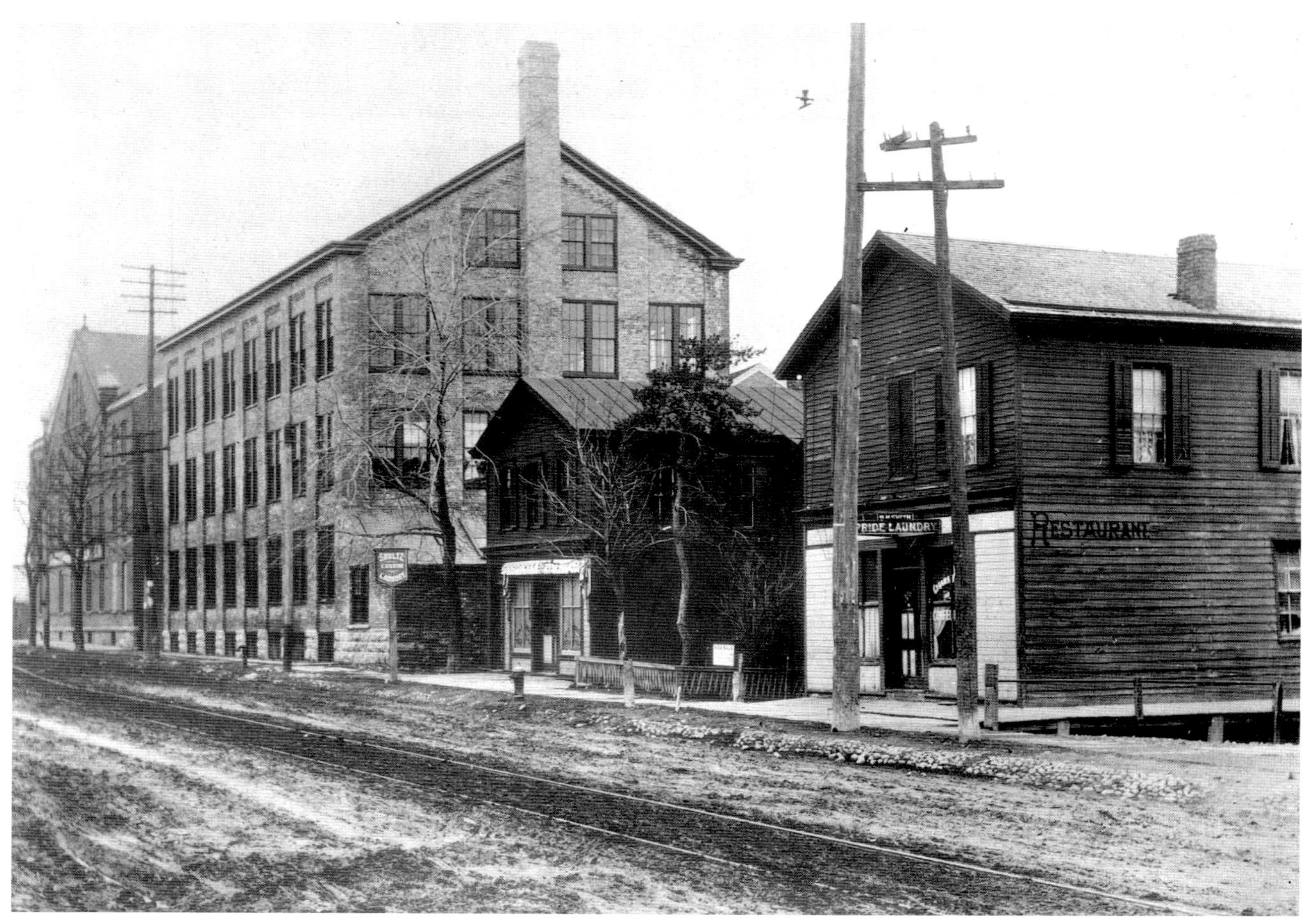

1.6. Fayette Street plant "B" unit, 1883.

remained another fourteen years. At the same time the company renewed Clark's appointment as superintendent at a salary of $1200, with the proviso that "he furnish the necessary information on instruction to the manufacturing, glazing, and turning of ware to a person to be designated by the president." This person proved to be Richard Pass (see fig. 12.1), a Staffordshire-trained potter, who replaced Clark in June. Clark had already announced his intention to leave the company as soon as his successor was in place. He went to Boston to start a business of his own.

Pass, who had been working in America since 1863, came from Trenton with a wealth of experience. His arrival at Onondaga Pottery did much to keep up the company's momentum and to improve its operations. Some months before he took on his new job, the national *Crockery and Glass Journal* in a report titled "Pottery in Syracuse" had noted in its 23 January 1875 number that the pottery employed sixty hands, mostly from Staffordshire, . . . all skilled in their various departments and among them a skillful designer who is constantly employed in producing new and elegant shapes. The works have nearly doubled their sales, having during the last year sold about as many goods as they did in the previous two years combined. The works have been enlarged by the addition of new buildings and a branch railroad has been constructed . . . for receiving and discharging freight.

The "new buildings" mentioned in this report amounted only to additions to the old Farrar/Empire facility. Five years later, in 1880, the company purchased two lots adjacent to this site for the considerable sum of $7500. In August of that year, Gere and Oliver visited potteries elsewhere in the country for ideas. In September, they began work on a $16,000 building that included a workshop, two kilns, an engine and boiler large enough to run the clay shop with steam, and a storehouse. Built in brick close to the canal, this three-story "B" unit reached completion in January 1883 (figs. 1.6 and 1.7). Its two new kilns made the pottery a substantial four-oven business. This advance was soon undercut, however. The new building had been in operation for less than a year when, in December, fire destroyed the old Farrar/Empire works.

Richard Pass had not lived to see the opening of the new facility. He died on 15 July 1880 at age fifty-six. Of Pass's several contributions to the success of the company, perhaps the most important in the long run was the introduction of his son, James, to the pottery's spirit of constant improvement. In time, after experience elsewhere, James returned to the pottery and led it to develop the first successful, factory-produced, American vitreous china. Richard began a

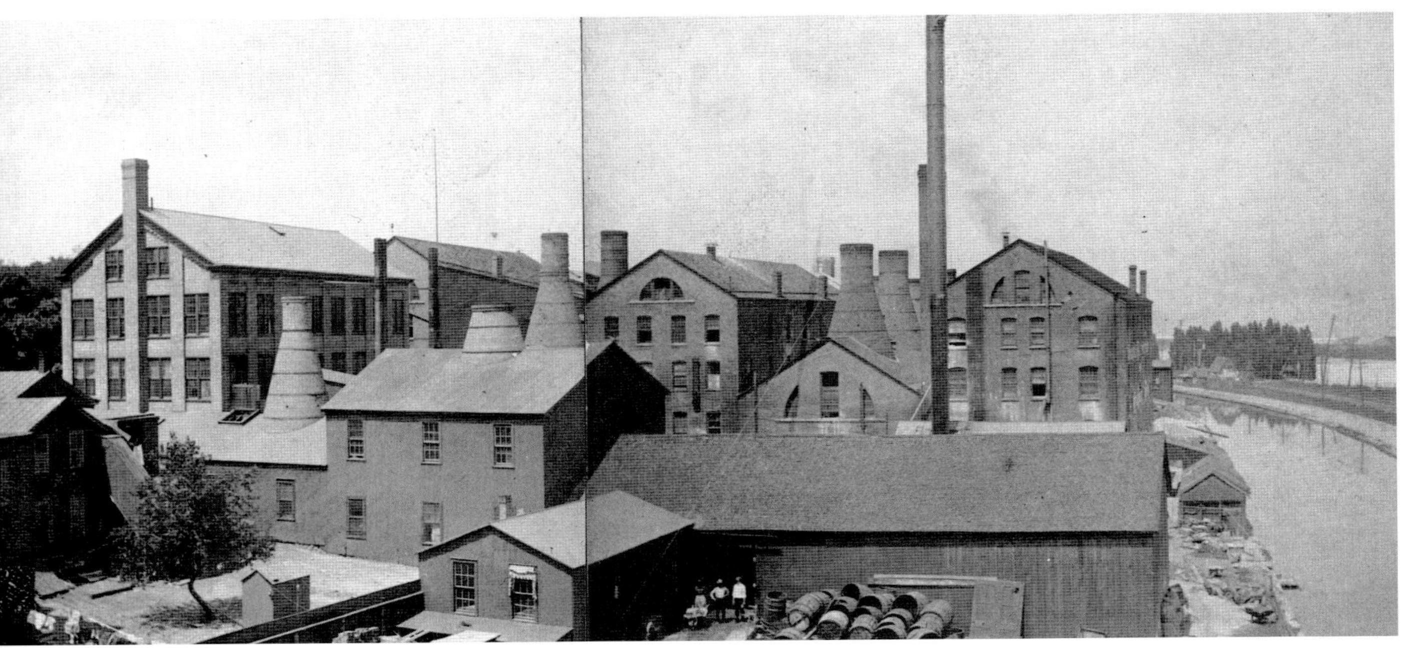

1.7. Canal side of pottery after 1890.

continuity of Passes that included his grandson Richard Pass and that carried the pottery to national distinction in the twentieth century.

At Richard Pass's death, the company promoted its modeler, Thomas H. Copeland, to take his place as superintendent of manufacturing. Copeland, an English-trained potter who had worked for the pottery since 1871, was in all likelihood the "skillful designer" mentioned in the *Crockery and Glass Journal.* He was the first in a line of uncommonly able shape designers who contributed highly successful designs to the company for more than a century.

In its first two decades, the Onondaga Pottery Company had limited its manufactures to a class of ceramics known in the trade as generalware. In the pottery's case this amounted to standard tableware and toilet sets. It left other ceramic products such as sanitary ware, door furniture (knobs and plates), druggist's ware, terra-cotta, tile, art pottery, and electrical insulation porcelain to specialist makers. Like all other makers of generalware, the new company judged the quality of its product by hardness and durability as well as by the attractiveness of its shape and decoration.

Pottery comes in four grades of hardness; they are, in increasing order: common earthenware (including redware), white earthenware, stoneware (Farrar's chief product), and, sharing the highest grade of hardness, true vitreous porcelain and china. The harder the ware, the more expensive it is to manufacture, because it requires purer raw materials, more fuel for higher temperature firing, and more carefully controlled conditions at each stage of making. In its earliest years, Onondaga Pottery concentrated on three grades of earthenware: CC ware, white granite ware, and improved earthenware.

The cheapest grade of earthenware was universally known as CC ware. The initials stood for "cream-colored" (also variously called corn-colored, common-colored, creamware, and, borrowing a name from Wedgwood in England, Queen's ware). This undecorated ware was used extensively for ordinary household service. After forming and firing, CC ware emerged from the kiln dense but still porous (nonvitreous), and opaque. Covered with transparent glazes, it varied a good deal from maker to maker in degree of color, finish, density, porosity, and weight. CC ware was only as good as its glaze. A glaze that did not "fit," that is, did not adhere thoroughly to the clay body during and after firing, would in time craze. The web of minute cracks in the glaze disfigured the surface even before it discolored the clay body underneath. Although Onondaga Pottery continued for some years to make CC table and toilet items, it never marked them with a backstamp. It sent this grade of ware to hardware and general stores and other outlets until the mid-1890s, though in continually diminishing amounts. CC sales provided income that allowed the development of better ware.

The next grade, called white granite at Onondaga Pottery (abbreviated to WG in the company's price lists), and also called "ironstone" in the industry, was a solid, serviceable ware with a bluish tint. Its creamy tone was masked by a blue stain mingled with the body clay. Because it was fired to a higher temperature, it was more durable than CC ware. Nonvitreous, it, too, depended on its glaze to provide an impermeable surface. White granite came to constitute the most widely used grade of ware in the United States during the last half of the nineteenth century. The quality of American granite ware in general was claimed to surpass that of common English white wares.[14] Onondaga Pottery's production of white granite toilet, dinner, and tea sets and other articles, both plain and decorated, accounted for a large part of its output during its first several years.[15]

The industry called the third grade of progressively improved, hardest-of-all, earthenware "semi-porcelain" or "semi-vitreous," though in fact it was neither porcelain nor vitreous. (It was, however, often "nearly vitreous.") These misnomers for this improved, hardest type earthenware persisted in the trade for many years, even at Onondaga Pottery. This ware might better have been called "improved earthenware," as it will be throughout the rest of this book.

The pottery made a constant effort to improve its clay bodies. All clay bodies are plastic when wet and then rocklike and permanent when fired. Only a clay body with ingredients precisely formulated to make true china or porcelain, and fired at the right high temperature, can harden, tighten and fuse its components into a glass-like, vitreous or "mature" ware. This ware has great compressive strength, denseness, and impermeability to liquids. Unlike earthenware, it possesses chemical inertness and insolubility, resistance to abrasion, and a very large and easily controlled variety of color and texture. To make vitreous ware successfully, a pottery needed the right clay bodies, the right glazes, the right kiln conditions, and the right balance of other things, none of them easily attained. Nonetheless, in the late 1880s, Onondaga Pottery would lead the industry toward a practical realization of a truly vitreous, translucent American ware.

The success of any pottery depended on the expert execution of every step in the manufacturing process. The best raw materials "batched" according to the best formulas, the arts of forming, firing, glazing, and decorating—all these things needed to work in concert to obtain good ware of whatever quality desired. An able superintendent needed to know all the steps, and like a good conductor of music, needed to lead his players to work together to realize an aesthetically pleasing product. By the late 1870s, Americans were becoming sensitive to the aesthetics of tableware.

In seeking an increasingly refined lifestyle during the mid–nineteenth century, many American middle- and working-class families took their lead from a number of books on etiquette and household management published in the decades following the Civil War. These books, including a standard, much-reprinted treatise by Catherine Beecher and her sister, Harriet Beecher Stowe, codified, among other aspects of domestic life, the rituals of "proper" dining, which included good manners at the table and the use of appropriate items of tableware. As the variety of foods available to households increased in the decades after the Civil War, and manners became more elaborate, so too did the array of tableware required for a "proper" meal, or even a tea. Eating as social behavior became transformed. By 1888, the word "eating" itself had been relegated to inelegant usage. In that year an American anthropologist made this clear when he said that brutes feed, the best barbarians eat, but only the cultured man dines. "Dining is no longer a meal," he added, "but an institution."[16]

This new propriety borrowed part of its form and display from upper-class and aristocratic table furnishings of eighteenth- and early nineteenth-century Europe, but it adapted them to America's egalitarian spirit. Industrial manufacture brought forth in quantity and at relatively

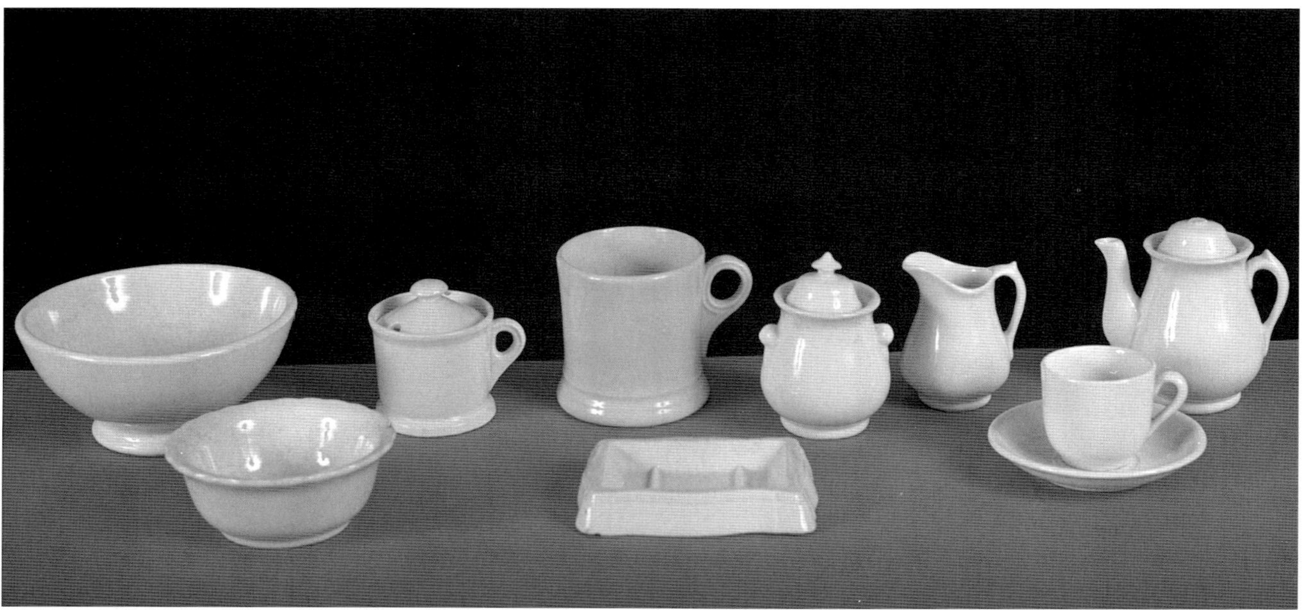

1.8. Examples of White Granite ware, c. 1873–97. *Left to right:* oyster bowl; scalloped nappie; handled covered mustard; coffee mug; soap dish; childrens' set: sugar, creamer, tea cup, saucer, and beverage pot.

low prices items of tableware of a kind that had formerly been rare and costly. In what has sometimes been called a "china craze," countless households began to acquire finer ware in a large variety of specialized pieces. This "good china" became a marker of social status, and it was not limited to the well-to-do middle class. Working-class families acquired elaborate place settings and accessories of relatively inexpensive ware.[17]

Dining in the nation's earliest years had for most people been a matter of eating from red-ware, yellowware, wooden bowls, and pewter plates. By the 1880s, however, "sets of china" (usu-ally earthenware) had become common. They included plates of several sizes; dishes (platters), also of different sizes; tea, coffee, and after-dinner (AD) cups with saucers; sauce boats; pickle dishes; oyster nappies (a kind of bowl); oatmeal bowls; cake plates; comports, soup tureens and ladles; bone dishes; fruit saucers; "ice creams" and punch bowls; shirred egg dishes; butter pats; tea sets; and sugar and cream sets, among numerous other items. All this was in a spirit of "more is better."

Americans of an earlier era would have been at a loss to name many of the items of ware popular in the 1880s or to explain their uses (fig. 1.8). This demand for variety and finer quality in household tableware propelled American generalware manufacturing. And finer ceramic ware, with equivalently fine glass and silver, called out for display. The purchase of "a set of china" often amounted to a once-in-a-lifetime event. Its owners kept their "good" service in safety (but for all to see) in glass-doored china cabinets, a furniture type that grew tremen-dously in popularity in the last decades of the century.

The increase over a generation in the variety of foods Americans consumed had also encouraged a proliferation of items of ceramic tableware. Home canning, which swept the country in the 1870s, made fruits and vegetables available year-round (in cooked form), rather than only during harvest season. Newly built networks of rail lines transported spring produce from the South to northern towns and cities still in winter. They carried to the populous East grain from the cultivated lands west of the Mississippi and beef from the vast grazing lands beyond.[18]

While the opening and spread of railroad routes westward in the generation after the Civil War helped to keep America's larder full, it also intensified the demand for a sturdier white ware. In fast-growing towns along the routes, new hotels, most of which operated on a room-and-board plan, opened dining rooms. Hotels began to vie with one another in securing durable and attractive service for their tables. They sought tableware that could withstand steady usage and the hard wear of hotel kitchens. Most household whiteware, both European and American made, had proved not durable enough. Hotels persisted in seeking better ware, and in increasing the variety of items needed to set a "proper" table. They constituted a new market for tableware, one that in the late 1890s would further propel the industry. They sparked a distinction that would grow between fine dinnerware for domestic use and sturdier ware for hotels and restaurants.

The tableware industry used a language of its own. The term "item" denoted a single category of ware, such as a cup, or a bowl. "Shape" meant a group of items that shared a common three-dimensional design. "Flatware" referred to items onto which foods were placed, such as plates and dishes; "hollowware" referred to deeper items into which foods went, such as jugs, cups, cream pitchers, and sugar bowls. The term "dish" meant oblong flatware: platters were called "dishes," oblong casseroles were called "covered dishes."

The most reliable guides to what shapes and items Onondaga Pottery manufactured in its early years are its published price lists. They show that it offered a large variety of items of both table and toilet ware and that it steadily added more shapes to its product line. Three price lists published before 1890 have survived to document the extent of this early earthenware production. In the first of these, the *Revised Price List* of 1879 (whose title leaves no doubt that there had been at least one earlier list, no copy of which seems to have survived), Onondaga Pottery offered fifty-six item categories of "White Granite, & C.C. Table and Toilet Ware." The CC ware was cheaper, but available only in a limited selection. Beginning in 1873, the pottery had given its first designated shape of distinctive design the name ***Cable,*** to distinguish it from more common ware. The name as well as the moulded cable ornamentation of the ware had originated in England (see fig. 1.5). Other American manufacturers had also borrowed the shape. Onondaga Pottery's ***Cable*** came in such items as butter dishes, casseroles, covered dishes, tureens for oysters, sauces, soups, jugs, sugar bowls, teapots, tea and coffee cups, chamber pots, ewers, and soap dishes. The company made ***Empire*** (named after its predecessor) and ***Cable*** items in CC as well as the better white granite ware.

To late twentieth-century eyes accustomed to a standard five-piece place setting with very few accessories, the number and kinds of items offered by these lists to grace tables seems excessive. One wonders whether there could really have been any regular use of bone dishes, comports, or seven plate sizes, but the answer to this questions seems to be: yes! Even at full tide, however, in the 1880s and 1890s, the American inclination for elaborate displays of tableware settings and accessories probably never reached quite the variety available to the well-to-do in Victorian England. There was something more pragmatic and democratic about the American attitude toward dining at home and in hotel dining rooms, an expectation that the various array of items ought not to differ a great deal from house to house, regardless of class. Like other American potteries, Onondaga made its best ware not for an exclusive, well-to-do customer, but for the tables of ordinary Americans.

The 1879 price list also offered fancier items made only in white granite, such as bread trays, individual butter plates, comports on footed pedestals, egg cups, ice cream dishes, pickle dishes, sauce boats, soup ladles, and toy tea sets for children "of sufficient size to use." It offered better quality toilet sets in white granite. A toilet set (see plate 2) typically consisted of large and small pitchers (called ewers), a wide basin (wash bowl), a mug, a brush jar, a covered chamber pot, a slop jar (to hold used water), and a soap dish with a drain, enough items to make toilet ware a significant part of generalware production in the industry. Toilet sets made up a significant part of the company's business in its early years, declining in importance around the turn of the century as modern bathrooms with running water, lavatories, and sinks increasingly became a standard feature of American homes and hotels. Until then, however, a set of toilet ware had been expected in nearly every middle-class bedroom, often arranged on a wash stand.

Items exclusive to CC ware were bed pans, bird baths and feeding cups (for bird cages), butter or molasses pails, soap slabs, pie plates, blanc mange or jelly molds, round cake cups, and a "tankard" jug. Some of these items had descended directly from traditional colonial American folk pottery. A good many of them would disappear from the company's list as the years passed.

Two other price lists published by the company carry no dates, but fall between 1879 and 1886 and differ in several ways from the 1879 list. They progressively dropped the number of categories offered in CC ware, from twenty-eight groups of items in 1879 to twenty-four, and finally to eighteen. They stopped the ***Empire*** line (except for a jug) and offered in its place a newly named shape: ***Square.*** The gradual phasing out of CC ware coincided with the company's successful introduction of its improved earthenware.

In its first dozen years, Onondaga Pottery had become a significant competitor of the older and larger potteries in New Jersey and Ohio. It had put its roots down in the community by building a substantial new factory building. Its ware had gained a reputation for consistent good quality. Having begun in 1871 with a single kiln, in 1883 it operated four. Onondaga Pottery, now earning a profit every year, had reached a point where it could set its sights higher and claim a place in the first rank of American tableware manufacturers.

Two

The James Pass Era Begins
1885–1890

His life was typical of the sentiment so well expressed by Emerson,
"Every great institution is the lengthening shadow of one man."

— E. L. Torbert, 1947

IN 1947, in an article written just after Onondaga Pottery's seventy-fifth anniversary, the trade magazine *Ceramic Industry* looked at the history of the firm and found in it three broad periods of development, each having lasted for approximately twenty-five years.

First there was the preliminary stage in which the company took over the doubtful assets of the Empire Pottery in 1871, gradually found itself through continuous experimentation, and arrived in about 1900 at the point where real expansion could be considered. Next came the stage of consolidation and extension of these gains, a period lasting roughly to 1921. . . . The third stage began with [a second] expansion and led . . . [to 1946].[1]

But another, less abstract, three-part division would be: the era of James Pass, what came before, and all that followed. This remarkable man did more than any other person to achieve national leadership for the company and to place it in the first rank of American fine china manufacturing. What he accomplished as a potter, practical scientist, manager, executive, and humane leader left so lasting an impact that for more than a generation after his death Onondaga Pottery still seemed to be, in many respects, "his" company.

Already a skilled potter at age nineteen, Pass had come to the pottery in 1875 when his father, Richard, succeeded Lyman Clark as superintendent. At age twenty, he worked as a foreman for two dollars a day, six days a week. This was a good wage for a young man when boys earned fifty cents a day and $2.75 was top for a modeler, one of the best paid jobs. The men who made the ware worked for piece rates, earning in a typical week amounts ranging widely from three to thirty-six dollars according to how much work was available. In the mid–1870s, the company's weekly payroll averaged about $450 for forty or so employees, at least four of whom were women. Pass soon mastered every job in the factory.

Convinced that he could deal best with the problems of ceramic manufacturing through the principles of scientific research, he enrolled in an evening course in analytical chemistry at Syracuse University. This began a lifetime of constant experimentation with clay and glaze formulas. In 1879, at age twenty-three, he left Syracuse to seek broader opportunities. He went first to Beaver Falls, Pennsylvania, to manage a pottery for the Economite Society. When the Mayer brothers took over its operation in 1881, he was no longer needed. A small inheritance following his father's death in 1880 allowed him in 1881 to invest in what turned out to be an ill-fated venture for making pottery near the clay deposits at Cape Girardeau, Missouri. He returned to Trenton, where he found work at one or more of the potteries, perhaps ending at Ott & Brewer Pottery. He apparently tried unsuccessfully to persuade the potteries to establish a laboratory for cooperative research in fundamental problems of ceramic manufacturing.

Looking for a man to take over Richard Pass's old post as superintendent of manufacturing of Onondaga Pottery, George Oliver consulted John Brewer, who recommended James Pass. Oliver hired him in June 1884. He already knew him, of course. Oliver encouraged Pass from the start, making research part of his job. This move accelerated the era of "continuous experimentation" mentioned by *Ceramic Industry*. It added, "Onondaga's star began to rise."

Having gained a broad range of practical experience, being blessed with an analytic mind, possessing natural, often inspiring, leadership skills, and proving himself indefatigable as a worker, Pass, at age twenty-eight, was one of the best-equipped men of his generation to run a pottery. To make the challenge more interesting, the pottery he now superintended was still a maverick. Far from Trenton, even farther from East Liverpool, in both of which places competing potteries stood shoulder to shoulder, Onondaga Pottery was rapidly becoming the most significant manufacturer of ceramic generalware in upstate New York and New England. Perhaps because of its isolation from the rest of the trade, it had already developed an innovative spirit, one that Pass intensified. The company's steady growth and its success for some decades —long after his death—remain inseparably linked to his leadership.

In 1921, Mark Haley, who had joined the pottery in 1891 following his graduation from Syracuse University's College of Fine Arts (where he had studied sculpture and modeling in clay with Hiram Gutsell), and who in a nearly forty-year career designed many shapes for the pottery, described Pass's approach to his job.

> Mr. Pass's days were rather full, with supervising all construction and repair, instructing the modeler and carpenter and later the glaze maker, managing the clay shop, preparing the clays and the glaze with Tom Newton in the slip cellar. He bossed the kilnmen and looked after the dippers, in several instances dipping whole kilns of ware himself, and then fired both biscuit and glost kilns. He found time to improve the gold mixtures for the decorators, construct a pugmill of his own design, and so thoroughly understood kiln building as to erect them with the assistance of local brick masons—conducting all of the operations of installation or replacement so that not a minute of working time was lost and making an occasional all-night session a matter of routine.[2]

A number of valuable recollections of Pass have survived (and some are quoted below and in succeeding chapters), but other evidence comes from the man himself. Between 1884 and 1913 he kept a set of pocket diaries. He used them not only as a schedule of events but also to document production statistics, record his experiments, and keep notes for his annual reports. For a while he jotted his personal expenses in them. The diaries shed much light on the factory's operations. As a day-by-day record of work, they also served as a repository for such "secret" matters as formulas for clay bodies, glazes, and dating codes. When he began as superintendent, these formulas were considered industry-wide to be personal rather than company property, though this would change by the end of his career. Pass's diaries are a unique body of documentation in the history of the American ceramic industry.[3]

In December 1883, six months before Pass assumed his duties as superintendent, the devastating fire at the old Farrar/Empire factory complex had left two badly damaged kilns and little else. Meeting in January, the company's directors voted to build two new kilns, salvaging as much as possible from the old ones. This would return the pottery to its four-kiln status. But they also considered whether they might better find a buyer for the whole property. They even considered organizing a Rockingham redware manufacturing operation at the site. The directors were unsure of their direction; Pass would find it for them.

One of his tasks in his new job was to assess loss from the fire. By January 1885 he had repaired the old kilns. Only then could he set to work to improve the company's clay bodies. He brought to this task the knowledge that he had gained from trials in mixing clays in Missouri and elsewhere. His diaries for the period show that he made changes in the CC and white granite body formulas almost weekly, substituting new materials and altering their proportions.

He also set himself the larger goal of developing an industrially viable true china body, not a superior earthenware but a vitreous, translucent, durable product with a glaze that would "fit" (not craze or crawl). This was no small order. The goal had been at the heart of the art of ceramics since ancient times and had been realized in Europe, but had remained unfulfilled in America's industrial potteries.[4] By 1886, he had introduced a new clay body that was not true china, but a highly superior, white, hard-fired improved earthenware. Because the term "china" in America had come to embrace any reasonably good quality ceramic tableware, even earthenware, Pass called his product "O.P.China." To distance it from the Trenton potteries' reputation for craze-prone products, Onondaga Pottery boldly guaranteed it "not to craze." Pass remained confident that he would in time develop a true china; O.P.China was a step in that direction.

To distinguish his improved earthenware from the pottery's white granite ware, Pass devised a new backstamp in 1885 bearing the pottery's initials, "O.P.Co." over the word "China," separated by a broken line. Used only on ware made with the new body, the "broken-line" O.P.China backstamp heralded a new era for the company. O.P.China did not, however, immediately supplant the cheaper CC and white granite ware. Pass's diaries and other company records show that he kept up with the continuing demand for CC goods until 1895 and for white granite until 1897.

In 1884, at about the time it hired Pass, the pottery began to put a new emphasis on decoration. Ware could be embellished in two ways: with moulded ornament and with applied surface decoration. Designs carved into a mould left impressions standing in relief as an integral part of the ware. Hand-applied surface decorations (over or under the glaze) included (alone or in combination) transfer-printing, hand-painting, banding, or lining. Industrywide, the surface decoration of most American generalware up to the 1890s was at best competent. Most often it amounted to little more than crudely executed, unimaginative adaptations of English, French, and German decorations.

O.P./China (1885–90)

Surface decoration was meant to be not only attractive in itself but also to moderate the stark whiteness of most white ware. As the art critic Clarence Cook wrote in 1878, white ware "took the yellow out of butter, made the milk look blue, cast suspicion on the tea, [and] took all the sparkle out of the sugar.[5] Company records indicate that as early as 1876 the pottery had arranged for the surface decoration of some special ware, but most of its decorations were of the moulded type. That the company believed more needed to be done in the realm of surface decoration is clear from a stockholders' meeting in March 1883 at which Mills P. Pharis was "appointed to put the decorating shop in suitable condition to rent." Just what the decorating shop may have been remains uncertain.

Then, in August 1884, a young ceramic decorator, Elmer Walter, left Boston and traveled in New York along the Erie Canal as far west as Buffalo looking for a place to start his own shop. He settled on Syracuse with its fast-growing Onondaga Pottery. His Boston China Decorating Works came into being on West Fayette Street directly across from the pottery in a rented building to which he added a wood-burning muffle kiln. This placed him but a short walk away from his major customer. He hired a printer and trained a few young women to transfer and finish the decorations.

Walter had brought with him expertise in transfer printing. This process, in which an impression of a copper-plate engraving was transferred to the ware, developed in England in the 1750s and revolutionized applied decoration in the pottery industry. In the 1880s, decorations of this kind tended toward floral motifs, usually printed in black or brown, to which color was applied by hand. Walter and his workers also hand-lined or banded ware with colors and sometimes embellished it with gold. Pottery workers transported ware to and from his shop across Fayette Street. Walter was free to decorate ware from other potteries, but the great bulk of his business, if not all, came from his close neighbor. This arrangement of an independent ceramic decorating shop serving one or more potteries was already well established in New Jersey and Ohio.[6]

In March 1886, disaster struck when fire destroyed the building that housed Walter's business. Only the kiln remained standing. Oliver had already begun to think about setting up a decorating shop in the pottery to eliminate the carting of ware back and forth across the street. Following the fire, Oliver offered Walter shop space on the pottery's third floor. Walter moved in June 1886, becoming a company employee and making Onondaga Pottery one of the first in America to operate its own in-house decorating shop.[7]

In 1921, looking back over many years, Walter recalled that he had devoted the first year of his operation to decorating white granite toilet sets in the *Syracuse* shape. He then began decorating "dinner ware, of the old square shape." As production increased, Walter supervised a small staff of decorators, most of them women. As the department grew, he relinquished his leadership to others and returned to decorating ware himself, with notable achievements.

The constant experimentation with formulas for clay bodies and glazes, not only at Onondaga Pottery but elsewhere in the industry, had made it increasingly important for purposes of quality control to know when a piece of ware had been made. In August 1886, Pass initiated a system for imprinting ware with a date code. He adopted a series of four separate alphabets, run in sequence, followed by three sets of numerals to indicate to the month when an item had been made. (The first letter was a Roman capital, i.e., an "A" for August 1886, "B" for September 1886, etc.; See appendix A for list). Workers imprinted the date codes on the bottoms of the ware while it was still wet. With an adjacent trade mark backstamp, this meant that most of Onondaga Pottery's O.P.China and white granite ware carried two discrete but related bits of information on its underside, one indecipherable by the public and the other seemingly only a trade mark, though to a knowing observer it identified the piece's clay body and sometimes more.

In 1887, the company introduced the **Doris** shape, offering its first full line of earthenware table service in the O.P.China body (fig. 2.1). The customer could now have good quality table-

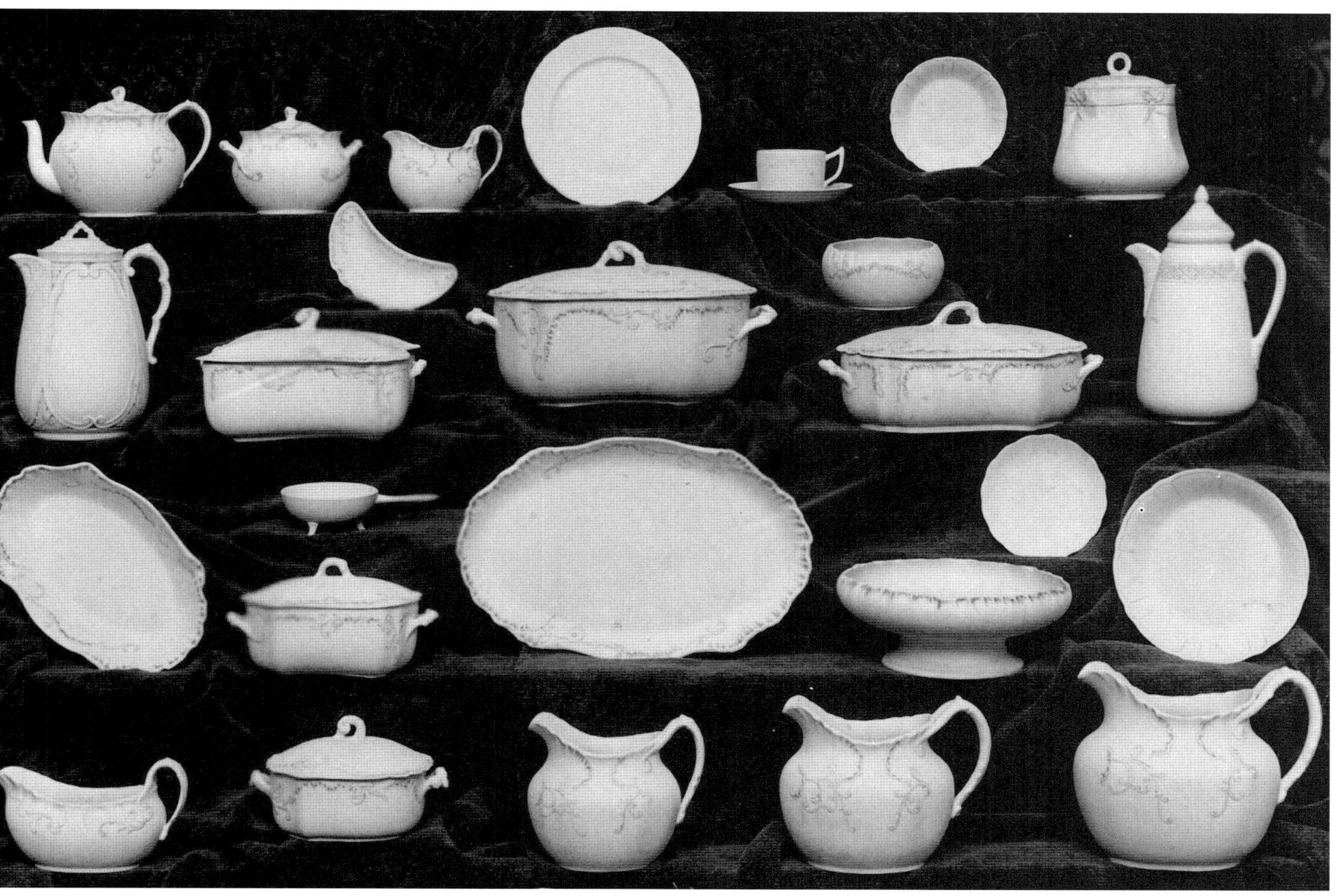

2.1. "**Doris** Shape in the Celebrated O.P. China. This is a ware designed to fill the demand for something between Earthenware and China. It is made very light, the body is very dense and hard, the hollow-ware pieces are translucent. Decorations in the best style." Caption for catalogue photograph, 1893.

ware, with plates, cups and saucers, and other items harmonious in style with casseroles and jugs. Either plain or decorated, often embellished with gold, **Doris** sets were made up from a selection of over fifty items.

O.P./China (1890–95)

In 1890, five years after his introduction of O.P.China, Pass made such a significant change in the formulation of its body that he modified its backstamp, substituting a solid line for the broken line to distinguish this superior product. The clay body was even harder, still more impervious to stains, and state-of-the-art within the industry, though not yet vitreous china. When the company began to advertise this product in trade journals, business picked up considerably. "We are working to capacity," Oliver wrote to *American Pottery and Glassware Reporter* in 1889, adding in his best promotional style,

> The Onondaga Pottery Company [workers] . . . are making a line of goods second to none in this country. The most painstaking care is evident in every department . . . and nothing but the best materials that money will buy can be found about their works. Their goods are standard wherever known. For several weeks past they have run their decorating department nights to keep pace with their orders for decorated goods. Their decorations are very fine, tasty and artistic, the colors harmonious and the designs pleasing. This company contemplates putting up two more kilns in the near future.[8]

Pass and Oliver had also determined that the company needed a new toilet ware shape. In 1887 the pottery brought out a short-lived Grecian-styled, unnamed shape (fig. 2.2). The design apparently went too far in the direction of formality and simplicity, for it sold poorly. The compact form of this relatively unadorned classical-revival design gave it less tendency to break. Although it offered an alternative to Victorian ornateness, a well-entrenched preference for overadornment still dominated the market. Not until the early twentieth century, when the Arts and Crafts movement and early modernism began to advocate greater simplicity, would something so unadorned have an appeal. The simple "Grecian" shape came too soon, and was dropped.

In 1888, the pottery brought out its **Juno** shape, and it had a very different reception. This was a richly embellished design for both table and toilet ware, made from the *O.P.China* body (fig. 2.3; See also plate 2). Six years earlier, in 1882, the company had begun to run a weekly unillustrated advertisement in *American Pottery and Glassware Reporter* for "China, White Granite, C.C. & Decorated Ware." The **Juno** shape required more. On 21 November 1889, that journal observed, "The Onondaga Pottery Co. are getting there with both feet in the way of trade and have orders in large number and volume. They have just brought out a new toilet set, which they have named "the Juno," and it does credit to the name, being a beauty in every respect." Soon after, on 12 December 1889, the *Reporter* published the company's first illustrated advertisement to the trade, introducing the "new 'Juno' toilet set." Its caption read: "Send order for some decorated sets. Our line of tableware, both plain and decorated in O.P.China, is one of the most desirable in the world" (fig. 2.4). The company waited until 1892 to publish its first *O.P.China Price List*. In this list it illustrated its solid-line trademark, and included as tableware shapes **Fork Handle Square, Doris,** and **Juno,** and as toilet ware, **Juno** and **Oneida**. It claimed for its product: "warranted not to craze, all hollow pieces translucent."

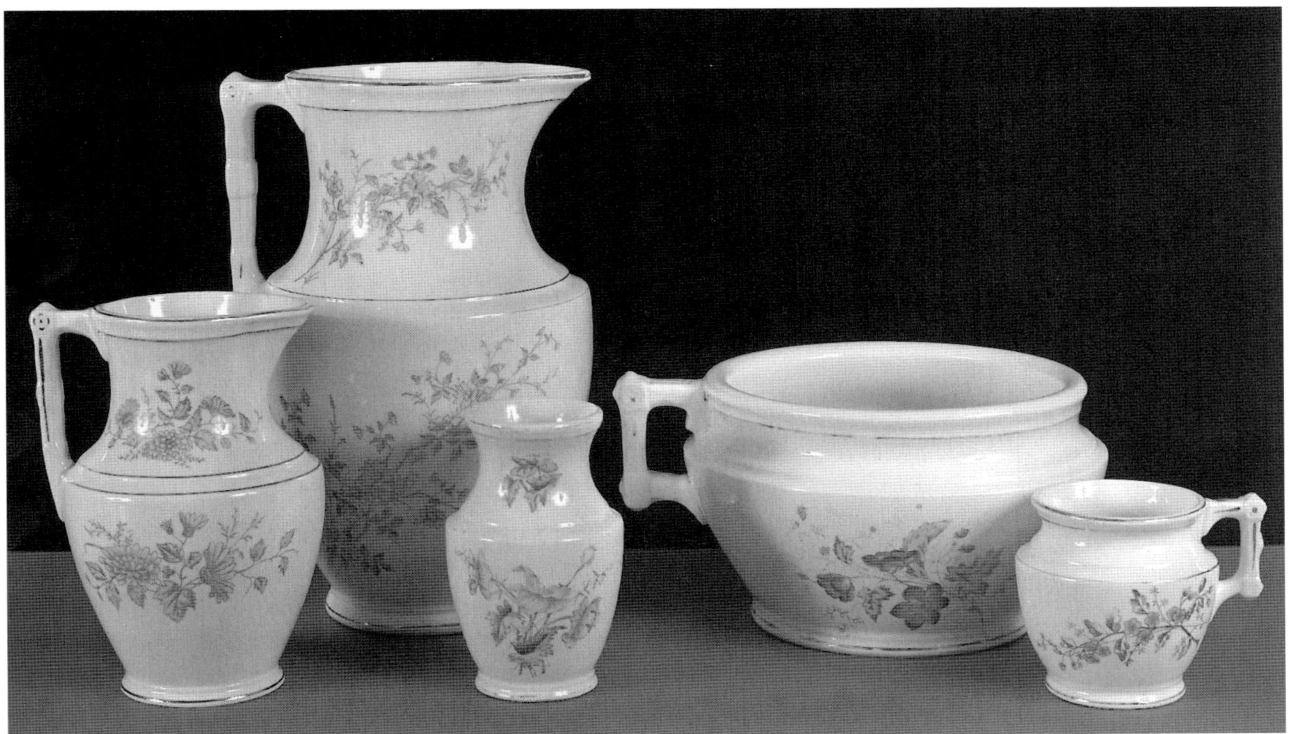

2.2. "Grecian" shape toilet ware, hand-filled, transfer printed, and gold lined, 1887.

Although Pass's achievements in these years were crucial to the success of the company, he depended also on the vision and leadership of a few others. While he ran the company's manufacturing on a day-by-day basis, improved its products, and developed new ones, Oliver, as general manager, promoted the ware, directed sales, and supervised business operations. Pass and Oliver together oversaw the growth of the company's facilities. In 1889 they began construction of the C unit, a three-story brick factory building similar to and connected to the B unit of 1883. With its completion in 1890, the pottery increased its capacity to six kilns.

Oliver received praise from the directors year after year for his leadership. Then in its 16 January 1890 number, *American Pottery and Glassware Reporter* announced: "Mr. Geo. W. Oliver expects to retire from his position with the Onondaga Pottery Co. at an early date. The company is in a very flourishing condition and on his retirement he contemplates an extended trip in search of rest and health through the West, with possibly later a European tour." On 21 January, at his last official function representing the company, he was elected second vice president of the U.S. Potters' Association at its annual meeting in Washington, D.C. He had been active in this seminal organization since its founding in 1875. He had served as its secretary since 1879 and had worked steadily for the standardization of raw materials throughout the industry. After

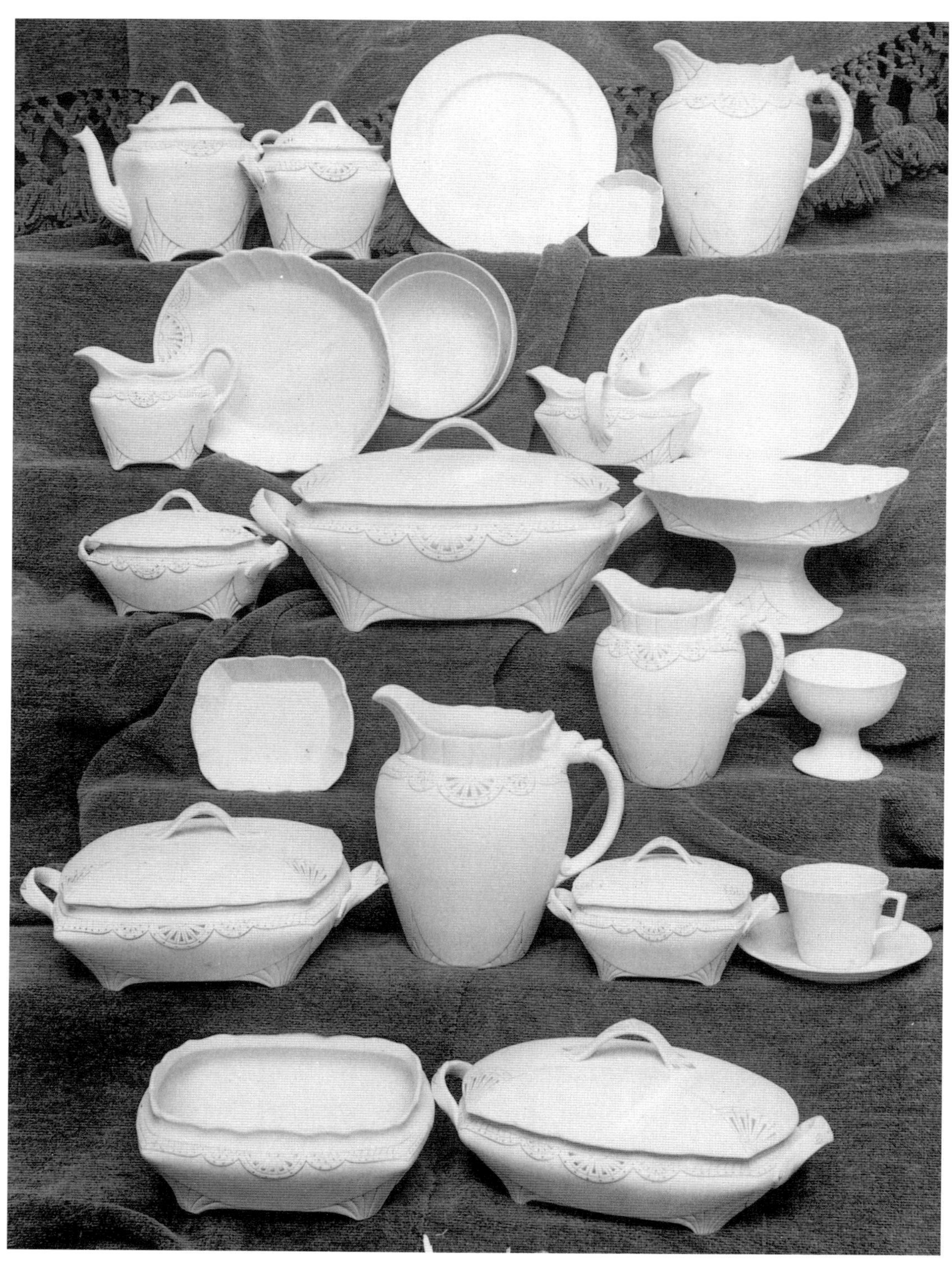

2.3. *Juno* for household ware in the "Celebrated O.P.Co. China." Catalogue photograph, 1893.

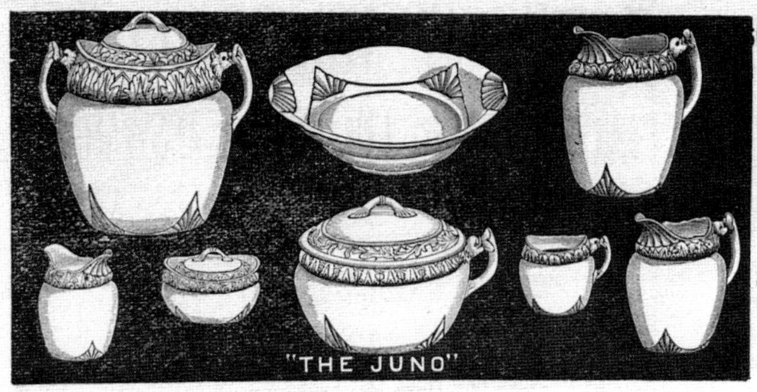

2.4. Advertisement for *Juno* toilet ware beginning in December 1889 issue of *Pottery* and *Glassware Reporter*.

his retirement, he continued to live in Syracuse until his death in 1909. Francis "Frank" Alexander succeeded Oliver as general manager.

The life of a pottery worker in the 1870s and 1880s was arduous, even in the "modern" B and C units. As was true throughout the industry, skilled workers had to pay for things that made their work possible, such as power, light, and tools. Sidney Lodder, who in 1937 recalled his early days at the pottery, remembered these conditions. He had begun to work at the pottery fifty years earlier as an unskilled young man in the dipping room, earning $3.50 for a week of sixty hours. He did menial chores in the clay shop before he began his training as a presser.

All the power was steam and man. . . . In the old [Empire] shop which burned, I have been told that it was all hand jiggers and all hand tools and one had to go to the clay bank for his clay so when they moved into the new building and had steam jiggers [jiggermen were charged for the steam] and the clay hoisted to the floor on which we worked, . . . a charge of 50 cents [was made] to jiggermen and 25 cents to pressers for clay hoisting and we had to wedge our own clay, get our own boards, carry our scrap to the cellar, carry out our dirt, move all our moulds and get drinking water from whatever pump we could find around the neighborhood that worked, or hire these chores done. Also, we had to furnish our own light when it was needed, which was a hand lamp, until the insurance company stopped their use. Then the Company bought large brass Rochester burner hanging lamps which they sold [to us] for $1.55.

Since then we have changed from hand tools to pull down [machines] and now everything is brought to and taken from the jiggerman, so all a jiggerman has to do now is make ware. That is why we can produce the increased quantity of ware with the small number of jiggermen. The same applied to the pressing method which has been replaced with casting and everything brought to the worker and taken away from him.[9]

In 1887, when Lodder joined the company, the working environment of all potteries was remarkably unhealthy. Clay dust was ever-present and damaging to lungs. The work load was heavy and the work week long. Kilnmen, working in exceptional heat, consumed great quantities of beer, which they rationalized as more sanitary than water (it "dissolved the dirt out of the system").[10] In East Liverpool, "Englishmen were notorious beer drinkers." The same was undoubtedly true in Trenton. It was not uncommon for men paid for piecework, as nearly all were in Trenton, to spend overlong lunch hours at the bar. Because they worked on Saturdays, they sometimes gained a two-day weekend by spending their Mondays ("blue Mondays") at the local tavern.[11]

Pass worked to improve this state of affairs. By 1903, the company had established a well-equipped clubhouse where workers could socialize away from taverns. But major improvements in environmental conditions occurred only in the twentieth century, when electric power made possible effective ventilating and vacuuming systems, and natural gas replaced coal as fuel for the kilns. Onondaga Pottery would be in the forefront of these developments also.

Late in the 1880s, the pottery was still a relatively small producer within the industry. It had gained an enviable reputation for the quality of its ware, however, especially its *O.P.China*, the first ware made by the company that could be considered to be of reasonably good quality by twentieth century standards. Pass's experimental, developmental approach had appeared elsewhere in the industry, and it would become a hallmark of progressive potteries in the 1890s, but Onondaga Pottery was now positioned to take a lead in a number of crucial areas and to hold it for a generation and more. First among these was the search for a true china.

Three

The Birth of American China
1891–1896

Of the many inventions that arouse our wonder not the least amazing is the discovery,
made over one thousand years ago by the Chinese, that the rude stone and clay of the
mountainside could be converted into that white, translucent, gem-like substance
which the Western world called porcelain.

—R. L. Hobson, 1908

Pass set himself the goal of creating a true china body, one that was thin, translucent, and completely vitreous. What was this elusive "china"? How did it differ from the historical porcelain of T'ang Dynasty China (618–907 A.D.) that had found its way through the early trade routes to the Western world where the Italians named it *porcellanna* ("Venus shell")? How did it differ from modern porcelain, made since the eighteenth century in Germany and France, and from bone china, made since 1800 in England?

Porcelain, ancient and modern, had been coveted and collected by the wealthy for many generations. One of Pass's contemporaries, the British historian of ceramics R. L. Hobson, wrote in 1908:

The collector's alphabet begins with the distinction between pottery and porcelain . . . [then] the distinction between true and artificial porcelain, properly called hard-paste and soft-paste. . . . In composition the main distinction lies in the nature of the fluxing [melting] material. . . . True porcelain consists of two natural feldspathic substances; a non-fusible clay (called by the Chinese *kaolin*) combined with a fusible stone (called *petuntse*), the latter melting in the kiln to a glassy material which holds the former in suspension and gives the porcelain its translucent and vitreous character. One is the bones, the other the flesh of the porcelain body. Over this body is a skim of glaze formed of pure *petuntse,* sometimes softened with a little lime. . . . In the case of artificial, or soft paste porcelain, the body is formed of a natural clay suspended in a fluxing material artificially prepared . . . (a glass or frit made of sand, lime, flint, bone ash, soda, etc.), the ingredients differing at almost every factory and producing a variety of wares of diverse tone [color], hardness and translucency. The glaze, too, varied, but as a rule it consisted of a soft and fusible glass largely composed of lead.[1]

As a rule, hard-paste porcelains are fired to a higher temperature than are soft pastes. After firing, both porcelains are quite hard, but the hard paste is hardest of all.

From late medieval times onward, countless potters in Europe had aspired to discover the nature of hard-paste Asian porcelain manufacture. In working toward this goal, a few of them found ingenious ways to produce a largely satisfactory imitation; they developed soft-paste porcelains and fine decorating techniques to serve them. Then, in 1709, Johann Frederick Boettger of Dresden found a source of kaolin in Germany and in working with it realized that he had discovered the essential ingredient of Chinese hard-paste porcelain. His discovery made possible the famous Meissen porcelains. Although he was virtually kept a prisoner by royal authority, his secret soon leaked out. In 1768, the French discovered kaolin beds of their own near Limoges. This enabled the potters at Sèvres near Paris to make some of the most beautiful hard-paste porcelains the world had ever seen. Like the Meissen potteries, those at Sèvres were state-supported on the grounds that their splendid products added to the glory of the state.[2]

The English also made fine ceramic ware, but with different formulas. They did so without the state support that made possible the magnificent wares of Meissen and Sèvres. English fine china came from private entrepreneurship and was directed to a broader market. The availability of kaolin clay in Cornwall hastened the growth of the English potteries in the eighteenth century. In 1800, Spode at Stoke-on-Trent developed modern English bone china, a vitreous ceramic ware with a distinctive whiteness achieved through the addition of finely ground animal bone to the clay body formula. Different from both hard- and soft-paste porcelains, it combined the durability of the former with the beautiful, softer glaze characteristic of the latter. This vitreous bone ware, and others like it, produced by Royal Worcester and other potteries, was called "china" rather than porcelain.[3]

Displays of high quality English and French wares in trade fairs in the 1870s inspired American manufacturers, Pass included, to try to imitate and better them. This goal drove some American potteries to the expense of importing English china clays to add to native clays. Others searched for a purely American formula, for economy as well as for national pride. The Carolina mountains soon yielded an exceptionally white kaolin called "uniker," although its properties in firing differed from those of Cornwall kaolin. This Carolina kaolin was for many years difficult to obtain in uniform quality, and it tended to create more problems than it solved.[4]

Rather than attempting to make porcelain or bone china from European recipes, Pass worked to transform his earthenware clay body into a new kind of vitreous ware, the equivalent of European china and porcelain in its intrinsic beauty as ceramic art.[5] Although the terms "china" (in its true sense as vitrified ware) and "porcelain" had often been used synonomously (and continue to be), Pass and others in ceramic manufacturing distinguished between them because of their different ways of production. In making porcelain, the ware is typically first-fired at a relatively low temperature in a biscuit kiln to a state hard enough to be handled without breaking and to accept embellishment by carving or by painting with ceramic pigments. It is then dipped in glaze and fired to a very high temperature where the thin glaze and clay fuse together. A fractured edge appears vitreous and shell-like throughout; it is difficult to determine where the glaze ends and clay body begins. China, as we will see, reverses this sequence of firing.

Porcelain is hard and cold to touch, glistening and translucent when held up to light. The edge of the foot rim is free of glaze, close and compact in texture, slightly brown from firing. Thin porcelain is fragile, however, and unable to withstand everyday use in homes, hotels, or restaurants. Pass sought something that approached porcelain's aesthetic appeal but that had greater durability. This required both a different formulation of materials and a new process of making. When it was finally achieved, with Onondaga Pottery leading the way, this "American china" would have no close counterpart in Europe.

In 1930, Professor Charles Binns, one of the great figures of American ceramics and a professional colleague of James Pass in the industry, commented on the origins of this new product in a paper, "Turns of the Potter's Wheel in America," written jointly with Pass's son, Richard, for the British Ceramic Society. Perhaps to avoid offending old associates, the two authors credited a number of potteries with contributions to the development of American china before they identified James Pass and his pottery as the first to succeed in producing a marketable product.

> It was in the late 'eighties and early 'nineties that the American potters, by trial and error, finally perfected and made commercially practical a white, translucent, non-porous ware, stronger and more durable than any tableware yet produced. This product, which has come to be known as American china, was the first outstanding contribution of the American potter to the progress of his trade. Today this ware is used in most of the hotels and restaurants of the United States, where its good appearance and unusual durability secure its popularity.

> It is not possible to assign credit for the first success of American china to any one person or factory. In fact, like the progress of most improvements, one reacted upon another—a little here, a little there—until the present quality was reached. The Greenwood and Maddock potteries in Trenton, the Knowles, Taylor and Knowles Company and the Sebring family in Ohio, and the Onondaga Pottery Company in Syracuse, New York, competed closely in the quality of the ware, but the last named, after the usual beginning in CC and semi-porcelain, succeeded, under the direction of the late James Pass, in manufacturing a thin, translucent china decorated with underglaze colors.

> The essential ingredients in the composition of American china are the same as those now used for hard porcelain with the exception that the former invariably contains appreciable quantities of lime, or lime with magnesia, introduced in the carbonate form, whereas this is not universally true of porcelain. The glazes of the two types of product, however, are quite different. Porcelain glaze contains the body ingredients only, though in different proportions. China glaze has been developed and refined from the earthenware type. The porcelain glaze is fired with the body mass, and of course, at the same high temperature. The china glaze is placed on the already matured body and fired at a temperature which, while lower than that applied to porcelain, is yet much higher than that used for the cheaper glazes.[6]

"American china," as Pass and his company came to call it, was a product of the industrial age's alliance with science. Pass's research in clay body chemistry and his continual, systematic experimentation with firings would have been impossible in an artisan pottery. American china was a product of and for a democratic industrial society. Pass intended from the outset to manufacture it in quantity for sale at affordable prices. His pottery soon became its largest producer.

Pass's diaries record to the day—27 July 1886—the first formula he mixed as he systematically began his search for a durable high quality china. He marked his series of experiments "JP" and numbered them in sequence. In a trade notorious for its secrets, where potters sometimes carried their recipes to the grave, Pass devised a code of ciphers with which to write his formulas:

1 - •   6 - ∟

2 - | 7 - ⊔

3 - ⟩   8 - ⊔⌐

4 - Ɲ   9 - ▢

5 - ⅝   0 - ○

He then created unmarked "blind" measures and scales for use in the factory to prevent his workers from knowing the exact proportions of the ingredients they used. This was a common practice in the industry. He mixed and tested more than fifty formulas. His last was JP-52, mixed in 1893. Each trial involved more than adjusting the percentages of the raw materials in each recipe. The characteristic behavior of each kiln, no two of which ever produced precisely the same results, added another complex variable.

High on Pass's list of requirements for a china clay body were sources of raw materials of consistent quality. This was no simple task at a time when American mining and clay extraction was still a chancy business. Suppliers did not always ship the same clays from a single location. New sources of clays were still being discovered. Working as a chemist in his laboratory at the pottery, Pass sought empirical results by testing every newly arrived batch of clay before he further refined and used it. He recorded what each mix produced after firing in a specific kiln location.

In 1930, Richard Pass recalled that his father, hunting for American materials to use in his new American product, learned of the remarkably pure North Carolina kaolin (uniker) discovered in 1888 at Hog Rock mine, near Dillsboro. He went on to report that his father had the clay selected from the mine according to his specifications and then had it washed especially for Onondaga Pottery. In his experiments, Pass mixed the Hog Rock clay with more than a dozen variations of the other basic ingredients: flint, silica, and ball clay. Adjusting the proportions by only a single percentage and substituting one variety of flint or silica for another made significant differences in the fired product. He ground the raw materials finely and sieved them to remove impurities such as mica. He fired this china formula first to a high temperature (2250° F) in the biscuit kiln and then, after cooling and decorating, fired it again at a lower temperature (2050° F) in the glost kiln. This reversal of porcelain's firing process made it possible to decorate the ware with colors that would burn off at higher temperatures.

He glazed the feet of this ware to prevent it from scratching table surfaces or, when stacked, itself. Both would have been potential problems for hotel dining rooms and kitchens. To do this, he placed the pieces of ware onto pins set in the walls of the saggers (ceramic boxes in which ware was fired) in a way that prevented the glazed items from touching each other or any other surface. Three small marks left by the pins on the underside of the ware were ground smooth and polished after firing. This was not a new practice. Onondaga Pottery had always pinned its glazed earthenware.

As a standard for his china, Pass used the French porcelains of Sèvres. He noted in his diaries in 1888 that one or another of these trial formulas "when fired at . . . intense heat equals the French body. Both W.G. & . . . [O.P.China] glaze craze on this body immediately"; "good color"; "tough but cracks on moulds"; etc. He experimented continually with variants of his Hog Rock recipe for the next two years.

Pass had good reason to take French porcelains as his standard. In 1853, porcelain decorated in France (at Limoges) and imported to the United States by the new firm of Haviland Brothers & Co. had been a great success at the New York Crystal Palace exposition. A rage for "Haviland China" followed in America despite its relatively high cost. Part of the cost reflected the long-established 40 percent tariff on imported ceramic goods, and part reflected the costs of the ware's decoration. Haviland's decoration far surpassed any done in America in its sophistication, execution, and originality. In the late 1880s, however, Pass's primary concern was to find a china clay body of a quality on a par with that of Haviland porcelain, though of a different composition and differently made. Until he attained that, improvements in decoration would be a secondary concern.[7]

In 1888, he found what he had been looking for. On 11 April, when he reached JP-20 and changed his kaolin to "C. Ground" (probably meaning "Carolina ground"), he noted next to the formula in his diary, "best in the world," and "equal to bone china." American china was born that day. The Hog Rock clay had made a crucial difference. In October 1888 he broke the sequence of his experiments at JP-28, inserting the variant formula JP-20A, and specifying the kaolin as "Carolina."

Years later the pottery claimed 1888 as the birth year of its celebrated *Imperial Geddo,* Pass's first true chinaware. Although he may have used his new body with a few *Juno* items, it took him nearly four years following his first successful trial to bring *Imperial Geddo* to market. Pleased as he was with the formula, he continued to experiment with it intermittently, reaching JP-35. In 1892 he appears to have resurrected JP-20A, noting it as JP-20B. He followed this immediately with his JP-36 formula; this one fully satisfied him. At some point, probably in 1889 or 1890, he had set his chief modeler, Hamlet Bourne, the task of creating a new shape for the purpose of introducing this American china to the public.

The introduction of a new shape required major commitments of time, the making of new tools, additional space for the storage of many dozens of moulds, the design of patterns for decoration, and much faith that the shape would sell. In introducing a wholly new clay body, Pass also needed to learn through experience its behavior during firing in each of the kilns. Although other progressive potteries were also moving ahead in the search for a commer-

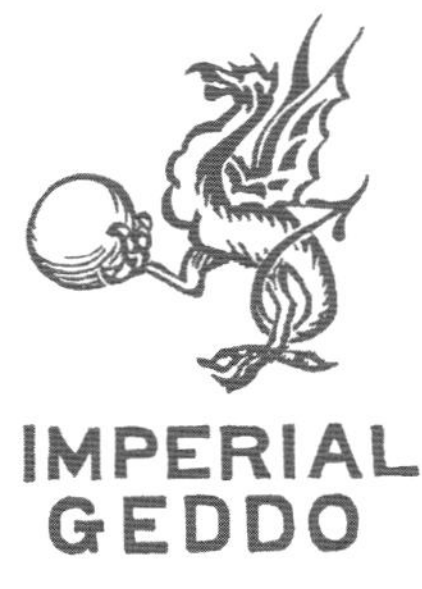

Imperial Geddo
(1892–94)

cially viable American china, Pass seems to have been right in his belief that Onondaga Pottery, far from the two major centers of the industry, had achieved this goal first.[8]

Pass introduced his **Imperial Geddo** American china to the public in 1891. For this delicate, white, translucent vitreous ware, he devised a Chinese dragon, "rampant," for use as a backstamp. While some of the pottery's shapes called for more than fifty items, **Imperial Geddo** called for only about two dozen. Bourne's new shape fell close to the category of art ware. **Imperial Geddo** (fig. 3.1) offered such fancy accessories as an orange bowl, a vase (plate 3), a rose bowl, two biscuit jars, three tête-à-tête tea sets, a boudoir set, and a smoking set rather

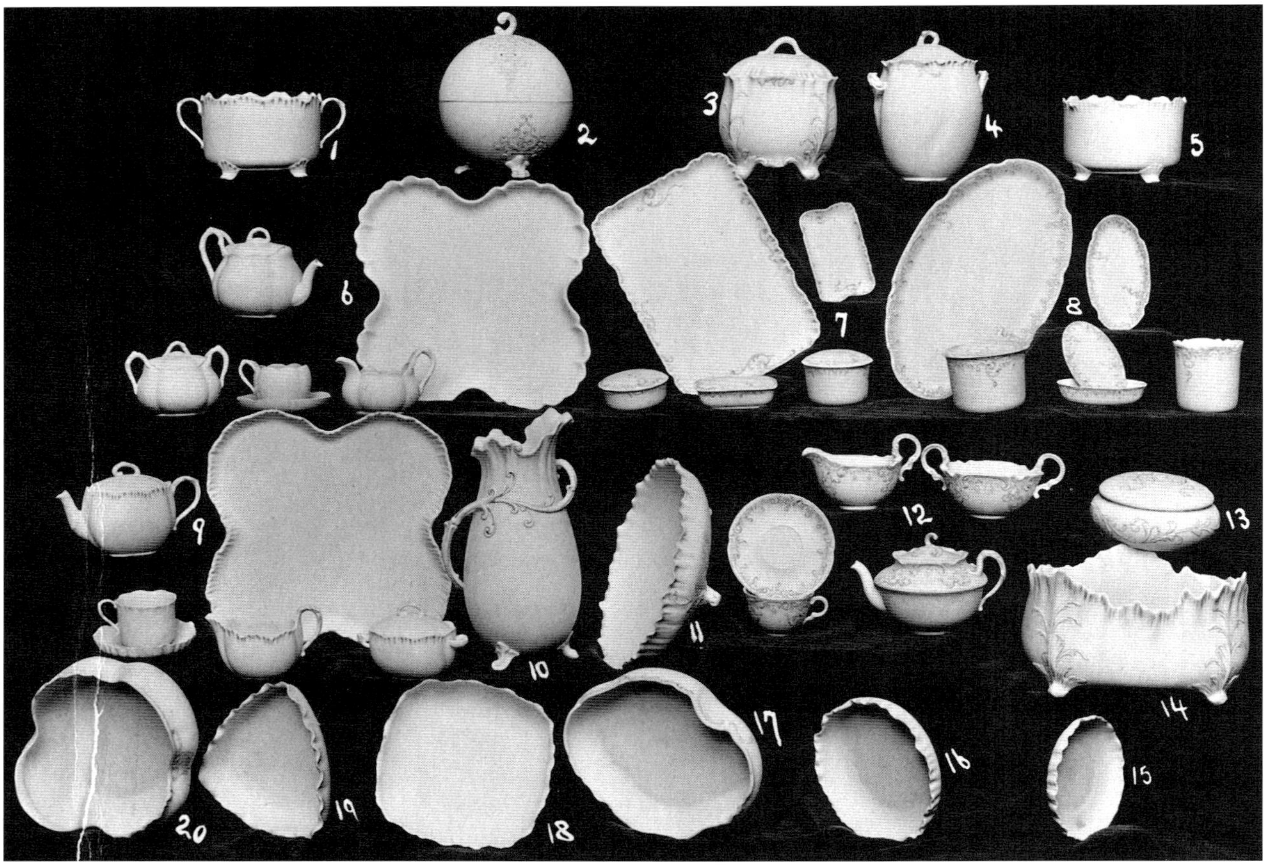

3.1. **Imperial Geddo**, "A fine line of White china for Amateur decoration. The same goods are sold in very rich decorations from sample." Caption for catalogue photograph, 1893. Shown: (1) nut bowl (handled); (2) globe rose jar; (3) biscuit jar (**Pharis**); (4) biscuit jar (**Moro**); (5) nut bowl (unhandled); (6) tea set (**Passo**), more commonly called "tête-à-tête" set; tea pot, sugar, cream, two cups and saucers on a tray; (7) boudoir set: puff box, square match box, pomade box, and pin tray on a square tray; (8) smoking set: oval match box, tobacco box, oval pin tray, and cigar vase on an oval tray; (9) tea set (**Alexo**); (10) vase (see plate 3); (11) flower tray; (12) tea set (**Moro**); (13) bonbon dish; (14) orange bowl; (15) ice cream (same as **Alexo** tea saucer); (16) round jelly; (17) fruit dish; (18) square jelly; (19) triangle jelly; (20) salad.

than a complete dinner service. Most pieces were richly hand decorated (probably by Elmer Walter) with a variety of special treatments, including raised paste embellished in gold (gilded slip decoration) over a pale Worcester matte, a solid rich cobalt blue, and china painting. The company also offered **Imperial Geddo** ware undecorated as "a Fine Ware of White China for amateur decoration."

The backstamp, signifying both the shape and the clay body, was the first to omit the company's name. Alexander had coined the shape's appellation, taking Geddes as the root and adding a final "o" to make it seem exotic. "Imperial" summoned up both China and England, and their associations with fine ceramic ware. He named the three subsidiary tête-à-tête sets *Passo, Alexo,* and *Moro.* Pass imprinted every piece of **Imperial Geddo** with his cipher for JP-36 (except for three dozen pieces that he marked by mistake with the cipher for JP-20. Only ware marked with this distinctive backstamp warrants the name **Imperial Geddo.** The cipher is visible when a piece of ware is held up to strong light. The potters knew the JP-36 body as "No. 1 china clay."[9] This was the first American china body to reach the market. Pass had made his point—that Onondaga Pottery could produce a grade of translucent true china unsurpassed in America, and one worthy of comparison with European porcelain and bone china.

Compared to its other shapes, the pottery produced very little of **Imperial Geddo.** Pass mixed only a few batches of No. 1 china clay between 1891 and 1894. In dollar value this expensive (to make) decorated ware amounted to 6 percent of the company's total output in these years. When he ran out of the North Carolina kaolin needed to make this body, he thereafter made the same forms with No. 2. china clay and called the ware simply **Imperial. Imperial Geddo** was the finest ceramic ware Pass ever created, and its elegance drew attention to the pottery's achievement in a way that a more ordinary shape would not have done, but the need to make china tableware suited to the much larger market for household use now occupied him.

Imperial Geddo called for gold decoration, as did some other shapes. Pass recorded in his diaries for 1889 his work to improve the formula for gold used to decorate this ware. Though many potteries used a cheaper, inferior product called "bright gold" (also called liquid gold), Walter had brought to the pottery skill in the use of 24 carat coin gold to embellish finer ware. He and his decorators used gold for lining, stippling, tracing moulded decorations, and other embellishments. After it was low-fired in a hardening-on kiln, coin gold required burnishing with a cloth to obtain a lustrous sheen. Coin gold decoration was an expensive process. Every speck on rejected and broken ware, and even on wiping cloths, was reclaimed and sent to a smelter to be reconstituted for reuse. Walter occasionally employed other hand-decorating techniques: *groundlay,* achieved by dusting dry pigments on a surface prepared with a special oil; *enameling,* by applying a thick glaze (called enamel) as raised dots of color to serve as accents in a pattern; and *raised paste* work (see plate 3), by building a decoration on glazed ware with a thick paste made of clay components and, when fired, decorating it with flat gold.

Pass's abilities in ceramic research did not go unnoticed. Soon after his first successful Hog Rock clay experiments, he had an opportunity to put them to further use in the Syracuse community. In 1889, Albert Seymour, an electrical engineer and an officer of Syracuse's power company, approached Pass about the need to develop and manufacture high quality porcelain electrical insulators. Pass's diaries document his experiments for this project. He soon decided that for Onondaga Pottery to enter the field would unduly drain its resources, and so, with the blessing of the pottery's officers, in 1890 the two men formed a partnership, Pass & Seymour, to manufacture insulators.

Pass was then about to marry and had saved money to purchase a home, but his bride-to-be, Adelaide Salisbury, agreed that they should invest the house money in the new business. It was a wise choice. This venture met with remarkable success and made Pass's fortune. By 1913, the year of his death, Pass & Seymour employed 350 people in Solvay, a town adjacent to Geddes. The two companies remained closely linked by virtue of their similar (ceramic) products and common ownership for decades.

In 1955, after a long, influential, and highly productive life with the company, William L. Huber, the retired dean of the pottery's sales force, sat for an interview with Jeanne Corry, editor of *Syracuse China News*.[10] His often humorous recollections, even given the benefits of hindsight and an understandable tendency to exaggerate and oversimplify, provide a valuable firsthand account of life at the pottery in the 1890s. He had been present as Onondaga Pottery began its transition from the manufacture of white earthenware to vitreous china.

Huber had gone to work as an errand boy at age thirteen for S. P. Pierce and Son, Syracuse's leading wholesale and retail crockery dealers. He advanced to a sales position, where he gained experience selling German, English, and French ceramic wares. General Manager Alexander recognized Huber's talents in sales and his acutely perceptive understanding of the complex role of dining in American life. He hired the young man in August 1891 to represent the pottery as a commercial traveler (salesman) and packed him off immediately to the West on an eight-week trip by rail to a "territory where few had ever heard the company's name and knew nothing about the product."[11] He opened up the Western markets and devoted his early career to expanding and developing them.

Half a year later, on 2 March 1892, the pottery experienced a major crisis. The company's president, Mills Pharis, one of its founders and chief backers, died suddenly. Although the pottery had increased its plant and production, had a superior earthenware product, was preparing to introduce **Imperial Geddo,** and was by any reckoning an innovative, usually profitable, and nationally respected manufacturer of ceramic ware, the implications of Pharis's demise put the company in jeopardy. He had apparently personally underwritten notes to fund the construction of the pottery's C unit. Because his estate was found not liable for the debt, the company was left to assume it. Three days after Pharis's death, Frank Alexander died suddenly. At an emergency meeting, the company's board appointed Pass to act as general manager concurrent with his responsibilities as superintendent. They elected E. B. Judson as the firm's new president. He had been a director only since 1890. Soon after, shaken by the sudden financial crisis, the board lost confidence and voted to liquidate the company.

More than sixty years after the fact, Huber offered his recollection of what Pass and others had told him about what happened on that fateful day. According to him, following the board's meeting, Judson told Pass, "I guess you are going to have to look for a new job. We decided to liquidate." Pass asked him for an opportunity to propose to the directors that they reconsider their decision and allow him to reform the business side of the company as he had reformed its products. Judson convened a meeting for him with two stockholders who had insisted on liquidation. They seemed resistant to granting Pass time to rescue the company. After Pass departed, Judson, who had great faith in Pass's abilities, said to them, "Well, gentlemen, that is all there is to the Onondaga Pottery Company. If you don't keep him you will have nothing." Understanding that the company they were about to liquidate was worth less without Pass than it was with him, and that he was the key to increasing its worth, they changed their minds, met with the other directors, and set Pass the task of single-handedly saving the pottery. So, at least, is the story as related by Huber.[12]

Had Pass been concerned only about his own welfare, he need have made no effort to rescue the company. He could easily have shifted his efforts entirely to Pass & Seymour. He might even have found backers to begin another pottery. He would have been a prize catch for any other ceramic manufacturer in the country. But his experiments toward bringing his American china body into mass production were proving successful, and, perhaps just as important, he had "bonded" with Onondaga Pottery—with its workers, its reputation, and its promise. He did not want to see his increasingly skilled workforce dismissed and dispersed. He set out to rescue the suddenly crippled company.

Huber recalled that "Mr. Pass wrote me a letter when I was at Galesburg, Illinois [on a sales trip]. . . . 'I would like to see you as soon as you arrive . . . to talk over our future hopes here.'" Pass trusted Huber's judgment on what the market would bear and, more important, how it could be reshaped. Together with Judson, they reconsidered the pottery's products, worked out new sales strategies, and in a sense created a working partnership that shaped much of the company's future. Huber remembered that "things began to hum and James Pass began to build the foundation for the Onondaga Pottery as it exists today. He was the architect; he was the builder. I became the hod carrier—the fellow who carried the bricks to him so that he could build."[13] At the time, in 1892, Huber was only twenty-four; Pass was thirty-six.

Pass told Huber that the pottery was now capable of making a durable thin china tableware and that, with **Imperial Geddo** already under way, he had begun to make a line of vitreous dinnerware for the dining room of Syracuse's new Yates Hotel. Designed by the Syracuse architect Archimedes Russell, the Yates opened on 17 September 1892. It was a first-class hotel. An observer described the hotel's dining room as the first west of New York where "one gets a hot plate every time; where they don't put 'Bass' [ale] on ice; where they don't ice claret, and where one can get a steel knife to cut steak with. They also have black pepper, and the napkins aren't starched."[14] The Yates operated on the American plan; its dining room was open as a restaurant to the public. It had little serious local competition in this respect. The age of stylish restaurants independent of hotels had barely begun and would not make a real impact on American dining until after the turn of the century. It would have been unthinkable to have fine china from any maker other than Syracuse's Onondaga Pottery grace the tables of the Yates's splendid Renaissance-revival dining room. Even the best earthenware service was out of the question.

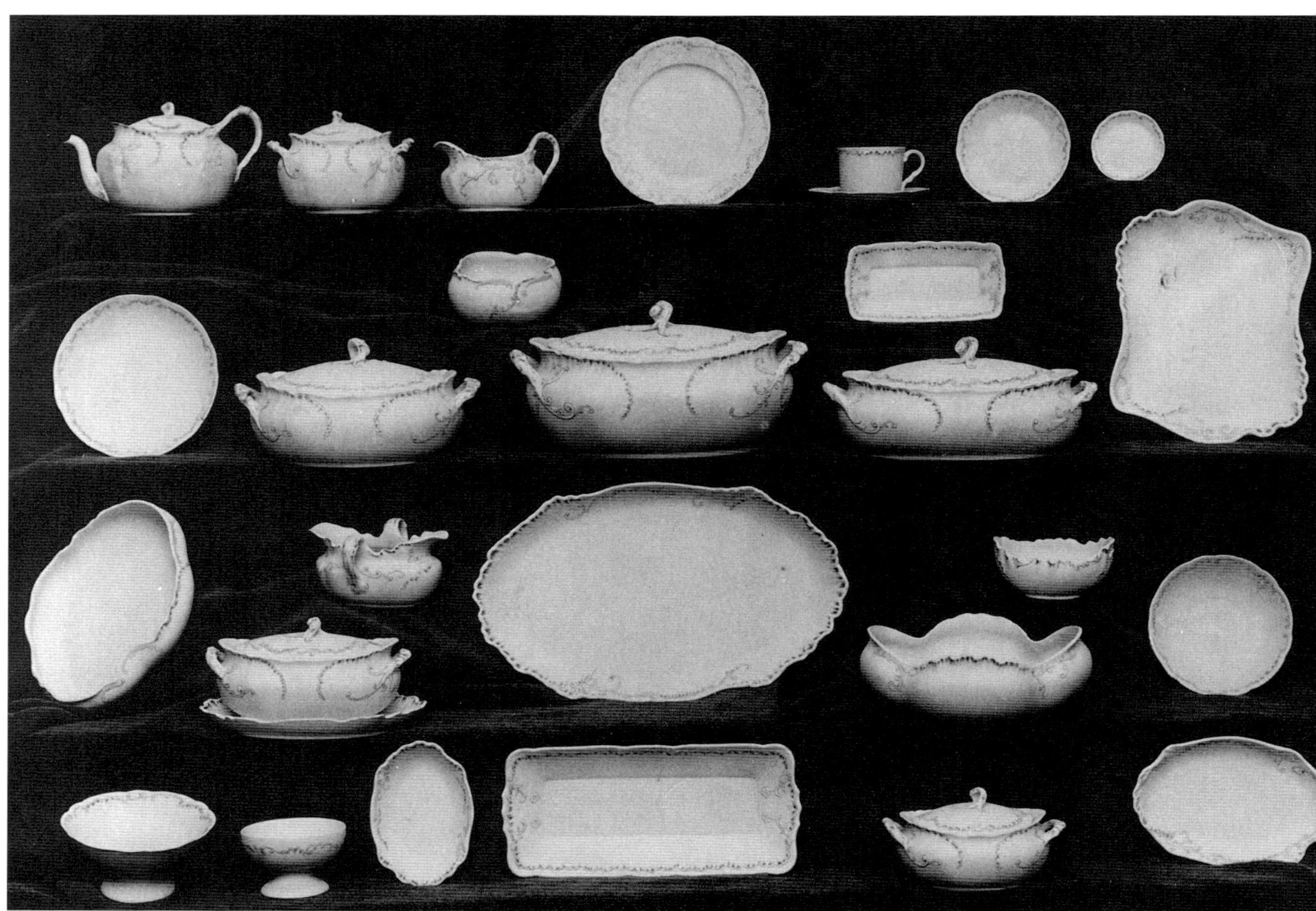

3.2. *Marmora* "A Fine Dining Room Service for Family use. Very choice decorations." Caption for catalogue photograph, 1893.

When the pottery ran out of its supply of the Carolina kaolin used in the JP-36 formula, Pass switched to an English substitute from Cornwall. He recorded in his diary for 1893 (possibly after the fact) that he had made the Yates service with his JP-52 body, known to his potters as "No. 2 china clay". This was the last recorded formula in his china body tests and one that he marked "Yates Hotel."

The single contemporary report of the Yates Hotel service stated only that "all the beautiful china, which bears the coat of arms of the hotel, and which, we venture to say, compares favorably with any used by hotels in this country, is from the Onondaga Pottery Co., who are among the foremost manufacturers of pottery in America."[15] Oddly, no examples of this ware seem to have survived.

Pass and his young designer, Haley, also began in 1891 to develop *Marmora* (fig. 3.2), the first dinnerware shape for the new china body; the company introduced it in 1893. Sidney Lodder, in recalling this time, said that "Mr. Pass made what was known as #2 china [clay] . . . [for] our regular dinnerware, which I think was the *Marmora* shape."[16] At the same time, Haley modeled the *Iona* shape, "an entirely new and very neat dinner service in white granite," but it never went into production, probably because of the financial restraints imposed by the recession of 1893.

Marmora was very much in tune with the fashion of the times as defined by the continuing popularity of Haviland porcelain. This French rococo shape nevertheless presented problems from the start. In use, the inside-fitting covers of its hollowware too frequently slipped out of users' hands, falling into the casseroles and cracking their bottoms. The company promoted *Marmora* as "Fine Dining Room Service for Family Use," targeting women as its most promising market, which they were. Its historical importance rests in its priority as the first large-volume production American china dinnerware shape to arrive on the market.

J. H. Brewer, president of Trenton's Ott & Brewer Pottery in Trenton and a founder of the U.S. Potter's Association, undoubtedly had *Marmora* ware before him when he wrote to Pass on 1 May 1894, "The samples came duly to hand and they are very fine indeed. What kind of china do you call it? It is not French—neither is it ordinary Trenton china. It is so much whiter, it stands pinning [in firing] just the same, however. These articles will grace my collection . . . for no cabinet would be complete without your china."[17]

In 1892 the pottery introduced a backstamp depicting a stylized Chinese dragon (couchant) with the words "China. O.P.Co." The dragon was another nod to the popular association of "china" with the Chinese. The dragon mark went on all ware made of the new No. 2 china body including *Imperial.* In 1892, the pottery marked its first *Marmora* ware with this China dragon backstamp.

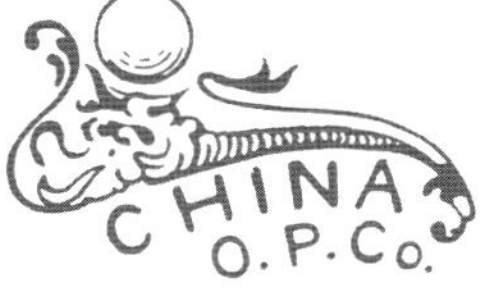

China Dragon (1892–95)

Huber described the mark as "a grotesque thing [that] looked like a worm with an egg in its mouth," and insisted to Pass that the company could not sell its American china with that trademark. "We've got to have something [else]—a given name for our ware." He came up with three suggestions: Syronda, Syrpass, and Syracuse China. The first, combining Syracuse and Onondaga, was awkward. Pass preferred not to lend his own name. He chose Syracuse China.[18]

In a simplification of what must have been a complex series of decisions, Huber claimed that about this time, Pass called him into his office and said, "Either we make china or earthenware. I can't fire them together because the glaze on the earthenware gets too hard in firing and crazes." Huber, who would be required to make major alterations in his sales strategies with the elimination of earthenware, responded, "By all means, china. . . . The earthenware market is flooded with cheap goods but the china market is ready." Pass knew that he would need to phase out earthenware products gradually, because he had an established market for them. He needed to build up his china business before he could drop the old lines. With this decision made, they set out to plan new shapes for Syracuse China. In pricing their envisioned new product they used as a guide the French Haviland tariff book's list and discounted it 40 per cent—the tariff rate—to come in under French prices for the American market.[19]

In 1892, the company published four price lists, one each for *White Granite, O.P.China, Fancy China,* and *Vitrified China,* all offering ware "warranted not to craze." To supplement these lists,

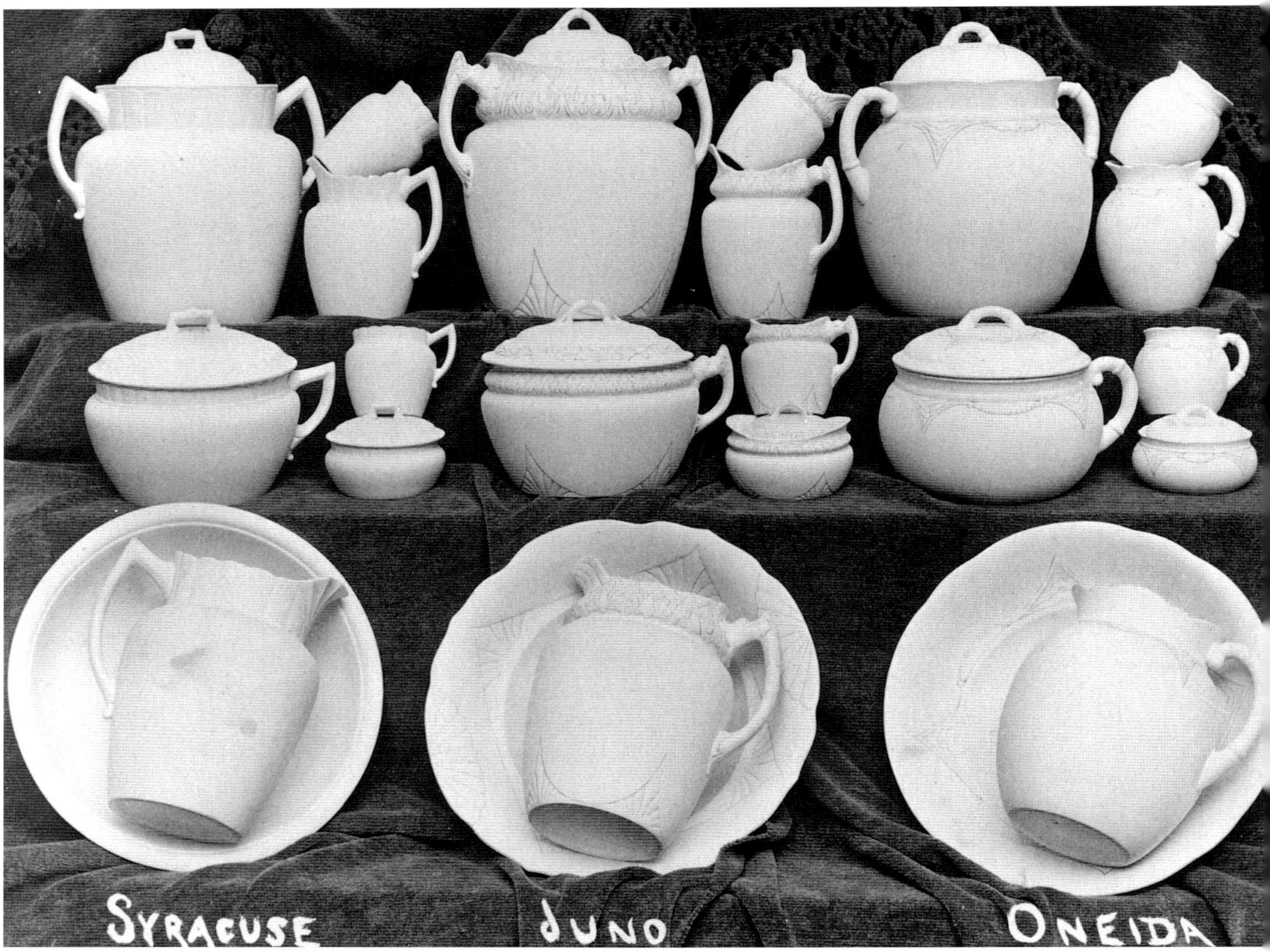

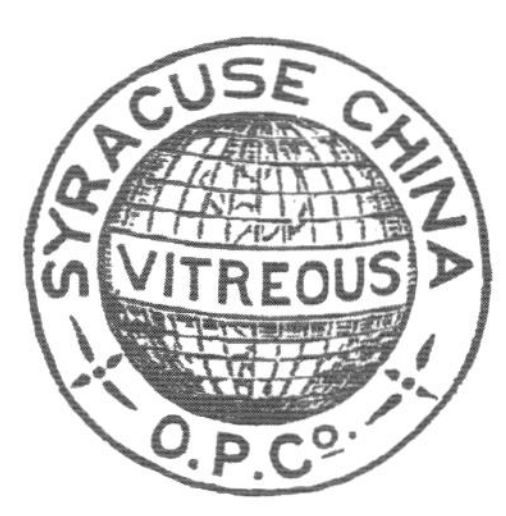

3.3. *Syracuse, Juno,* and *Oneida* toilet sets. Catalogue photograph, 1893.

it commissioned group photographs of the **Imperial Geddo, Iona,** and **Marmora** shapes (see figs. 3.1 and 3.2), and another arrangement of the three toilet sets, **Syracuse, Juno,** and **Oneida** (fig. 3.3). It also commissioned a set of high quality color lithographs illustrating its most popular applied decorations on the **Syracuse, Juno, Oneida,** and **Marmora** shapes (plate 1). With this new emphasis on the pictorial advertising of the ware, the company entered the world of twentieth-century marketing.

In 1892, the pottery created a distinctive trademark for its Syracuse China, depicting the Western Hemisphere on a globe banded with the word "vitreous" and encircled with the words "Syracuse China/O.P.Co."[20] The trademark first appeared not as a backstamp on ware but as an illustration on a 1892 **Marmora** price list. It evidently was not backstamped on ware until 1895,

Western Hemisphere
(1895–97)

when it replaced the "China dragon" to mark the **Marmora** and **Imperial** shapes and, later, some **Plymouth** items. The Syracuse China archives copy of this 1892 **Marmora** price list is annotated, "Carefully preserve, no duplicate, Pass", leaving the impression that it may be a proof of a list that was never issued. Furthermore, the illustrations of **Marmora** items show the China dragon trademark instead of the globe. This suggests that Pass designed the Western Hemisphere backstamp in 1892 but, again because of the recession, had to delay its use until 1895. In 1896 the Western Hemisphere backstamp appeared on the first price list for *Vitrified China*.

If the use of the New York State seal as a backstamp in 1873 had anticipated the nation's centennial with a statement of pride of place, then it may be that Pass wanted a trademark depicting the Western Hemisphere to exploit the widespread observance in 1892–93 of the four-hundredth anniversary of Columbus's great discovery. No evidence exists on this point, but, by intent or otherwise, the trademark in its own time suggested that the company's vision now extended far beyond its home state.

At the World's Columbian Exposition in Chicago in 1893, the pottery carried off a medal and a diploma for a display of **Imperial Geddo.** The citation described it as ware "of considerable merit in general quality of the ware and style of decorations, exhibiting considerable progress."[21]

This was the only award won by an American pottery for "translucent china," though Knowles, Taylor, and Knowles of East Liverpool received an award for bone china. If Onondaga Pottery exhibited a dinnerware shape in the "semi-porcelain, white granite" category, it was most likely **Juno** in the improved O.P.China body, which the company was promoting in trade journal advertisements at the time. The judges and critics noted the exceptional strength of the pottery's ware as well as the fact that although it had been fired on pins in the glost kiln, it had a hard glaze and was perfectly straight. They commended the translucency and beauty of **Imperial Geddo.**

In reporting on the exposition, a writer for the *British Pottery Gazette* on 1 August 1893 gave a thoughtful and discriminating critique of the china display, though his assumption that Onondaga Pottery needed imported clay to make its china was erroneous. The review supports the pottery's later claim that it was the first to manufacture a true fine china tableware of comparable quality to those of Europe.

Foremost amongst the firms engaged in making goods I find the Onondaga Pottery Company, of Syracuse, N.Y., who are stepping to the front with samples of a fine china of French character; the clay they use is, I imagine, imported, for I have seen no native clay that is capable of giving the results this company achieved (there is a large exhibit of potters' materials here, but the china clay [kaolin] shown is about the color of our Dorset blue clay). The china is good and clear, fairly well potted, and the glaze quite satisfactory. The Onondaga exhibit consists for the most part of dinner ware in fancy shapes—all, like the paste, of French motif and tastefully gilt; this company has apparently freed itself entirely from the seductions of liquid gold, the bane of cheap pottery, whether here or at home.[22]

On 23 December 1893, the *Syracuse Journal* reported with a little puffery that a

> medal and diploma are the trophies which the Onondaga Pottery Company brings from the competition with the world's products at the Columbian Exposition at Chicago. These are recognitions of the superior character in all respects of the fine table china produced at this pottery. This company is the pioneer in the manufacture of these goods in the United States, and the results of intelligent, painstaking, and persistent efforts are in a splendid success. Whoever visits the works located in the former village of Geddes, now the Tenth ward of the city of Syracuse, will be surprised and gratified at what is being done there in the manufacture of fine china ware. For beauty of design and form, for elegance of finish and decoration, and for strength and durability, this ware is not surpassed. It is tested with the celebrated Haviland ware, and outranks it. At the Exposition, experts were surprised not only at the beauty and symmetry of this china, but were astonished at its enduring quality. The reason for this success is in the material used, the blending of the components, and the thorough processes of manufacture.

In January 1893, when the company began to illustrate its ware in *Crockery and Glass Journal* advertisements, it first depicted shapes in its semi-vitreous O.P.China ware, an **Oneida** toilet set and a group of **Doris** items. It then stated about its china that the "manufacturers of the Celebrated Onondaga China" promise that "these goods are absolutely free from crazage [*sic*] and very light in weight, the hollow pieces showing a translucency. Large line of choice decorations, both in Dinner, Toilet and Fancy pieces."

Pass's accomplishments in the 1890s seem even more remarkable viewed against the broader context of the times. Business in the American ceramic industry nationwide had been poor in the early 1890s, and it worsened in 1893 following the late spring onset of a worldwide financial panic. The situation took on crisis proportions as congress moved toward a reduction of duties on foreign ceramic imports (the Wilson-Gorman Bill). One Syracuse newspaper predicted that if the duty was removed, "with the English potteries paying only half the wages American workman were receiving, the English would take the American market, which they already half possessed. Attendant results if Americans met British manufacturing standards would mean lower wages to American workmen, inferior American products, or more likely, no American manufacture at all."[23]

During the height of the troubles in 1893, good fortune brought the company a large order to make the ceramic part of the Premier Egg Cup, a china egg coddler with a "German silver" screwtop cover (fig. 3.4). This order kept the pottery open, but doubt remained that the plant would ever get back into full operation.

Referring to the Wilson bill, the *Syracuse Journal* on 13 December 1893 editorialized, "With the splendid success of such an industry as that built up by the Onondaga Pottery Company, and its future filled with encouragement, it is a burning shame that the policy leading to such results should be overthrown, and that future blighted." The Trenton potteries responded to the economic collapse of 1893 by cutting labor costs and promising to cut further if the tariff were lowered. Pass explained to the *Journal:*

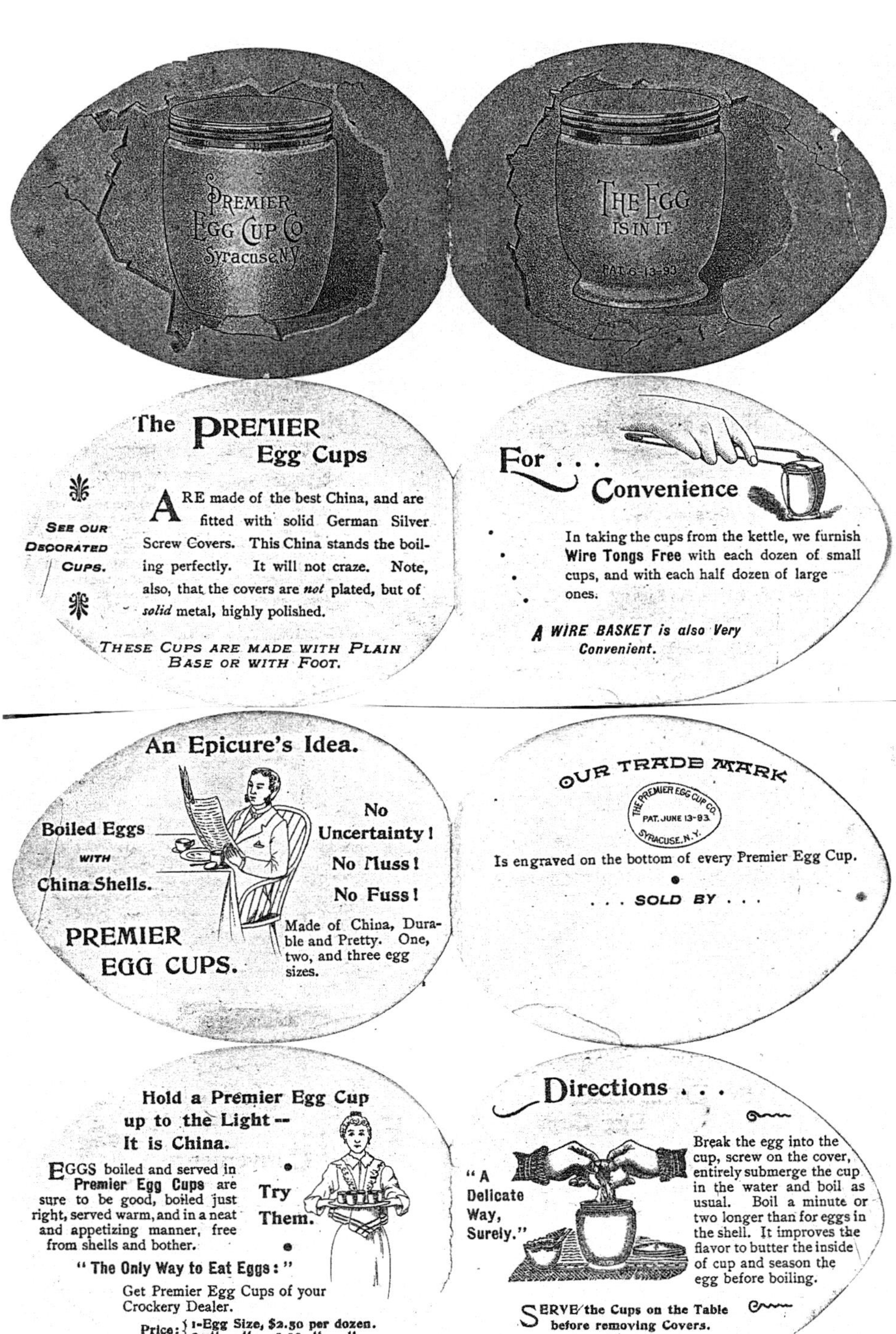

3.4. *Premier Egg Cup* flyer, 1893.

The United States Potters' Association, which embraces all the pottery manufactories in this country, at their annual meeting . . . decided upon a reduction in wages. The business of the past few years has been unremunerative . . . even under the present tariff laws. . . . It was decided to make a reduction of wages independent of the present tariff legislation. And, should the [Wilson] bill pass, a still further reduction of twenty per cent will be made in all departments.

Faced with such major wage reductions, on 25 January 1894, the Trenton pottery workers went on strike. Onondaga Pottery's men struck in sympathy. Pass, who was liked and trusted by his workers, went into a huddle with them soon after they walked out, and although nothing was resolved, and no punches were pulled on either side, mutual respect survived. Nearly six months later, a few days before the strike ended, Pass addressed a meeting of his company's directors. In his capacity as general manager, he reviewed the situation and sounded a brighter note than the industry as a whole. Although buffeted by recession, a strike, and reduction of the tariff, the pottery was advancing. Only two and one-half years had passed since the crisis brought about by the near-simultaneous deaths of Pharis and Alexander.

The sales for the year past are $121,305, for the previous year $137,334. Considering the condition of business for the past year and the fact that the strike has limited our production, the showing is very satisfactory. It is with pleasure that we are able to state that for the first time in the history of this company, <u>there is no indebtedness</u>, excepting to the Stockholders. It is only a little more than two years ago [at the time of Pharis's death] that we were owing some $46,000. This has all been paid, without any reduction of the assets except such as is shown by this year's inventory amounting to about $10,000, and reduction of stock may be viewed with satisfaction when we consider the prospects of the future. . . . A considerable sum has been expended in extraordinary repairs, all of the kilns . . . have been thoroughly repaired at an expense exceeding $1200, $1000 has been expended in tearing down all the dilapidated old sheds and building a large new one that gives us ample room for outdoor storage, . . . new fences entirely enclosing the premises, cleaning up the site of the old Pottery and putting the barn in thorough repair and raising it.

The product has been kept up to its usual high standard and the kind words (even praise) frequently received from our customers testifies to the appreciation our efforts in this direction receive. We have also been awarded the Medal and diploma of the <u>World's Fair</u> for <u>Translucent China</u> which is a very comprehensive term applying to our new product, . . . the Syracuse China.

On January 25th [1894] commenced the strike, which has kept nearly all the potteries of the country idle up to this date. There are rumors of a settlement having been reached but we have no official information confirming the rumors. . . . Early in the contest and with a view to keeping our men employed a proposition was made[:] . . . they go to work at the new scale of wages and this company guarantees to rebate to them the amount of any concession that might be obtained as a result of the strike so that from the day they commenced work they would receive the rate of wages that is settled on as the standard for the pottery workers of the country, but this proposition was rejected.

As soon as it became evident that the contest would be a long one we began to look about for machinery to simplify our methods of work and with this in view . . . [We (Pass and company officials)] visited the Pottery districts of England for the purpose of studying their methods and while there bought two machines for jiggering [pull downs] of a type which in that district have entirely superseded the skilled hand labor, effecting a great economy in cost of making ware, so much so that in this particular class of work in England they pay less than one third the price we have been paying. These machines, [along] with improved methods of drying the ware, are now in operation [at Onondaga Pottery] with good promise of success. Should they prove as successful as now seems possible, other improved methods of similar character will be adopted and we hope in a short time to have the factory in full operation whether the strike is settled or not.[24]

Settled on 16 July 1894, four days after Pass had made his report to the company's directors, the strike would be unique in the company's history. The good will generated in the Pass era persisted far into the twentieth century. As early as 1888, Onondaga Pottery workers had formed an Employees Mutual Benefit Association. By the mid-1890s, it was widely known in the industry that the pottery actively developed its own men and women by training local beginners and having them work with and learn from the highly skilled English and other journeymen whom Pass had managed to attract to Syracuse. The company guaranteed and paid for fifty weeks of work a year—a situation not common among other American potteries in those days. The workers did not join a modern union until 1972.

With the introduction of the pull-down jiggers that Pass obtained in England, it became clear that the Industrial Revolution's machine age, rolling forward relentlessly, had begun to make an impact on American ceramic manufacturing even outside the major centers of the industry. Yet despite the machines, or perhaps because of them, the force of skilled and less skilled workers at Onondaga Pottery continued to grow.

James Pass had saved the pottery, given it direction, and established its reputation as a maker of fine true china. He knew that neither he nor the pottery could rest on its laurels, however, for American competition would soon catch up. He forged ahead to find new and better ways to produce American china and to create a new market for it.

In 1894 the company revised its straight-line O.P.China backstamp for its best earthenware with the addition of the words "semi-vitreous." This reflected one final change in the O.P.China earthenware body. The new mark went on the last of the *Juno* and *Doris* items. In the same year the company issued one more earthenware shape, ***Seville,*** a casserole which it illustrated in 1895 on the cover page of its final *O.P.China Price List* (fig. 3.5). This shape proved to be short-lived because the company dropped all earthenware shapes in 1897. Instead of generating a price list of its own, in that year the company included these items in an industry-wide *Standard White Granite Price List,* published partly as a concession to the industry's troubled financial health. (Pass had recently been elected second vice president of the White Granite Association.) Planning to phase out earthenware entirely, Pass saw no need to promote it.

He now dealt with the problems of tooling up and training his men to make a greater volume of American china. With this plan in mind, in 1895 he introduced the new Syracuse China

O. P. CO.
CHINA.
SEMI-VITREOUS.

O.P./China Semi-Vitreous
(1894–97)

Plymouth shape (plate 4). Haley had designed
it. Superior to *Marmora* in many ways, the
Plymouth shape was the most carefully considered and consistently designed product the pottery had yet produced. In form and decoration
it resembled the popular Haviland French
Baroque designs of the day. The strong foothold
Haviland had gained with its imported porcelain tableware forty years earlier had grown
steadily since the Civil War. By 1896 it offered
several new shapes with which American potteries had to compete. *Plymouth* presented a successful challenge to Haviland's dominance of
the market.

The *Plymouth* hollowware forms were wide
and low, curving elegantly to fluted edges and
scalloped rims. The comports, casseroles, and
tureens stood on graceful pedestals. The over-
hanging covers fit well. Huber had shown Pass
examples of Haviland's new *Ranson* shape (introduced in 1893) as an ideal product, and Pass
had set Haley to work on a similar concept. His
Plymouth shape reflected the exuberance of
Haviland's *Ranson,* echoing and refining its
rocaille moulded decorations, but with its own
integrity of design. *Plymouth* sold beyond the
company's most optimistic hopes. It sustained
its popularity into the 1920s. It made the late
1890s a more confident and optimistic time at
the pottery than the first half of the decade had
been. It confirmed the trust that the company's
directors and workers alike had placed in Pass
in the difficult years of 1892–94. It established
Syracuse China as the standard for industrially
produced American china.

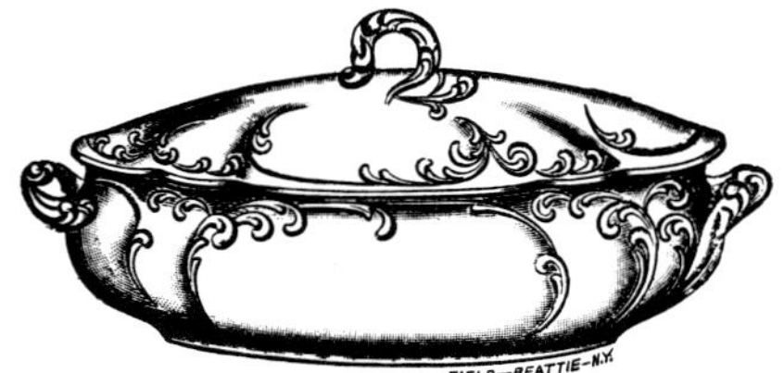

3.5. Page of 1895 price list showing *Seville* casserole
and Semi-Vitreous trademark.

Four

A New Beginning
1896–1900

We have at our place eight kilns; six are in use and two have never had a fire in them, because we can't get men to man them.

—James Pass, 1900

Between 1894 and 1897, Pass made at least three trips to European potteries. In June 1896, two years after his preliminary visit abroad with company officials, the company's board of directors authorized his absence for a trip from mid-July to mid-September for "the purpose of visiting European potteries." He went to England, France, and perhaps Germany. Already acquainted with many imported European ceramic products, his aim now was to assess the increasing use of machines in the manufacture of European china and porcelain. The strikes in 1894 had made it clear that labor unrest could be a major stumbling block in his plan to produce fine china. He realized ever more sharply that for Onondaga Pottery to turn out quality china in competitive quantity, he needed to concentrate on increasing manufacturing efficiency and productivity. Machines could help, but only to a modest degree in the 1890s. Pull-down machines for jiggering had begun to replace some of the labor-intensive methods that the pottery had used since it had begun to make earthenware, but machines in the 1890s could help only in a few operations. The making of first-class American china depended on highly skilled hand workers.[1]

Pass had instilled in his pottery an outlook that welcomed advances in technology, whether in manufacturing, plant design, or any other aspect of the business. In this way, Onondaga differed from the Trenton tableware potteries, where labor and management alike resisted the introduction of machinery. The East Liverpool potteries, with their Ohio "western spirit" of ingenuity and innovation, had led the way in this respect in the 1870s and 80s. Trenton's resistance came from a belief in the superiority of skilled hands over machines, as well as from the under-capitalization of its potteries. When one Trenton manufacturer said in 1895 that "no machinery [for tableware making] had ever been invented to work automatically and none can work without the guiding hand of the potter," he was expressing a view widely held throughout the industry. But at Onondaga Pottery and elsewhere among more future-oriented manufacturers, success in manufacturing seemed certain to depend on advances in machine technology. Pass and others like him believed that improved machines would free skilled labor to spend more time doing those things in china-making that can only be done by hand.[2]

Beyond learning firsthand about European manufacturing methods, Pass studied advances in ceramic technology and chemistry relating to the problems of forming and firing. On his return, he adjusted the composition of his china body, making it possible for a jiggerman to get up to 95 percent of good ware off his moulds. Nothing like this performance had been possible before. He almost certainly also modified his mould-making and ware-drying techniques to achieve such a significant yield.

To fire different wares at different temperatures in the same kiln presented a major challenge. It was one thing to fire the china clay body protected in the kiln by saggers of semi-vitreous ware, and another to fire a kiln loaded only with china. Periodic bottle kilns of the type used in the industry at the time did not, in any event, fire evenly throughout. In firing earthenware and china together, he had placed saggers of different kinds of ware in rings around the kiln according to the temperature reached in that part of the oven. As the company developed its true chinaware, Pass placed it only in the hottest sections of the biscuit kiln. With experimentation in placement and firing, china gradually came to occupy the entire kiln. Pass believed that Onondaga was the first pottery in America to successfully fire kilns filled exclusively with china tableware.

For many years, selectors at the pottery had inspected and graded decorated and undecorated earthenware, sorting out imperfect pieces by category. They separated first quality ware from seconds. Some customers in the trade preferred to buy unsorted R.O.K. (run of the kiln) ware as it came from the kiln. In the late 1880s, Oliver told an inquiring customer that Onondaga Pottery discarded pieces that other potteries classified and sold as thirds. As Pass began to fire china, he retrained the selectors to make even finer distinctions in grading.[3]

While in France, Pass had taken a close look at a decorating process that had been developed there in the 1870s. In the trade it was called decalcomania. The word came from the French *decalcomanie* (*de calquer*, to trace; *manie*, by hand). When Pass decided to adopt this process at Onondaga Pottery he made a return trip in 1896 to negotiate the hiring of three Frenchmen skilled in its steps: a lithographer, a stippling specialist, and a pressman. The lithographer, André Regand, remained with the pottery until 1903. When Pass installed in-house lithographic printing equipment in 1897, he believed that "the Onondaga Pottery Company produced the first ceramic lithography in the New World."[4]

Decalcomania was a method of affixing pictures and designs printed in multiple ceramic colors on a specially prepared duplex paper to a smooth ceramic surface. The color printing eliminated the need for application of colors by hand. The lithographic process made possible finer and more elaborate and intricate patterns with crisp outlines.

Decalcomania decoration caught on so rapidly that Pass soon had twenty-four women crowded into the decorating room applying decals to glazed ware. This overglaze decoration offered bright, clear, sharply delineated colors. It produced uniform, high-quality multicolored decorations at a lower cost than did hand-painting. Furthermore, unlike transfer prints from etched copper plates, which had to be used fresh from the press before their pigments dried, the sheets of decals printed from lithographic stones could be stored in quantity for future use, even many years later. Decalcomania did not replace transfer printing, however; the latter remained in use for one-color decorations and patterns that called for the freer techniques of hand-filling.

On his travels, Pass also investigated the French process of giving a rolled edge to plates, saucers, and other pieces to lessen chipping. Years later Huber took credit for introducing Pass to this distinctive feature of some French tableware, though Pass surely already knew of it. Huber's contribution most likely was that of persuading Pass that the round edge would sweep the rapidly growing market for hotel dining ware in America. Huber claimed that he first encountered the shape in New York at the Knickerbocker Hotel, where his place setting included a four-inch French round edge bread-and-butter plate. "I took the butter off the plate . . . slipped the plate in the folds of my *New York Times* and walked out. . . . I asked [Pass] if he thought we could make it. 'That's very simple, very simple,' he said, 'we can make that. Is there any volume [of business] in it?' 'You bet there is,' I said, 'that's hotel ware.'"[5] Pass set out to develop the product for the American hotel and restaurant market.

The advantage of the round edge was that it resisted chipping. The edge worked from the principle that when a round object is struck, the energy radiates, lessening its effect on the point of impact. The edge of the rim had to be perfectly half-round, however; from this roundness came its strength. In jiggering the edge, the potter had to keep the center of the roll exactly between the tool and the mould. Once the rim was formed, the finisher, who handled it next, needed to preserve the edge and avoid trimming, flattening, over-finishing, or otherwise altering it. To succeed, the rim of a round edge plate could have no flat facet.

A slightly elevated angle to the rim of rolled-edge ware (a feature Pass adapted) added to its strength. An article in *Food Profits* in 1927 pointed out: "A flat rim will chip much more readily than the ones which have the dished-up rims. In flat plates, it is often observed that the chip shears off downward. . . . When two flat rim plates collide (as for instance happens when they are slid across a dish table) the result is a solid impact. Plates with dished-up rims, however, deflect part of the force upward, giving a glancing blow."[6]

Pass must have understood this intuitively. He showed Haley how to make the round edge and assigned to him the task of designing what would become the most important shape in the emerging hotel ware industry. Haley initially designed more than sixty items for this shape; he would more than double that number in years to come. The pottery named the shape simply **Round Edge** (fig 4.1). (In the shop, the men always called it "rolled edge.") In mid–twentieth century the company renamed it first **Rolled Edge,** then **Rolledge.** Pass made another trip to Europe specifically to secure an elliptical jigger to form a round edge on oval ware.

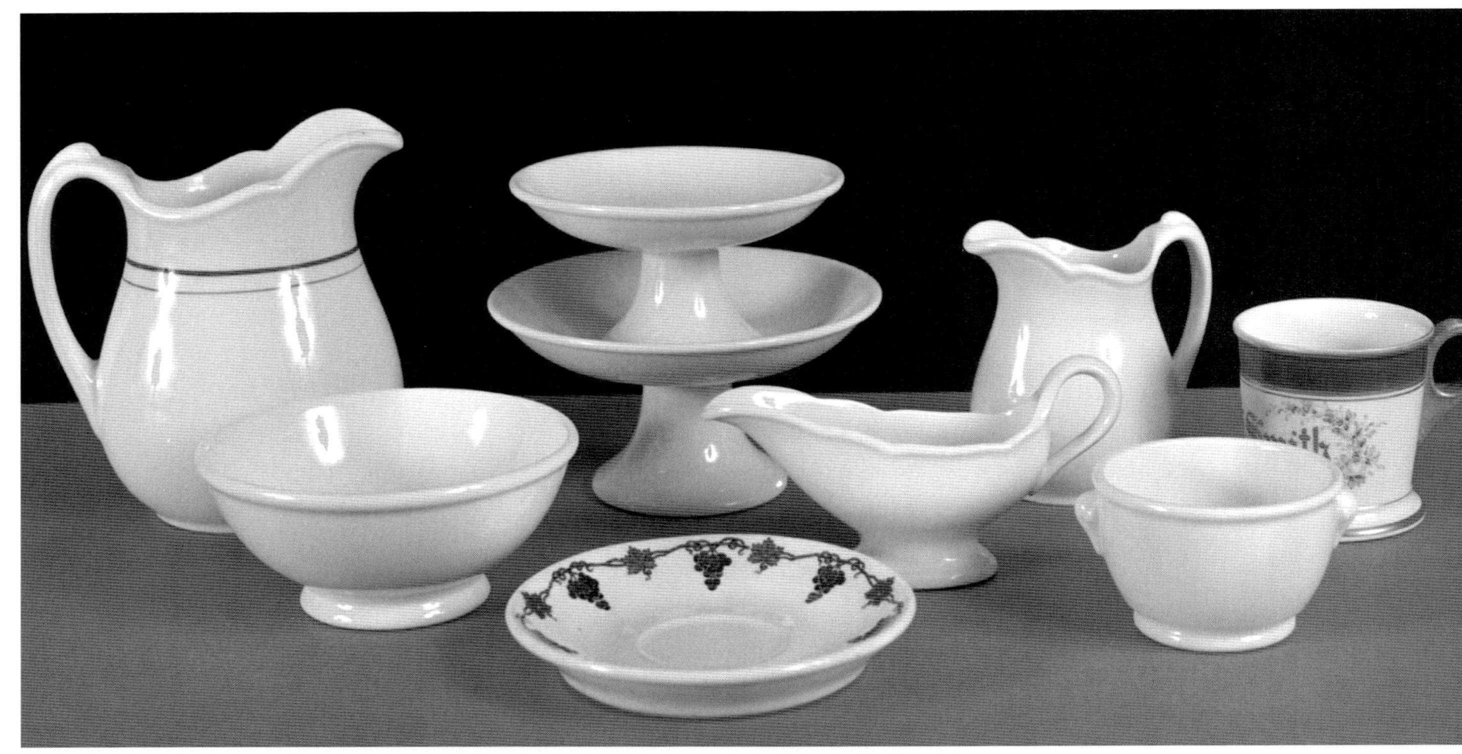

4.1. Some **Round Edge** items. *Left to right:* jug, oyster bowl, high foot comports, saucer, gravy boat, jug, handled sugar, and coffee (or shaving) mug.

With work on items for the new **Round Edge** shape under way, Pass sent Huber and his men on the road with prototypes to test the market. On 14 August 1897, surprised by the rapid response in some quarters, Pass wrote to E. M. Euniac, his New York sales agent:

> For heaven sake let up on that rolled edge plate. . . . You were instructed to ascertain how it probably would be received in case we decided to manufacture it, and not to take orders for it. Now here is Putts & Johnson China Co. wondering why we are selling goods that we cannot furnish [yet]. Orders from Philadelphia and New York have been duly received. You are getting hold of the right kind of people. . . . We are just beginning to fill our orders for the new [decalcomania] decorations, and the goods come through very nicely. It is a success sure. We will consider this trip a missionary [one].[7]

Once **Round Edge** was ready to market, Huber went on the road with Pass's directions to give it away, if he had to, to prove its worth. Huber noted in his memoirs that hotel dining rooms across the country then tended to use semi-vitreous ware from Trenton's Greenwood Pottery more than from any other.[8] With a thinner, harder, round-edged china, Onondaga Pottery had an opportunity to claim much of Greenwood's hotel business nationally.[9]

Traveling with samples of ***Round Edge,*** Huber first demonstrated its superiority to the manager of the Cadillac Hotel in Detroit by striking an Onondaga plate edgewise with a Greenwood plate. Only the Greenwood plate chipped. When, after further strikes, the Onondaga plate at last chipped, it did so out of sight under its round edge. Another strong selling point was that ***Round Edge*** stacked up perfectly, its foot resting snugly in the well of the plate below it. Authorized to send the hotel twenty dozen plates for trial, gratis, Huber did so, and then repeated this "giveaway" with ten other hotels across the Midwest. Orders soon started coming in. As ***Round Edge*** gained in popularity in the hotel market, it established Onondaga Pottery's reputation in that field. Soon the company's other shapes also began to appear in dining rooms and restaurants.

The demand for Onondaga Pottery's American china grew rapidly not only in the hotel dining business but in the domestic fine china market as well. The company could now claim without exaggeration that its ware was as good as any made in the hemisphere, and better than nearly all. It possessed a high quality body in attractive and practical shapes, well-glazed and tastefully decorated by the newest processes. At this point, the pottery's lack of skilled workers posed the chief hindrance to increasing china production to keep up with demand. Even for its earthenware production in the 1880s and early 1890s, the company had never had enough well-trained workers. For china production, which called for different and more exacting skills, the lack was severe.

Immigrant skilled labor continued to offer a way out of this problem. Just before the turn of the century, Pass hired William H. Swift, a Staffordshire-trained potter who had worked in France for Limoges. Swift introduced the highly advantageous method of casting ware (instead of pressing it), a process not previously used at the pottery. But even without the casting process, since the mid-1880s the pottery's clay pressers had excelled in making earthenware of high quality with the use of the "tommy stick" (a sponge on the end of a stick) and other simple tools. In his short tenure, Swift improved the craftsmanship of his fellow workers at the factory.

As the production of American china increased late in the 1890s, Pass invited another English-trained potter, Thomas Astbury, to introduce the method of throwing cup liners on a potter's wheel. (Pass had already, around 1893, introduced lathes for turning cups.) Like Swift, Astbury commanded much attention from other workers. He represented a level of expertise impossible to find even in the best locally trained potters. Astbury was followed by a number of less-experienced younger Staffordshire potters, four of whom were recruited directly from England. Most did not remain with the pottery for long. Coming from pottery traditions that still emphasized individual artisanlike production within the divisions of labor in a factory setting, they found it difficult to adapt to Pass's emphasis on team-oriented methods. Pass had instituted these methods in part to make up for his lack of skilled help. A few of the Englishmen who arrived at the end of the 1890s left to join the small Chittenango Pottery twenty miles or so east of Syracuse, a venture that did not long survive.[10]

One faithful newcomer from Staffordshire, Levi Eccles, worked as a dipper at Onondaga Pottery from 1900 until his death in 1921. How Eccles came to work at the pottery is known from a remarkable document preserved in the Syracuse China archives. It is a record of Pass's testimony at a hearing conducted in New York City on 4 April 1900 by the United States Immigration Service (parts of which are quoted below). Pass's responses to the questions of the members of the service's Board of Special Inquiry illuminate much about the pottery, its need for skilled workers, and his perception of its place within the American pottery industry.

Pass had authorized his engraver, the Englishman Tom Bryan, to recruit skilled compatriots in Staffordshire with a guarantee of work in Syracuse. The immigration hearing concerned four Englishmen, including Eccles, who had arrived from Liverpool four days earlier aboard the SS *Lucania*. They had been detained on arrival in New York after explaining the purpose of their journey and the circumstances that had led to it. Syracuse was their destination, they said. "About three weeks ago, in Staffordshire, England, Lewis Johnson, whom we all know, told us he had received a letter from Tom Bryan which authorized him to engage several potters . . . and that they would pay 'dippers' three pounds and twelve shillings and 'placers' two pounds and fourteen shillings per week if they would come."

Learning of their detainment, Pass traveled immediately to New York to appear before the Immigration Service's Board of Special Enquiry. His answers to the questions put to him by the board members shed much light on the problems that attended his commitment to manufacture true china. He now and then also offered a hint of his quick wit.

He said that his company produced its fine china in the same manner that English potteries made fine china, that the English manner differed from that used on the Continent, and that he had visited most of the principal potteries of England and the Continent. He explained some of the differences between manufacturing earthenware, for which he had barely sufficient labor, and china, for which he had an inadequately skilled work force and no way of obtaining a better one within the United States. He said,

> The method employed in placing the [china] ware in the [glost] kilns is very different from that used in similar operations in the manufacture of earthenware. Also, in the dipping, greater skill is required because, the ware, being thoroughly vitrified, does not absorb moisture as does earthenware and requires much greater skill in securing a uniform coating. . . . [Because] there is no China ware manufactured in the United States excepting at our factory, and possibly one or two other places where it is still in the experimental stage and is hardly on the market as a commercial item, there is nothing in the way of potters to be secured in this country but the ordinary earthenware workers. They . . . [might gain China making skills] after a period of time, but a certain nucleus of those already skilled must be . . . [present] around which to accumulate them. . . . [This would require] a period of several years.

Pass indicated that he hoped that the four detainees would act as the nucleus for the training of American apprentices. When the board asked what efforts he had made to find this nucleus in the United States, he responded that the company had advertised for even ordinary skilled earthenware workers and had not been able to secure them. "I advertised in the *Daily State Gazette* of Trenton twice . . . two weeks at each time, separated by an interval of four weeks, and secured not even an inquiry." He was asked, "Have you tried to advertise in Syracuse?" He replied, "You might as well advertise in the Sahara Desert for wheat—it doesn't grow there. We have every potter in Syracuse employed."

Asked how much of the pottery's production consisted of vitreous china, he said that six years earlier it had been probably less than 10 percent of output, but that, "today [1900] the thin chinaware would amount to 60 percent," (Pass referred to only thin dinnerware; all ware by this time was vitreous). When asked what other American potteries manufactured fine china, and how long they had been doing so, he answered, "I only know of two others: the Maddock Pottery Company, of Trenton, which is experimenting with it and have put out some goods, but a limited quantity. . . . The other is Bell Bros. of, I think, Findlay, Ohio. . . . It is difficult to say when they were experimenting or selling [but] it is within the last two years, I think, [that] any of it has appeared on the market." He said that Onondaga Pottery's production constituted "practically all" of American china production. "We are the one concern turning it out commercially."

Asked how much of a market existed for vitreous china tableware, Pass responded, "I think somewhere between seven and eight million dollars worth is imported, largely coming from the Continent, from France and Germany. Then the English china is high grade and sold at a high price, but only a limited amount." He indicated that if there were a sufficient supply of skilled labor in this country for making thin table china, the amount of fine ware imported would be reduced, though this would take time, for "you would have to work against the competition of [European] manufacturers already well entrenched. We have at our place eight kilns; six are in use and two have never had a fire in them, because we can't get men to man them."

Pass made it clear that he had not knowingly meant to violate Immigration Service rules, and that the pottery would in the future follow the required procedures. The board voted to admit the four men. They went to work at the pottery, but only Eccles stayed on, remaining for twenty-one years, until his death at age forty-four from pneumonia.

Committed exclusively to the production of Syracuse China, the company needed a distinctive new backstamp. In February 1897, a prototype mark entered use for one month. In this, "Syracuse China" appeared as a circle crossed horizontally through its center by "O.P.Co." In March 1897, the company replaced this mark with the more distinctive backstamp, a direct, straightforward statement of the product: "O.P.Co. Syracuse China" accompanied with an impressed date code. From 1896 until 1926, the pottery used only one clay body (Pass constantly improved his No. 2 china, which became known as the flint body) for all vitreous chinaware, no matter what shape or grade of thickness.

The pottery had first produced its American china as thin household tableware and decorative pieces. In 1940, Pass's son, Richard, recalled that initially it had been sold at retail almost entirely through jewelry stores. With a few exceptions such as the Yates, hotels and restaurants continued for a few years to buy white granite or O.P.Co. ware. He said, "Our hotel china was an outgrowth of our dinnerware china and not vice versa, as many assume." He added that, "In the later 1890s it began to have considerable acceptance in the hotel field because of its extreme strength, being far superior in that respect to any ware available up to that time. . . For hotel

O.P.Co./Syracuse/China
(Prototype, March 1897 only)

O.P.Co./Syracuse/China
(1897–1920)

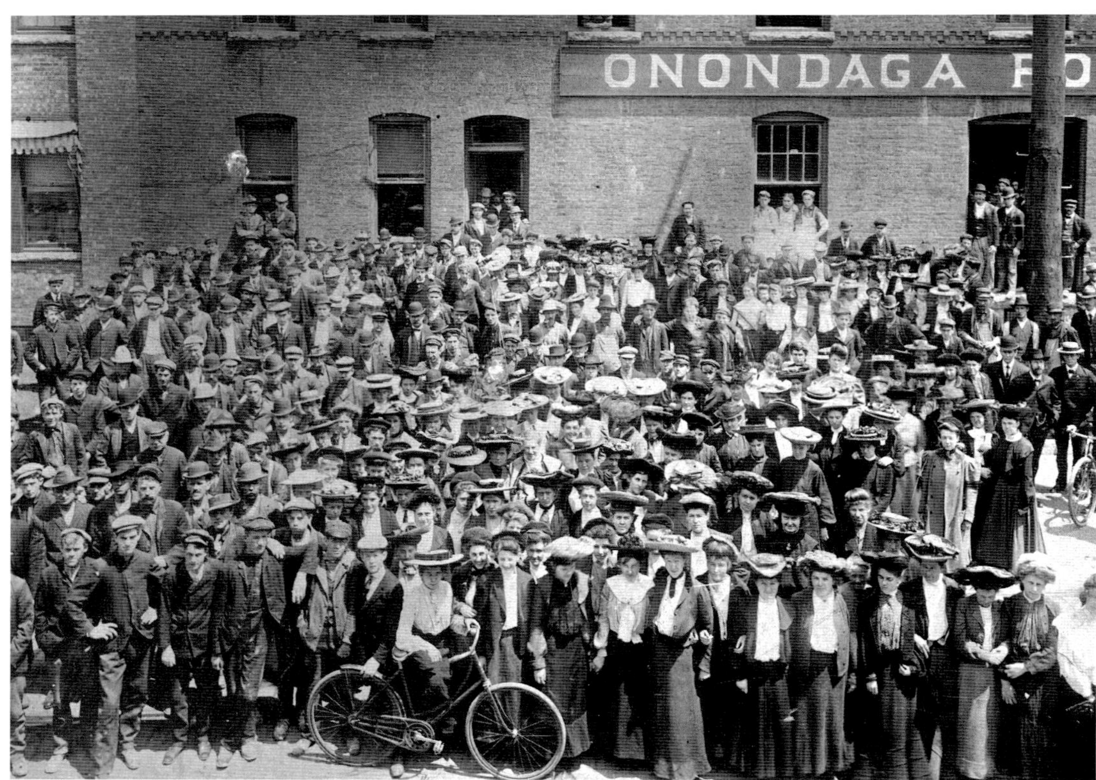

4.2. Pottery employees gathered in back of Fayette Street plant, c. 1900.

ware use we modeled new shapes and adopted the Rolled Edge for the first time on American pottery." As the new china body came to be the basis for all ware, so the new backstamp became the only one in use.

The company's price list for its American china, published in 1899, was titled *Hotel Ware*. It offered "China, White and Decorated Round Edge and Thin Hotel Ware." This amounted to **Round Edge** plus the **Plymouth** shape, some holdover **Imperial, Marmora,** and **Moro** items, plus some odd-shaped items. The company stated with confidence: "This Ware is Warranted to be the Strongest China on the Market."

In 1899, as the new century was about to dawn, the company had expanded once more, completing its D unit and doubling the factory's capacity by adding four kilns, bringing the total to eight. As Pass testified the next year, two of these kilns were yet to be used, but this situation would soon be remedied.

He had accomplished what he had set out to do, not only through his remarkable skills, vision, and determination, but also through the remarkable loyalty of his workers and sales force (fig. 4.2). He inspired their trust in many ways, but especially through his genuine concern for their welfare. In time his successors as president would amplify that concern as "industrial humanics," enabled to do so by the company's continuing prosperity. That prosperity now depended increasingly on the pottery's ability to stay in the forefront of its industry in design and decoration at a time when the arts of design were undergoing great change.

Five

American China for the Twentieth Century
1901–1913

Mr. Pass, the manager of the pottery, is ambitious to raise the art standard
of its china so that it will occupy a unique position in the American pottery
world and eventually abroad.

—*Keramic Studio,* 1902

As the new century opened, Mark Haley was at work on a number of new shapes. He designed them not only to take advantage of the properties of the china clay body but also in response to the spirit of design reform that was then making an impact on the decorative arts and domestic furnishings across the country. The departments that decorated the ware added new patterns in newer styles. Like all tableware manufacturers, the pottery followed what the trade demanded. It now demanded new departures in design. The long-standing unadventurousness of the American tableware industry in matters of decoration began, cautiously but steadily, to give way to the influence of worldwide design reform.

The new thought in design that influenced tableware decoration came chiefly from two sources: the Art Nouveau style and the Arts and Crafts movement. Art Nouveau had originated in France and Belgium late in the nineteenth century and soon made its way to American shores. Although it was short-lived as a decorative style, it represented a major departure from well-entrenched academic and historical modes of decoration. It was a breath of fresh air. Avoiding straight edges and perpendiculars, it emphasized instead sinuously undulating curvilinear lines, many derived from plant forms. In the 1890s, Art Nouveau represented "modern" design. It stood for freedom from all that seemed stodgy and tradition-bound.

The Arts and Crafts movement, which had began in England a generation earlier, had little effect in America until the late 1890s. It was not an identifiable style as much as an expression of faith in handcrafting as a positive influence in society. It sought to free design from the often impersonal, soulless, and monotonous constraints of machine mass production. Unlike Art Nouveau, the Arts and Crafts movement paid careful heed to the past, and to medieval times in particular. Like Art Nouveau, it found fault with the directions that design in the decorative arts and architecture had taken through most of the nineteenth century, but the movement looked backward as much as forward. It idealized the Middle Ages as an era when human creativity in honest crafts had been a force for the good in society, one that industrialization had swept aside. Both reform movements had developed richly in Europe, though chiefly at levels that left the ordinary middle-class home little touched. In America, these reforms in design had a much broader influence, especially in the home.

Early in the first decade of the new century, Syracuse became an important center of the American Arts and Crafts movement. It did so in good part because of the presence of two figures whose publications helped to shape the American acceptance of design reforms: the china painter and ceramist Adelaide Alsop Robineau and the furniture maker Gustav Stickley. In 1899, Robineau moved to Syracuse and began to publish the monthly *Keramic Studio* magazine, directed at first to American china painters. She was then one herself; her great achievements as a studio potter lay a few years ahead. Her magazine offered, among contributions by many noted ceramic artists, patterns for the use of china painters. China painting had gained national popularity, especially with women. How fortunate it was that Pass, whose pottery needed skilled china painters, had one of the field's leading professionals in his own community.[1]

In August 1901, Pass hired Robineau as a part-time consultant. For nearly two years, she earned fifty dollars a month for her services. These consisted chiefly of instructing the decorating department's ever-growing number of china painters, but she also guided the efforts of the company's pattern designers quietly into the spirit of reform. In addition, she created several decalcomania patterns. In the December 1901 exhibition of the New York Society of Keramic Artists in New York City, Robineau submitted some of her Onondaga Pottery decalcomania designs. In its February 1902 issue, *Keramic Studio* reported in an unsigned review:

> Under the direction of Mrs. Robineau, the [Onondaga] Pottery is making the experiment of introducing more artistic and original decorations on their printed ware. The effort is a commendable one, as a good deal of financial risk is involved in persuading the public to buy artistic designs, and in a big factory like the Onondaga Pottery where hundreds of girls are employed, the [danger of] loss from work poorly done is greatly increased in introducing this class of design. Mr. Pass, the manager of the pottery, is ambitious to raise the art standard of [its china], so that it will occupy a unique position in the American pottery world and eventually abroad.[2]

Although it was not unusual for potteries to call on the services of freelance designers, it was indeed unusual, and perhaps unique in the industry to that time, to invite a freelance artist to raise a manufacturer's "art standard." That a woman would be invited to do this was remarkable, and further testimony to Pass's vision.

In March 1903, *Keramic Studio* published a photograph illustrating several plates with prototype decal designs by Robineau. The pottery put two of these designs into production (plate 5). It considered one, "Violet Border," important enough to obtain for it the company's first known patent for a decoration, registered on 30 May 1904. Two years later, on 13 March 1906, the company patented another design, "Moss Rose," also called "Wild Rose Art Nouveau Border."[3] Records show that Robineau's "Moss Rose" pattern was ordered for the newly installed, Arts and Crafts–inspired Tabard Room at the Yates Hotel, a restaurant whose "Craftsman" furnishings came from Gustav Stickley's United Crafts.

Robineau's decal decorations are stylized treatments of floral subjects. She filled the outlines of her motifs with muted tones rather than bright colors, creating a flat-patterned, flowing, decorative border with a "hint of modern" feeling to it, while keeping the charm and romantic essence of ever-popular floral motifs. Reflecting both the flowing line of Art Nouveau style and the conventionalized simplicity of Arts and Crafts design, they departed from the chastely naturalistic Victorian floral patterns that had long been the dominant style in the decoration of American tableware. The newer style patterns did not drive out the old. They coexisted for many years.

In 1904, the company again entered ware in a World's Fair, this time in St. Louis, at the Louisiana Purchase Exposition. Although no documentation survives, it is reasonable to assume that the pottery entered its latest products. The award, "Grand Prize for Clays and Table Ware," probably referred to decorated **Puritan** and **Round Edge** ware.

In 1908 the company offered but did not copyright a third Robineau pattern, "Grape Border" (plate 5). By this time, Robineau was fully involved with her great work as a studio potter. The unique one-of-a-kind porcelain vases, bowls, and vessels that came from her studio in Syracuse, each wheel-thrown and hand-decorated by her, had placed her in a realm of ceramic art well removed from the industrial manufacture of china tableware.

Stickley had begun to manufacture Arts and Crafts–inspired furniture in Syracuse in 1899. He founded *The Craftsman* two years later. This magazine, with its broad subject appeal, spread the ideas of design reform across the nation. It drew on the writings of William Morris, John Ruskin, and others, including contemporary Americans, to reinforce Stickley's arguments for simplicity in design, honesty in materials, and using machines in the service of handcraft production rather than as producers themselves. In 1903, Stickley held a large, nationally important Arts and Crafts exhibition in the Craftsman Building, his downtown Syracuse headquarters.[4] The Craftsman Building housed Stickley's editorial offices, showrooms, a lecture hall, and a tea room. For the last, he ordered a service of Syracuse China decorated with his now-famous Jan Van Eyck–derived *Als Ik Kan* motto and logo. The exhibition drew together most of the elements of Syracuse's growing Arts and Crafts community.

As demand rose for customized patterns for the burgeoning hotel and restaurant trade, and some of the china-buying public shifted its taste in decoration from Victorian elegance to simpler styles, the pottery's decorating department developed a virtuosity in design far beyond what it had possessed in its early years. As early as 1897, before Robineau's participation as teacher and designer, the pottery had hired Louis A. Cowan, an able East Liverpool decorator, to head the department. He followed a series of decorating shop foremen (Charlie Lee and a Mr. Copson among them). By this time, the pottery had three separate but coordinated departments concerned with the visual appeal of its products: lithography (decalcomania), decorating (engraving, filling, lining, banding, and free-hand painting), and design (shapes).

Cowan had come from a subcontracting tradition widespread in the pottery industry. It encouraged individual artisan practice within an independent decorating shop. Walter's short-lived Boston China Decorating Works had been part of the same tradition. In 1903, Pass reassigned Cowan to the clay shop. He left in 1905.[5]

Needing to find a replacement to head the department, Pass wrote for advice to his close colleague in the profession, Charles Binns, who was now head of the School of Clayworking at Alfred University.[6] Binns and Pass held each other in high regard. The two men had been founders of the American Ceramic Society, in 1898. Binns's role in this organization kept him in touch with nearly all American ceramic manufacturers and gave him a unique opportunity to know everyone in the profession. Three letters from Binns, now in the Syracuse China archives, illuminate their exchange. Binns wrote in response to Pass on 27 February:

> As to a man for you I fear they are hard to find. Do you know John Wigley of Trenton? He is head of the dec. dept. at the Empire Pottery. . . . He would be a good manager I think, but [I] have not a very high opinion of his artistic ability. He is an Englishman. Almost all the men are simply hack decorators as is shown by the poor stuff most of them turn out. I imagine that you want a good *executive* & that you can take care of the quality of the product yourself. At the same time you want a man of fair education and one who understands the processes. It is strange how scarce they are. Would you care to have me make some enquiries? . . . I wish you could have a brother of mine from Worcester associated with you. Together you would make the fur fly.[7]

Another statement in his letter is illuminating. Onondaga Pottery's head of decoration had long worked as a salaried employee rather than as a subcontractor, and, even as late as 1903, this departed from much of the industry's practice. "I do not know how you arrange your decorating business. Whether it is controlled by the head decorator and charged by him to the factory or whether he and his people are on salaries. The former is the Trenton method, I believe. At least I know that John M. Pope hires his own people, buys his own colors, etc. and charges the factory with the cost of decorating."[8]

Pass found Harry Aitken, a promising young English designer working in Trenton, and hired him in 1904 to replace Cowan as head of the decorating department. Aitken streamlined the color system as well as the process of printing engravings to be filled. Where before, six women had worked on each press, he cut this number in half. (Men operated the presses while women removed the printed sheets, inspected them, and hung them over dowels to keep them from sticking together.) In addition to instituting team methods of greater efficiency, Aitken added to the company's reputation for fine decoration through his talents as a watercolorist and painter in ceramic colors. He worked in a picturesque and painterly style, developing for the pottery several series of hand-painted decorations depicting such subjects as tall-masted sailing ships, scenes from Dickens, deep sea gardens, water lilies, and a group of forty "Nature Studies" depicting birds and flowers. He insisted on painting many of these decorations himself. They constitute some of the most accomplished hand-painted pictorial decorations of the era to come from an American pottery (plates 6–12).

Long freed from supervisory work, Elmer Walter worked for many years in the decorating department as its designer. In 1905, he created a series of designs very much influenced by the Art Nouveau style promoted by *Keramic Studio.* One of these, the lovely "Chelsea" border (plate 5), was the third surface decoration patented by the company. He also designed a "modern art" series and a "duo-tone" two-color transfer print process. In the same year, he designed "Blue Plum," which in time became the most sought-after (by collectors) decorated ware the company ever made.

Tradition has it that though Walter created the design and worked out the technique (new for Onondaga Pottery), Harlow Pierce was the only person who decorated "Blue Plum." To start, a print from a copper plate was transferred in the usual way to glazed ware. Pierce then shaded the print with a mouth-blown air-brush application of blue cobalt. This was the first time this technique had been used in the department. The ware was placed in saggers with a small amount of flow blue powder (cobalt oxide), which, during the firing, owing to the volatile nature of cobalt, caused the blue to flow and blend into the print. The ware was gilded in another operation. The "Blue Plum" pattern appeared on the **Puritan, Imperial,** and **Mayflower** shapes (plate 13). "Blue Plum" apparently was made for purchase only by employees at holiday times. Although the pottery illustrated it in 1913 in a small booklet directed to homemakers, *The World's Most Durable China,* there is no record that "Blue Plum" ware was ever sold to the general public.

The demand for customized decorations of different kinds in different styles grew by leaps and bounds. Electric lighting, the new mobility brought by automobiles, and the popularity of motion pictures, all helped to transform dining in American life. The middle class had more leisure time, could afford to eat out, and increasingly traveled to do so. Quality restaurants independent of hotels began to appear in great numbers. As automobiles became common, many restaurants located themselves outside city centers. Dining rooms in hotels (which by now were nearly all on the European plan, in which the price of a room did not include meals) served the public at large as well as hotel patrons.

The demand for customized decoration of fine dining service in the form of monograms, coats-of-arms, logos, or other symbols (called "crests" by Onondaga Pottery and sometimes "badges" by other potteries) increased (plate 14). These might be combined with lines, incorporated into borders of stock patterns, or stand alone. The customer could also have a unique, elaborate, one-of-a kind decoration. The possiblities were endless (plate 15). Where in the nineteenth century crests and other decorations often appeared as one-color transfer prints, colorful decalcomania enlivened the choices. Orders that had once come chiefly from restaurants and hotels now also came from country clubs, yacht clubs, colleges, fraternal orders, hospitals, and countless other organizations and institutions that served meals. Even lunch counters at dime stores, diners, and coffee shops wanted personalized ware of better quality. Railroad travel created an entire specialty of dining car service, in which Onondaga Pottery was a major supplier (see fig. A.20); customized china decoration graced steamboats on the Great Lakes as well as ocean liners (see fig. A.21). The pottery made countless custom commemorative designs; one of the first of these commemorated the launching of the yacht *Meteor* in 1902 (fig. 5.1).

The customization of china through distinctive patterns for dining venues is evident in the pottery's price lists. In 1901, it issued one for its new **Club Colonial** shape, "designed for fine club and hotel service," a graceful shape with widely scalloped rims. In 1902, the company published a full price list with a handsome Art Nouveau cover. Emphasizing hotel ware, it included about 430 items in **Plymouth, Club Colonial,** and **Round Edge,** with the addition of a new, lighter-weight, hotelware shape line, **Normandie.** Stylistically similar to **Club Colonial, Normandie** offered rectangular flatware in place of ovoid, and was limited to bakers, dishes (platters), fruit saucers, and individual butters. **Normandie** cups came in three sizes, interchangeable with **Club Colonial** ware, just as **Club Colonial** plates went with **Normandie** items (fig 5.2). Neither shape offered hollowware items such as casseroles and sugar bowls; indeed, hotel shapes were generally more limited in accessory items. American hotel dining rooms and

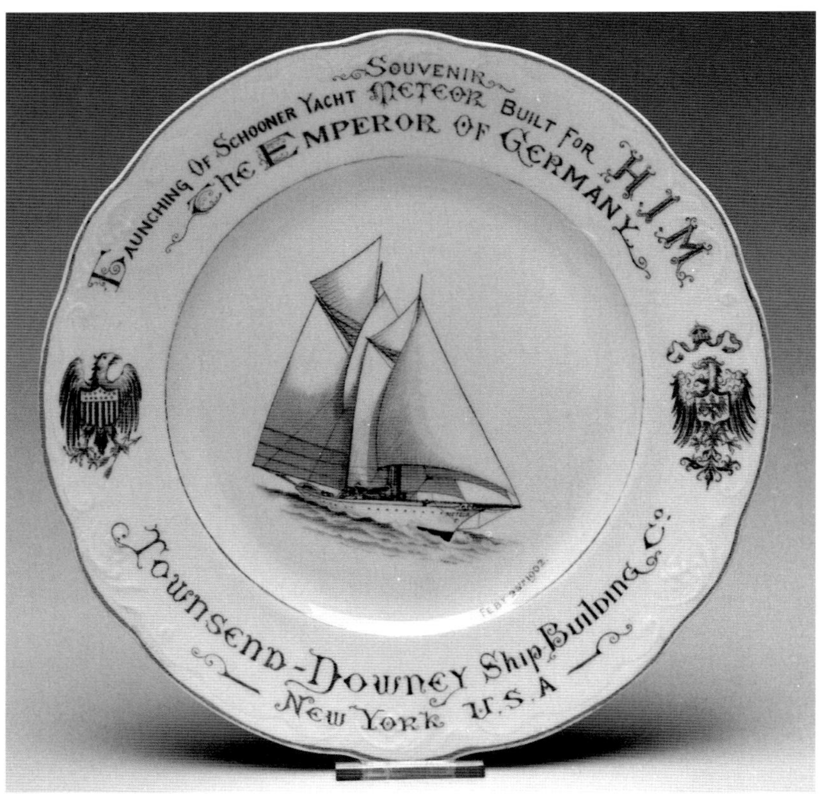

5.1. Souvenir plate to commemorate the launching of the schooner yacht *Meteor,* built for the emperor of Germany by the Townsend Downey Ship Building Company, New York, 1902.

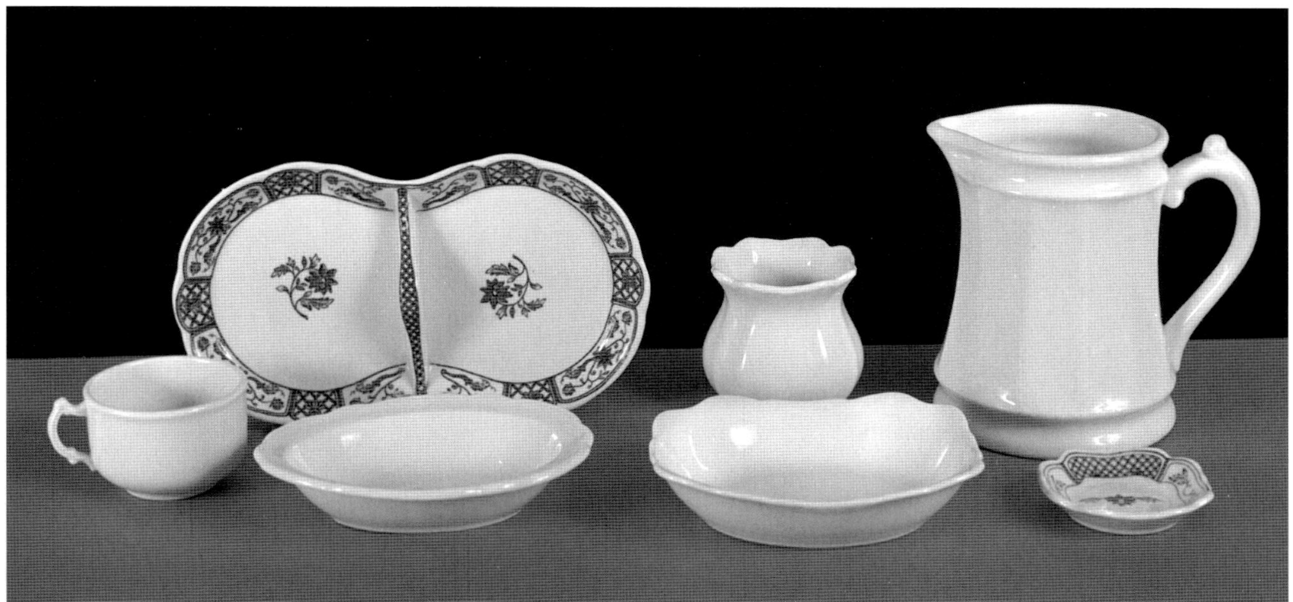

5.2. **Club Colonial** and **Normandie** ware, c. 1902. *Left to right:* **Normandie** cup, **Club Colonial** handled bread tray and baker, **Bag** sugar, **Normandie** baker, **Hall Boy** jug, and **Normandie** individual butter.

restaurants now served meals individually on plates and not, as before, "family style," in which each diner delved into bowls and platters of food to fill his or her plate. Customers wanting items of hollowware ordered them from the **Round Edge** or **Plymouth** lines.

By the early years of the twentieth century, the pottery offered a few hundred printed decorations, a great increase over the several dozen offered a decade earlier. Some, such as the new Art Nouveau and Arts and Crafts–inspired designs, were applied by decals; others were transferred engravings, hand-filled. Lining and banding were still hand operations; a few decorations were freely hand-painted. Reflecting these options, the company's lists gave prices (by the dozen) for undecorated ware followed by progressively higher prices for gold treatments and other decorations. These price scales reflected the number of colors and the amount of work required to accomplish the decoration. The 1902 price list doubled the list of 1899, from eight to sixteen decorating options and from two to three gold options.

Price lists of the time included a measurement and a trade size for each item. The two were quite different. The practice of using trade sizes had crossed the ocean with English potters and was perpetuated for generations even though over the years it lost its meaning. Old-time potters knew that a number-six plate (trade size) was approximately eight inches in diameter. Traditional trade sizes had developed inconsistently in England. Some indicated not a measurement but the number of individual items a potter could make from a premeasured amount of clay. Others indicated something quite different, as in the practice of numbering jugs by size from "6" to "66,"usually impressed on the bottoms of the ware (the larger jugs having smaller numbers), meaning the quantity that would pack into a standard wooden shipping barrel. Similarly, archaic terms persisted in the trade (even to this day), though their original meanings have been lost. The word "nappy" is still used to describe a type of thick low bowl that evolved from a straight-sided, rimless, shallow bowl once used in English public houses and taverns to catch the overflow of "nappy" (a strong foaming ale).[9]

The pottery's new price lists included a system initiated in the industry around 1902 of using code words to facilitate the placing of orders by telegraph. One short word designated the shape, size, and item desired. "Tedious" meant a **Puritan** 6-inch baker. The company's salesmen combed the dictionary for more than seven hundred code words to be used in describing almost any detail in the ordering of Onondaga Pottery's ware.

On 1 January 1904, the company issued a price list that featured eighty-four items in its new **Puritan** shape (plate 16). Haley had departed from past practice with this shape, removing the dainty moulded rococo decorations typical of his earlier **Plymouth** shape. He made the new shape thinner, now drawing a clear distinction between hotel ware and household dinnerware. The pottery felt so confident about the originality of his design that in September 1903 it took out six patents for the **Puritan** cup, saucer, plate, dish, covered dish, and teapot. Preceding those for Robineau's decal decorations by eight months, these are the pottery's first known design patents.

In a field as competitive as industrial ceramics, the borrowing or adaptation of designs from rivals had been the norm, though not a happy one. Many shapes became standard within the industry. Potteries bid against each other to fill orders for a shape that was virtually the same from pottery to pottery, and not always named differently. To deter wholesale duplication of its new shapes, Onondaga Pottery began to register a few of its original designs on a regular basis, year after year. The number of these patents was small, however, compared to the number of designs that the company generated. It remains unclear how effectively the patents protected the shapes.

As decades have passed, the patents have taken on a different kind of value. The patent application forms completed at the pottery, and now preserved in the Syracuse China Archives, document two things. They establish which of the designs the company especially valued for originality and they identify the designers of a number of patterns and shapes. Some designers, such as Elmer Walter, Harry Aitken, Augustus J. Koch, and John Wigley were regular employees. Others, like Robineau, worked as freelance artists. On occasion, works of well-known artists such as N. C. Wyeth, Adolph Dehn, Athos Menaboni, and Grandma Moses were reproduced as special orders.

Having finally discontinued earthenware production, Pass worked steadily to minimize the losses that occurred in the process of making china. No kiln ever offered up only perfect ware, and in his early years of making American china, he allowed only first-quality pieces to leave the factory. Reluctant to discard the seconds and selected seconds, he stored them in the attics of the factory buildings. In 1901, the year of the Pan-American Exposition in Buffalo, J. M. Adams, a dealer from that city, arranged to buy $14,000 worth of these "crooked" goods, anticipating a ready market from the huge crowds expected at this World's Fair. To move the stock, a block rigged from the canal-side top-floor windows of the pottery buildings lowered the ware to ground level in baskets. It was then carted to the canal, loaded into the boats waiting nearby, and hauled to Buffalo. This was cheap transportation for so much ware.

Adams's advertisement in the Buffalo papers brought customers flocking to his store with baby buggies, wheel barrows, and children's wagons to carry their purchases home. Huber estimated, perhaps with tongue in cheek, that over a thousand baby buggies could be seen outside the sale. Police were needed to manage the crowds that came to buy imperfect Syracuse China at bargain prices. This was a measure of the national stature that the pottery's product had attained in only a few years.

The idea of a company-owned store as an outlet for factory seconds may have arisen from the success of the Buffalo sale. Although some provision must have been made in the years after Adams's sale for disposing of this ware on the premises, the first real store opened in a new building across the street from the factory in 1939. Well before that, the pottery began marking selected seconds with an additional backstamp, usually rubber-stamped in red: S.S., short for "store selection."

Selecting was a difficult task. The fine discrimination required in selection was always ultimately a matter of human judgment, and one requiring sharp visual acuity, but in the course of a work day, a selector's vision tended to flag. A headache, a bad day, a change in light, all could influence the final call. The Onondaga Pottery selector essentially discriminated among firsts, seconds, and rejects, the last being destroyed. Obvious flaws such as cracks and crazed glazes were easy to see; crooked ware and imperfect shades of color were less easy to spot. Most selectors tended to be overselective.

Some S.S. ware sold at the factory store was made up into special sets. These at times incorporated "firsts" as a way to use up old inventories. This practice also resulted in oddities, where an overstock of decals normally used on one shape might be put on biscuit ware of another.

Around 1906 Onondaga Pottery published an elaborate catalogue with an accompanying price list. For the first time, the catalogue included photographic illustrations. Earlier, when the company had decided to make chinaware exclusively, it had assigned an "illustration number" to each item in its 1899 price list, beginning with No. 1, the **Plymouth** 3 plate. It seems, however, not to have published any comprehensive illustrations prior to this 1906 catalogue. The catalogue's handsome lithographed cover (9" x 12"), embellished with a gilded medallion, pictured an Indian warrior. The catalogue contained six chromolithographed illustrations. One featured the new **Puritan** shape with the "Renaissance Border" printed in colors, including gilded trim. Others showed colored crests and monograms, gold-encrusted borders, souvenir plates, and the pottery's newest decorations. Following this, numbered black-and-white cuts illustrated the company's entire line of ware, about 370 items, grouped on pages according to type. Instructions referred the customer to the company's already published price lists.

The price list set out for the first time the available decorations in overglaze and underglaze prints and decalcomania special gold treatments, and also Walter's two-color "duo tone" prints. Although the catalogue carries no date, the decorations it lists place it in or near 1906.

Responding to a need for a shape with pure forms and smooth surfaces, Mark Haley designed *Mayflower.* This streamlined "oriental" shape without moulded decoration presented ideal surfaces for engraved and decalcomania decorations. *Mayflower* would prove to be one of the most admired and emulated shapes in the industry, and eventually every major pottery made its own version of it. A half-page supplement giving prices for the new *Mayflower* shape was inserted in the center of the 1906 price list sometime after its publication but before 1914, when the next comprehensive catalogue came out, featuring the shape in a full-page display (fig. 5.3). The pottery also began promoting *Mayflower* in its first national magazine advertisements (plate 17).

In 1908, Pass asked Emil Schnepf, who had headed the lithography plant for several years, to write out a guide to decalcomania operations. Schnepf described the pottery's successful techniques for overglaze applications. In the course of writing this guide, he observed, curiously, that no one had yet found a way to make underglaze decals work successfully. They were desirable because, protected by a film of glaze, they would be more durable.

Schnepf's view that underglaze decals did not work was not quite the case, however. The pottery had filled its first order for underglaze decoration some years earlier, in December 1902, though its technique needed improvement. A. J. Koch, who in 1908 had been Schnepf's assistant for two years, had already begun experiments on his own to improve the underglaze decal process.

During Schnepf's absence while ill early in 1908, Koch, on 30 January, produced what he claimed to be the first truly successful underglaze decal. This was a portrait medallion of Bismarck, drawn by Koch. Years later, in 1939, Pass's son Richard, who was then second vice president of the company, recalled in a memo that "We created underglaze decalcomania. . . . That had never been done before either in this country or abroad. . . . I think no one will dispute the fact that the creation of underglaze decalcomania on American-type china is the outstanding technical contribution of the new world to the ancient art of tableware production."

Richard Pass assigned the date of 1908 to this achievement, probably in deference to a statement made by Koch, who became head of the lithography department when Schnepf left late in 1908. Six years later, in 1914, Koch wrote an account of the successful underglaze process he had devised, stating that "to date, no other lithographic plant or pottery produces sheets in this manner."[10] Koch, who had been trained in Germany, had realized that although colors needed to be applied more thickly to show up under glazes, heavy application of pigments tended to result in blurry, "wishy-washy" colors. He reported how he had begun to think through this problem, though he was understandably imprecise in this public description of a process whose particulars were a still closely guarded secret:

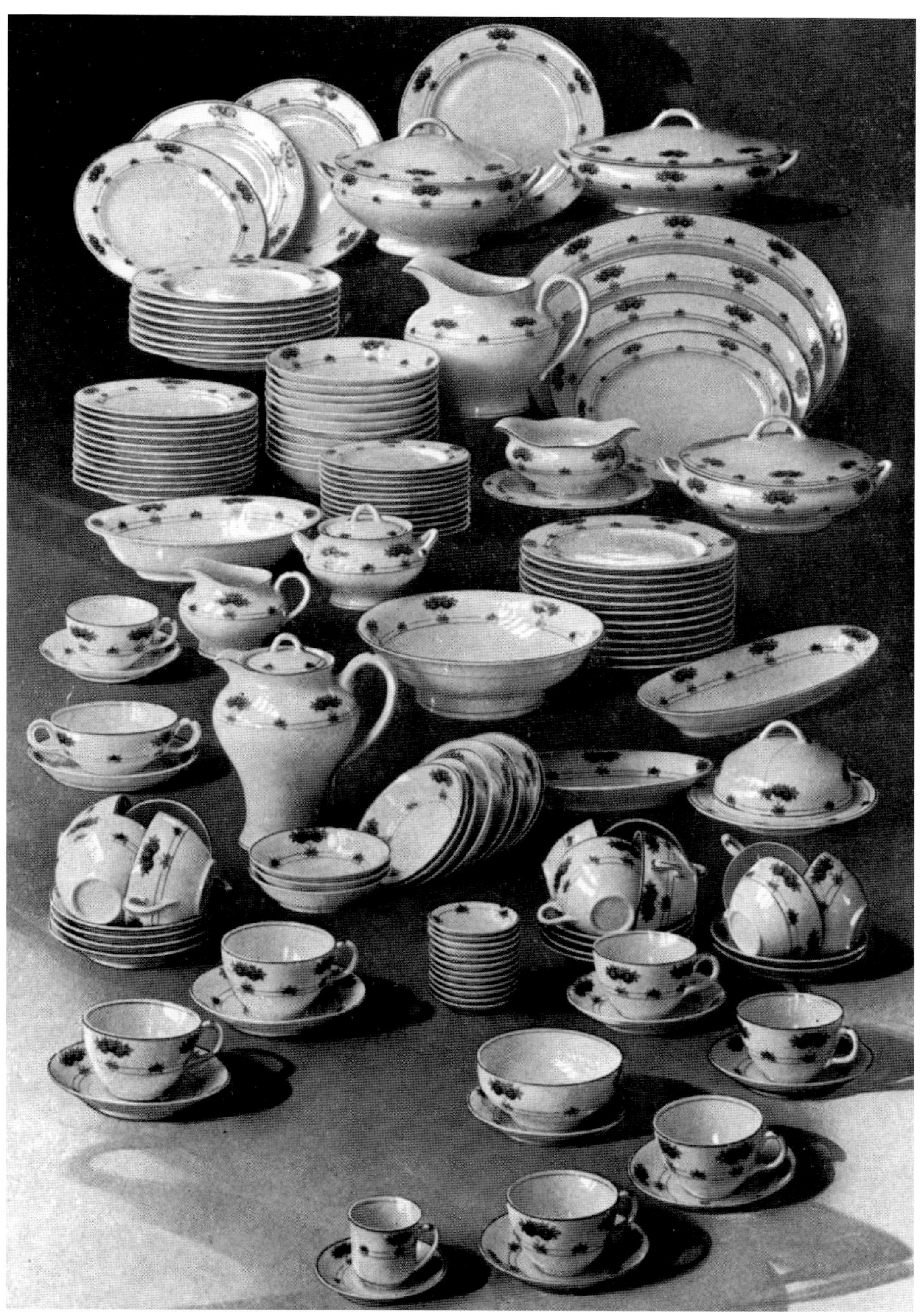

5.3. ***Mayflower*** line with "Canterbury" pattern, from the 1914 catalogue.

Personally, I was far from satisfied and I went to work in earnest to find out just how I could get a clean sharp print, and have more body to it. . . . One incident . . . started me thinking. In printing a certain steam press job, which was passed through the coloring machine, I noticed that a little surplus color which was left on the sheet, held fast to the work, and in those places made the work considerably stronger. This, then, gave me the idea of redusting my sheets, or running them through the coloring machine a second time after they laid long enough for the varnish to work through the color.

Koch reported that only by trial and error had he reformulated the varnish applied to the paper to accept the colors, etched the stone more strongly, and transferred the decal to the ware more solidly. His steam pressman increased the printing pressure and dampened the stone less. The sheets dried longer between colors. Other details and adjustments to the standard lithographic printing method proved critical, but Koch revealed nothing specific about them. One thing was definite, however: when it finally worked out, Koch's new and original process moved the company ahead by a giant step in its industry.

During the years of James Pass's successful development of a china clay body, Robineau's association with the pottery, Haley's modeling of the **Mayflower** shape, and Koch's success with underglaze decals, great changes occurred in American eating habits. These changes amounted to a simplification of the "properly" set table. Nutritionists (a new profession) argued persuasively that the Victorian diet had been excessive. They offered as a new model (in the spirit of the colonial revival of the times) a return to New England simplicity. "Boston" cooking schools and cookbooks appeared throughout the country. Onondaga Pottery's choice of name for its **Plymouth, Puritan, Mayflower,** and **Governor Winthrop** shapes responded to this aspect of the national mindset.

Other developments contributed to the drive toward simplicity. The "servant problem," meaning a steadily worsening shortage of domestic kitchen help to prepare and serve meals, and to clean up after them, added to the appeal of home menus with fewer courses. The new ideal became a three-course meal typically limited to a bowl of soup, followed by a single plate of meat (or fish), potato, and vegetable accompanied by bread on a smaller plate, perhaps a salad, and last, a single-dish dessert. A cup of coffee was the usual beverage. This was Spartan fare compared to Victorian plenty. At the same time, women's clubs preached the need for housewives to free themselves from the "tyranny of the kitchen." Although the arrival of labor-saving utensils and appliances helped toward this end, it was the simplification of the meal itself, and the reduction of the quantity of tableware needed for it, that offered the greatest relief.

New kinds of restaurants began to appear in larger cities. Some, exemplified by the Childs Restaurants chain, furthered the trend to simplicity by offering quick table service in sparely decorated, hygienically tiled rooms, making a virtue of relatively few choices of simple fare, efficiently served, using a limited selection of tableware items. The arrival of the coin-operated automat kept the number of items of tableware smaller still. Items such as bone dishes, covered casseroles, tureens, and others that had seemed essential in Victorian dining now disappeared from the American restaurant table to become curiosities of a bygone life.

The development of American china in the 1890s, and its ever-growing success in the first decade of the twentieth century, were James Pass's great contributions to Onondaga Pottery and to the history of American ceramics. What he had achieved had required all of his many talents, not to mention the unwavering support of the company's directors and the good will of his employees. His distance from Trenton probably helped, too, allowing him freedom from that center's dependence on the traditional ways of doing things and from the strife that often simmered between the workforce and the Trenton potteries' managers.

In the 1890s, Pass had kept a close eye on every operation in his company, but after his workforce began to expand rapidly, he increasingly depended on his department heads to get the job done and done well. He assembled and trained a remarkably able cadre to head and otherwise lead those departments. Walter, Haley, Huber, Cowan, Torbert, Schnepf, Koch, Aitken, and many others made vital contributions to the product's success.

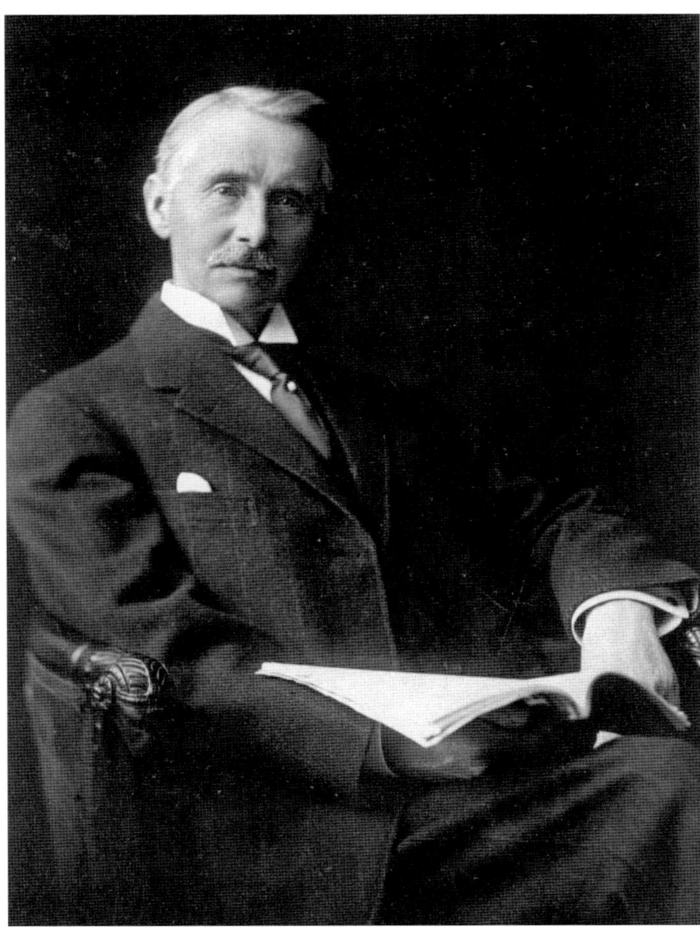

5.4. James Pass when he became president of Onondaga Pottery, 1910.

By 1910, when the company's directors appointed Pass (fig. 5.4) president of the firm, other potteries manufactured their own versions of American china. But Onondaga Pottery had been first, and the quality of its china for hotels and restaurants, as well as homes, remained largely unsurpassed as a result of Pass's constant striving to better it.

He was a remarkable man. His inventiveness extended beyond the development of clay and glaze formulas. When he found he needed a copier shortly after he arrived at the pottery in 1885, he invented a gelatin duplicator. When the underglaze decalcomania would not adhere to bisque ware in the early twentieth century, he designed and patented a tumbling machine that smoothed the biscuit surface enough to permit good application of decals. In the late 1890s, he studied the generation and utilization of X-rays and built an X-ray machine to look into his clays to learn the action of heat on their compositions. X-ray was a marvel of the decade, and his may well have been the first one in industrial use in the Northeast. He made nearly every part of this machine (crude by modern standards) with his own hands, from the small fixtures to the large induction coil and condensers. (The apparatus worked so well that local doctors brought patients with broken bones to be X-rayed.)

One of his last tasks was to oversee Haley's remodeling of the **Round Edge** shape, changing the contour of the underleaf of plates to a long smooth concave curve, lifting the leaf higher, and increasing the roll on its edge. He did this not so much to increase the already impressive strength of the plate in use as to increase its resistance to slumping in firing. The pottery bisque-fired its plates in stacks, called "bungs." Although the plates' leaf angles begin the same, they change in firing, sometimes tending to distort (slump, twist, warp, crook, or sag) under the intense heat. After firing, those on top are never quite the same size; the plates on the bottom of the bung are always a little flatter and wider than those above. The changed contour in **Round Edge** minimized this tendency and improved the yields. This late restudying of an old problem was typical of Pass's constant attention to the details of ceramic manufacturing. It helps explain why he, who could perform the job of any of his employees in the plant, commanded such respect in his profession.

Pass's death in 1913, at age fifty-six, of pneumonia, complicated by silicosis ("potters' lung disease") marked the end of a remarkable career. His son Richard wrote that his father's long hours at the pottery had taxed his strength. When James Pass died, the pottery employed more than seven hundred people. It was the largest producer of vitrified china in America. He left on his laboratory scales a composition for a china clay body that would return to the use of exclusively American materials. At every level the company felt deeply the "well-nigh irreparable loss" of the person who had transformed the firm from a struggling whiteware pottery into a major ceramic manufacturer. The achievements of the pottery's next two presidents over the next forty years would reflect much of his spirit.

Six

National Competition and Growth
1913–1932

It is significant that you are the only maker of china in America today with
the faith in your product and the vision of its future to use the modern miracle
of Merchandising—Advertising.

—M. W. Ayer & Son, advertising agent, 1921

Wɪᴛʜ the death of James Pass in 1913, the company had to rally quickly to fill his shoes.
His oldest son, Richard, in whom he had placed his hopes, was still in college. The post
of president went to his brother-in-law, Bert Eugene Salisbury, who could hardly have had a
closer association, personal as well as professional, with the company. Pass had hired him at
age twenty-one in 1891 to work at Pass & Seymour. Salisbury married Mills Pharis's daughter,
Mary, in 1895. Her mother, years after the death of her husband, was still one of the company's
largest stockholders.

Salisbury's father, Henry O., a building contractor in Geddes, had earlier worked as a foreman in the local salt industry; the family home stood across the Erie Canal in sight of the pottery. Educated locally, Salisbury attended Cazenovia Seminary to prepare for the ministry until
illness forced him to leave. When Pass hired him for Pass & Seymour, he required him to learn
every aspect of production from the ground up. He began by pressing insulators for a piece rate
of two-and-a-half cents a dozen and then learned to fire the kilns. At night he studied chemistry. In 1898, Pass appointed him superintendent of production. Salisbury was clearly a young
man on the rise.

In 1900, the Pass & Seymour Company, having outgrown its site on the canal, moved to
a newly built factory in nearby Solvay. When the steadily prospering company incorporated
the next year, Salisbury, at age thirty, became general manager and secretary of its Board of
Directors. In 1906, on the retirement of Albert Seymour, he became treasurer. In addition to
insulators, the firm now produced lamp sockets, switches, and fuse blocks, nearly all of which
contained ceramic parts. At its October meeting, two months after Pass died, the Onondaga
Pottery board of directors asked Salisbury to add the management of the pottery to his responsibilities, appointing him president. In October he became president of Pass & Seymour as

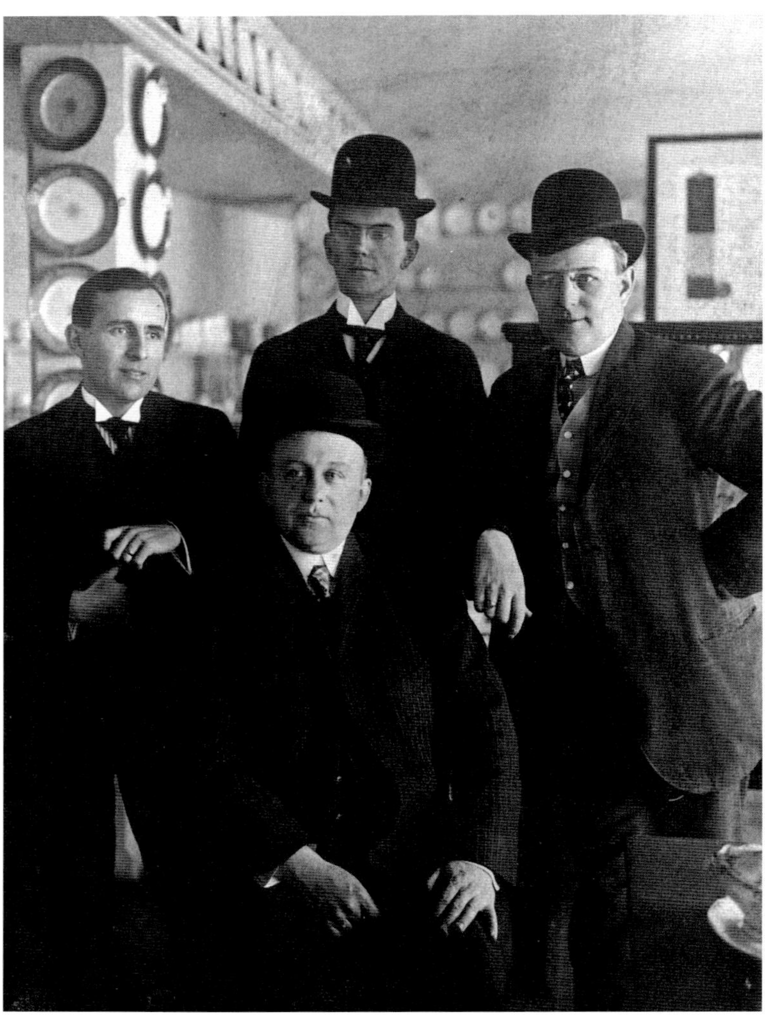

6.1. Salesmen visit New York showrooms in 1905. *Seated:* W. L. Huber; *from left:* E. M. Euniac, E. J. McMillan, and E. L. Torbert.

well. Until 1928, when Richard Pass took over the presidency of Pass & Seymour, Salisbury served as president of both companies.[1]

Salisbury took charge as enormous changes were beginning to reshape the pottery industry and American culture. William L. Huber, who was now the company's vice president for sales, had opened up the western markets, and E. L. Torbert, head of the company's sales department (fig. 6.1), had instituted new approaches to the promotion and marketing of Syracuse China, establishing ever more original and successful campaigns to sell china. Some of the early salesmen were J. Berkley Claiborne, E. A. Hinrichs, E. J. McMillan, William Lee Perry, Gordon Starr, C. B. Storms, T. F. Tunstall, and W. C. R. Williamson. The company took advantage of the interruption of imports from Europe during World War I to strengthen the market share of American producers.

Another important task undertaken by Torbert in 1915, as a member of the Transportation Committee of the U.S. Potters Association (with fifty members in 1917) and representing the newly formed eight-member American Vitrified China Association, was to work for a new classification for "hotel ware" so that it could be shipped at equitable rates in keeping with its durability rather than the more expensive rates that had been established in the nineteenth century for fragile "china" products (most of which came from Europe). He argued that the old rates gave the American industry unfair trade disadvantages. "It is cheaper to ship to Liverpool, England, than to East Liverpool, Ohio," he quipped. The struggle went on for years before it was resolved.[2]

In 1914, Salisbury's first full year on the job, the pottery intensified its already pioneering use of advertising. Under Huber's direction, it published an attractive forty-eight-page catalogue that seemed to mark the opening of a new era. For the first time, the catalogue designers illustrated the price list with photographs printed as halftones. A gold-embossed cover conveyed a level of self-confidence that would have seemed exaggerated a dozen years earlier. The title page stated in rhyme: "O.P.SYRACUSE CHINA Stands for Use, Leads the World from Syracuse," and then added, less poetically, "Elegance, Durable, Service, Economy, Thoroughly Vitrified, Nonabsorbent, Sanitary."

The catalogue included a page showing examples of custom crest decorations, a photograph of the factory with its twelve smoking bottle kilns, a substantial essay on "How Syracuse China Is Made," and a list of twenty-five categories of customers ranging from hotels, restaurants, railroad dining cars, and steamships, to hospitals, schools, and private homes. A second essay, "The Reason," beginning with, "Good table china makes a lasting impression," presented the advantages that could accrue from the purchase of a high-quality product. The catalogue included pictures of the plant's interiors showing manufacturing processes, illustrations of place settings of china with the latest decorations, and lists with illustrations of 705 items in 82 categories, from bakers and bouillon cups to teapots and ice tubs.

Following the list proper, a full page promoted the *Mayflower* dinnerware shape for home use, suggesting combinations for 100- and 112-piece dinner services. A 100-piece service consisted of twelve place settings and serving pieces; the 112 piece set included an extra plate size (see fig. 5.3). For hotel dining, the catalogue suggested compositions of items to make up orders for 500 dozen pieces of *Round Edge*. The catalogue introduced the new *Empire* shape, its name harkening back to the pottery's origins (fig. 6.2). A complete line of hotel ware, *Empire* was a thicker-gauge version of the *Mayflower* shape. Although it used many similar profiles and moulds as the popular dinnerware shape, it varied the handles, knobs, and other appurtenances to create a different-looking shape. *Empire* also offered new "room ware" items, utilitarian ware for hotel use outside the dining room such as dresser sets and cuspidors, as well as sick-room supplies. The age of the toilet set had passed.

This catalogue served the company's network of distributors and others who bought ware in quantity. Relatively few individual shoppers seeking fine dinnerware for household use would have consulted it. The household market had been served since the mid–nineteenth century

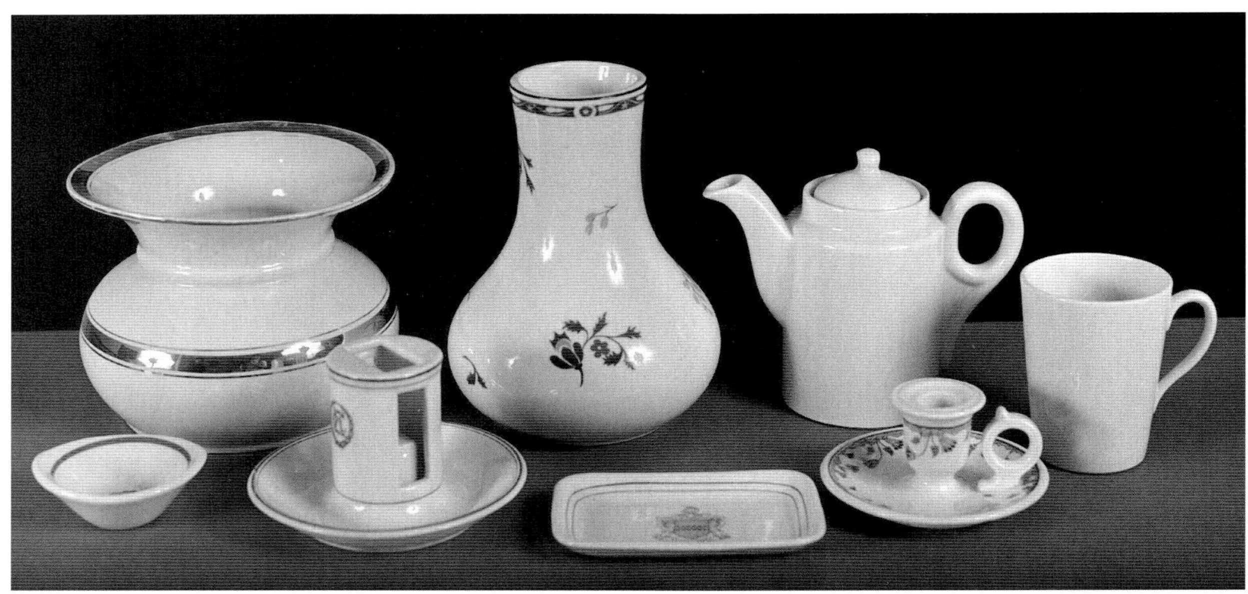

6.2. Some *Empire* ware, c. 1914. *Left to right:* individual jelly, cuspidor, barrel match stand, water bottle, pin tray, coffee pot, handled candlestick holder, soda mug.

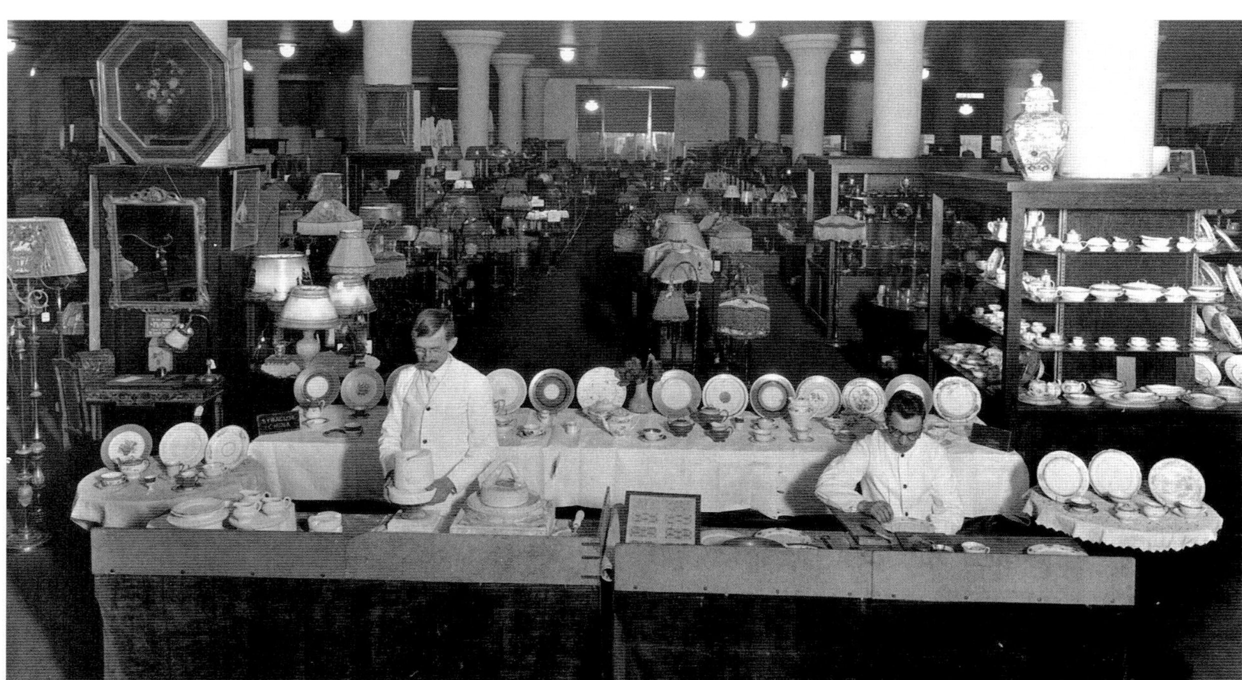

6.3. Sid Lodder (*left*) and Peter S. Lynch forming and decorating Syracuse china in a department store demonstration, 1925.

primarily by jewelry and other specialty shops that also handled
fine glassware and silverware. Now, however, urban department
stores, a late-nineteenth-century development in retailing, had
begun to sell fine china. The pottery's sales force recognized this
venue as one that would attract and hold the attention of a large
number of shoppers, who were mostly women.

In 1915, Torbert organized a promotional tour aimed at that au-
dience. For a period of six months, three highly skilled employees
traveled as a team to department stores in twenty-five American
cities from Boston to Dallas. At each store, they set up an extensive
display of Syracuse China and then demonstrated china-making
and decorating techniques. They repeated the tour in 1925 (fig.
6.3). Each store was a retail outlet for the company's dinnerware.
The stores ran promotions of their own. They sometimes spot-
lighted Syracuse China in memorably novel ways, as when, in
Cleveland in 1921 Emma Wiedel, a saleswoman in Kinney & Levan
Company's retail china department, dressed herself as the "Canterbury Girl" for the company's fancy costume ball (fig. 6.4). She
selected that pattern to represent her ideal of perfect merchandise.

In October 1915, the trade journal *China, Glass and Lamps,* reported on Torbert's traveling promotion:

> At the Jos. Horne Co's big store, Pittsburgh, Pa., throughout the
> past week a splendid display of samples of [Onondaga Pottery's] . . .
> fine ware was made and, in addition, patrons of the store were
> given an opportunity to see how the actual work is performed at
> the pottery. Included . . . were Sidney Lodder, an expert potter [and
> a senior one, for he had begun working at the pottery in 1887],
> Peter S. Lynch, who performed free hand lining and banding, and
> Miss Ellen Foley, a china decorator of more than ordinary ability.
> Thousands of people were entertained during the week. . . .
> Progress is a watchword among the company's officials, and it is a
> foregone conclusion that a considerable amount of foreign-made
> ware is being displaced by Syracuse China.

The pottery's opportunity to displace "a considerable amount"
of imported china was a result of the war in Europe. Soon after it
had begun in 1914, it had disrupted the pottery industry in France,
Germany, and England. By the time an armistice was reached in November 1918, the industries
in those countries were severely crippled. In 1915 and 1916, before America's entry into the war,
Onondaga Pottery and other American producers had a chance to make great headway against
imported fine porcelain and china.

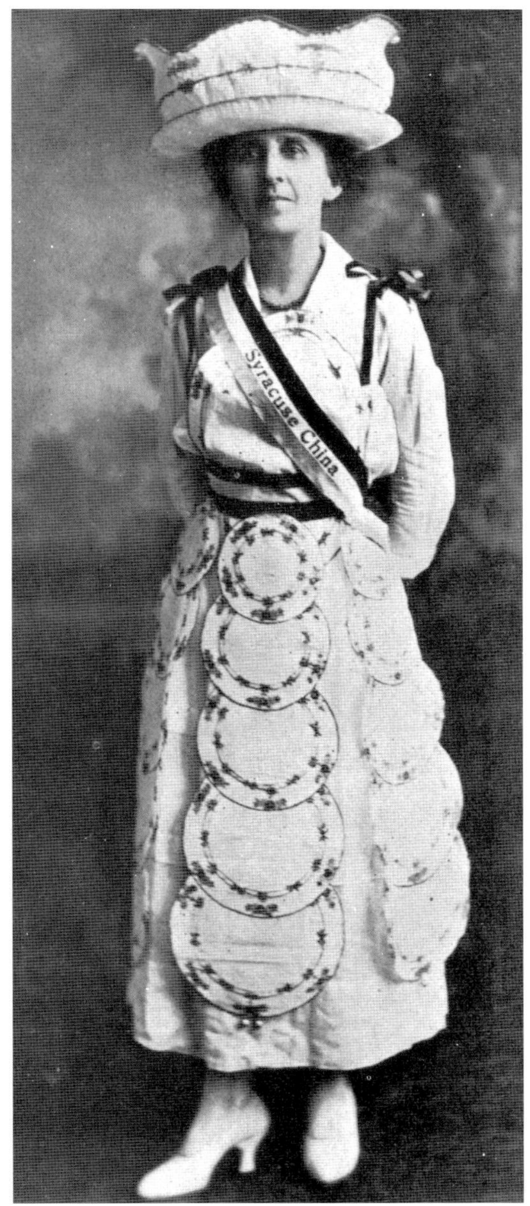

6.4. Emma Wiedel, the "Canterbury Girl,"
promoting Syracuse China in Cleveland, 1921.

The company's advertisements made no overt reference to the war. They went in the opposite direction. The October 1915 issue of *China, Glass, and Lamps* that reported the demonstrations by Lodder, Lynch, and Foley included a pictorial advertisement for Syracuse China that depicted a woman in Renaissance-revival attire pouring tea at a wicker table on a verandah, a large vase of flowers beside her, and greenery beyond. The design of her costume and her furniture, as well as what can be seen of her china, speak of Arts and Crafts simplicity in a peaceful setting.

The pottery had begun advertising in *Vogue, House and Garden, Country Life, The Craftsman,* and other magazines as early as 1913 (plate 17). A few years later, in June 1919, *Syracuse China News* explained to its employees:

> Three million homes scattered all over the country receive each month the magazines which carry the advertisements of Syracuse China. That means very nearly ten million people who read these magazines have Syracuse China brought to their attention. . . . Advertising makes it easier for the user to point with pride to his possession. . . . The widely advertised brand is the accepted standard of many people—advertising standardizes styles.

The company also issued handsomely designed and printed advertising pamphlets directed at specific markets. As early as 1911, it had prepared one of sixteen pages, titled *Individuality.* Addressed "to the sanitarium manager [and] to the hospital manager," the text emphasized the sanitary qualities of vitrified ware. It began, "If you believe in the highest degree of scientific sanitation, this booklet will surely interest you," and continued:

> It will surprise many people when we say that our kind of china is germ proof while some kinds of china and porcelain [i.e., so-called "semi-vitreous" "semi-porcelain"] may transmit disease germs. In hospitals and sanitariums it is of the utmost importance to have strictly sanitary china. By this we mean an article with a non-absorbent body, so that when nicked it can be thoroughly cleaned. When the body is not thoroughly vitrified it is not translucent, but often porous and becomes, especially in hospitals, a dangerous absorbent

The illustrations included plates with distinctive decorations from sixteen hospitals and sanitariums located throughout the country.

James Pass's older son, Richard, reported for work at the pottery on 1 June 1915, less than two weeks after graduating from Harvard University.[3] With full intentions of following in his father's footsteps, he had majored in chemistry and German. During school vacations, his father had brought him into the plant to work alongside some of the old timers in the clay shop, the kiln room, and the lab. There he heard stories about his father and his grandfather, about the old days and the old ways, as he learned about the new. With his father's sensitivity, Richard acquired a deep respect for the skills and dedication of these craftsmen, though he himself never became a potter. With something of his father's vision, he saw possibilities for the pottery in the fast-developing technological and other innovations of the time.

Richard and his father had spent long periods of time together on trips abroad to the pottery centers of the world. In 1911, they traveled to Germany with the idea of purchasing for Onondaga Pottery a company located in Sonnewitz, near Meissen. The plan was to manufacture hard-fired porcelain for export and to take advantage of the then-low tariff to import German-made ware for sale in the United States. This plan ultimately proved unfeasible.

Richard Pass started his career at the pottery by reporting to Dr. Edward Schramm in the laboratory. The first pages of his formula notebooks (maintained by him until 1946) indicate that he began by "cleaning the laboratory, examining the apparatus and arranging it." [4] On July 10, he entered "*Received my first pay check!* Makes me feel rather guilty. Salary $20.00 a week." In the middle of September he wrote, "Made a mess in lab—a hellish mess—an aggravating and expensive lesson."

One of his first important tasks was to find an improved method of refining the kaolin from the North Carolina Hog Rock mine property, to which the pottery had bought mining rights from 1914 to 1918, when German submarines endangered shipments from England. This was the same "uniker" clay that his father had used to develop **Imperial Geddo**. Pass was charged with finding a means of refining this clay more successfully and inexpensively than before, thereby making available a source of clay for a whiter body independent of the threats to wartime shipping. He uncovered and translated a German treatise describing a process developed by Count Schwerin to refine Zettlitz clay. Applying this method to experiments in the lab and in the refining plant at Fayette Street, he collected sufficient data and headed for North Carolina's Great Smoky Mountains. There he spent the next months on horseback exploring the Hog Rock deposits around Dillsboro, recruiting miners from among the mountain people, and constructing a mill from native timber. Even treated with his improved techniques, the clay contained too many impurities to be feasible as a reliable and economical clay source for the pottery, and Pass abandoned the project. In any event, the pottery's supply of English china clay lasted the duration of the war. In 1917, the United States entered the European war. Richard enlisted in the United States Naval Reserve Flying Corps.

In all, sixty men from Onondaga Pottery left for military service. Among the people hired and trained to replace them in the production departments were women who had never before worked outside their homes. Motivated partly by a sense of patriotic duty, they now also had an opportunity to earn good wages. They learned skills needed for the clay shop (but not for kiln setting or other jobs that required major muscle power). They joined the many women who had worked for years at the pottery as handlers, finishers, decorators, and selectors.

Both Onondaga Pottery and Pass & Seymour manufactured products for the war effort.[5] The pottery's production of hotel and household dinnerware fell sharply, partly because of shortages of raw materials and a wary market, but mainly because of the imposition of government restrictions on what could be made. In October 1918, the War Industries Board issued this decree:

> It is of primary importance in the present emergency that the country's resources be used to full advantage and that we husband our supply of materials, equipment and capital to aid in carrying on the war.
>
> *Schedule for Manufacturers of Vitrified China and Semi-Porcelain Ware:*
>
> No new moulds can be made during the war for any article not on the restricted list and no new decalcomania patterns or copper plate engravings to be added. No new lines to be placed

on the market even in cases where the decalcomania or copper plate engravings are made. Special designs already prepared by any pottery not to be duplicated by any other pottery during the war.[6]

Before the wartime restrictions went into effect, the pottery's October 1918 price list had featured over six hundred varieties of sixty item types in six major shapes: ***Round Edge, Club Colonial, Normandie, Empire, Plymouth,*** and ***Mayflower,*** with over four hundred different patterns of decoration. A month later, when the restrictions were in effect, the pottery cut back to about 120 varieties of thirty item types in ***Mayflower*** and ***Round Edge,*** the only two shapes it was allowed to make. The company used this cutback to discontinue shapes and items that had waned in popularity. In April 1919, when restrictions were lifted, the company's new price list showed the continuing effect of the cutback by offering only 454 varieties of sixty-four item categories. It discontinued about one third of its prewar offerings, including much if not all of the ***Club Colonial, Normandie, Plymouth, Imperial,*** and ***Marmora*** shapes. The time was right to clean house. Old-fashioned items such as chocolate pots, jellies and relishes, loving cups, and welsh rarebits finally disappeared.

Richard Pass returned from the war in 1919 and became assistant superintendent of manufacturing at age twenty-six. His diary notes, "handling all sorts of plant problems." His responsibility as superintendent was to respond rapidly and successfully to the constant flow of reports from the sales office. He improved the saggers and experimented with kiln construction, design, and repair, sometimes firing the kilns for thirty-six hours at a stretch.

In 1921, when it became clear that his father's old laboratory in the factory was outmoded, Pass supervised the design and construction of a new Federal-revival brick building to replace it on School Street across from the Fayette Street plant. It was a state-of-the-art laboratory, manned with the best personnel in the industry. Edward Schramm, the company's first official ceramic engineer, who had come to the pottery shortly after the death of James Pass to manage the laboratory, took charge of the new facility.

"War conditions are past; we are now entering a period of keen competition and the degree of success which we attain in the future will depend very largely upon the perfection of workmanship in the product we have to offer." This was the challenge voiced in March 1919 in the first issue of the *Syracuse China Bulletin*, renamed six months later *Syracuse China News*. This house organ was first edited by Robert Bryant, the company's comptroller, with the help of Henry Lane, a veteran employee, who had been hired by his cousin George Oliver in 1882. Lane, now foreman of the clay shop, had been one of James Pass's most trusted employees and friends. He was an early booster of the Potters Club, which group helped conceive the *Syracuse China News*.

In 1920, Maurice Olmstead, the newly hired employment manager, took charge of the paper. He had been a teacher, a high school principal, and an insurance agent before he took a night course in accounting with Bryant, who admired Omstead's ability and hired him. Manager of both employment and payroll, he became known as "Two Gun Olmstead." Olmstead and two women staff members figured the payroll. Armed with a loaded gun, he picked up the money from the bank, then delivered pay envelopes loaded with cash to the new Court Street plant. His job expanded as new laws and new employee benefit services expanded. In 1940, he became the company's first named personnel director and continued as head of payroll until 1946. He served three presidents before his retirement in 1959.

The stated purpose of the *Syracuse China News* was "to record items of interest to the 750 employees of the Company, to develop a closer social relationship among them, and to stimulate interest in the advantages of being associated with an organization which presents opportunities for satisfactory employment second to none." Issued regularly from 1919 to 1930, the *News* supplied much of the "glue" to hold the pottery "family" together.

The first number (March 1919) memorialized five employees who had died in military service. It reported on a company-sponsored English language class for twenty-six pupils conducted in the plant's lunchroom by teachers from the Syracuse public schools. In a lighter mood it reviewed a performance of "a romantic drama" by the Onondaga Pottery Drama Club, observing that "there wasn't a dry eye in the whole front row on several occasions." It added, jokingly, that "Mr. Salisbury was so exercised that he omitted his home gymnastics the next morning." Columns of local humor were headed, "Cracks, Crazes, and Gas," and "De Calco Manias." The "Decorating Department's Own Page" featured a humorous cartoon drawn by John T. Wigley depicting "working conditions" (fig 6.5).

The issue's editorial struck a more somber note. Referring to the recent wartime shortage of European-made ware, it recalled that

> hotels were forced to buy American ware or close their dining rooms. It may be safely said that ninety percent of the ware now used in hotels throughout the United States is made in America. . . . Now the time has come when they [hotels and restaurants] will decide whether they shall continue to use American china or go back to foreign products. All that can be said or done by the management or sales force is of no avail unless the man who makes the ware in the clay shop, those who fire it, or the people who apply the decorations will do their utmost to have their work fully acceptable. . . . The opportunity for each individual to do his or her share toward retaining this business for the home industry is greater now than ever before.

The *News* gave regular attention to women employees, often called "the girls." By 1921, women constituted a third of the

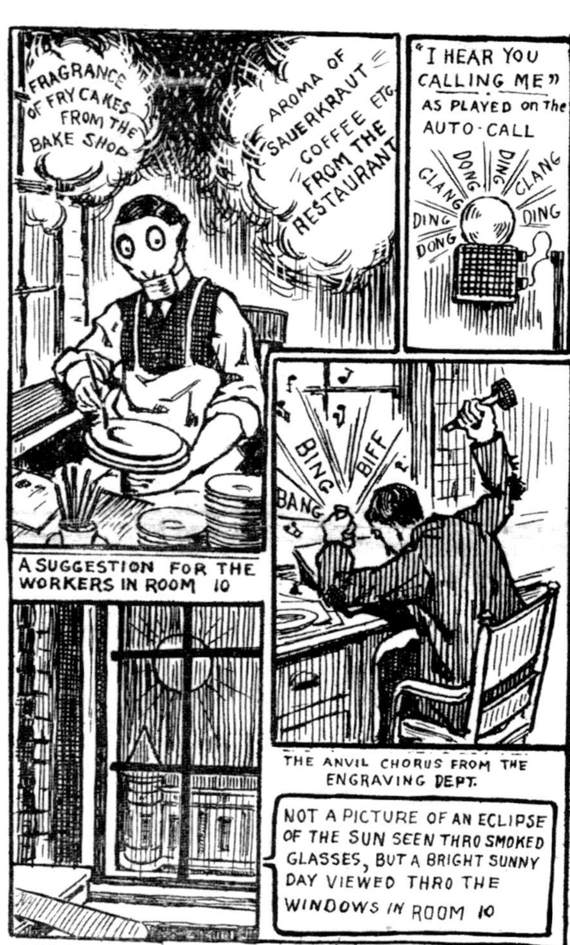

6.5. Decorating Department cartoon drawn by John Wigley for *Syracuse China News,* 1921.

workforce. As was true of most women workers, they were paid less than men, largely from the belief that men wage-earners were generally the sole support of their families and women were not (though there were notable exceptions to this). Many women workers who had joined the national Girls' Patriotic League during the war organized a branch of the Huntington Club in which James Pass's widow, Adelaide, took an active interest. Like the Potters Club for men, the Huntington Club brought women from all departments together for sports (basketball, softball, and swimming), classes, summer camping, house parties, Sunday hikes, dances, picnics, and play performances. The *News* reported baby and bridal showers, fishing trips, and vacation excursions. It regularly recounted the events of the Potters Club as well. The company provided a men's clubhouse equipped with four bowling alleys, a pool table, a game room, a library, and a large hall used for basketball, dances, and other entertainments. A family night once a week opened the facilities to spouses and children. The club's spring family picnics were legendary; more than two thousand people attended the one held in June 1919 at Long Branch Park in nearby Liverpool.

The cordial association of the company's officers with its employees, not only at picnics and family nights but also in the hospital and at funerals, continued James Pass's approach to management. In a movement led by his son Richard, it also reflected a growing national pattern of corporate concern for the welfare of its employees. In more recent years, such "paternalism," or welfare capitalism, has tended to be viewed cynically, but as late as the 1990s, many former employees of the company from the years between the World Wars, and later, reflected upon their lives at Onondaga Pottery with great affection for the company and its programs.

The *News* periodically noted the service anniversaries of long-time employees. In May 1919 it saluted William Shearer's forty-eight years and Joe Weiss's forty-six (he had begun at age thirteen). In an industry dominated by labor unions, the pottery had a reputation for being a good place to work. Wives followed husbands in joining the ranks of Onondaga Pottery employees, sons their fathers. Often three generations of a family were employed simultaneously. Cousins and nephews, sisters and aunts lined up for work. It was said that the best recommendation one could have upon applying for a job was having a relative employed with the company.

Even before the company had recovered fully from the war, it returned to its plans for the expansion of its manufacturing facilities. In 1917, on the eve of the country's entry into World War I, Onondaga Pottery had already completed a major construction project by adding the A unit (fig. 6.6). This filled in the space once occupied by the original Empire Pottery buildings that had burned in 1883. It increased production capacity by one-third. But even then, with sixteen kilns and seven-and-one-half acres of total floor space, the pottery had remained hard-pressed to meet the growing demands for Syracuse China, a demand intensified by the unavailability of imported ware. The making of both hotel ware and household dinnerware in the Fayette factory had been awkward at best. Making ware of any kind in a five-story factory building was increasingly inefficient. Ware constantly needed to be moved up or down a level. The A, B, and C units, each functioning to some degree as a separate entity, duplicated facilities and operations (fig. 6.7). The Fayette Street site had become crowded and offered no room for

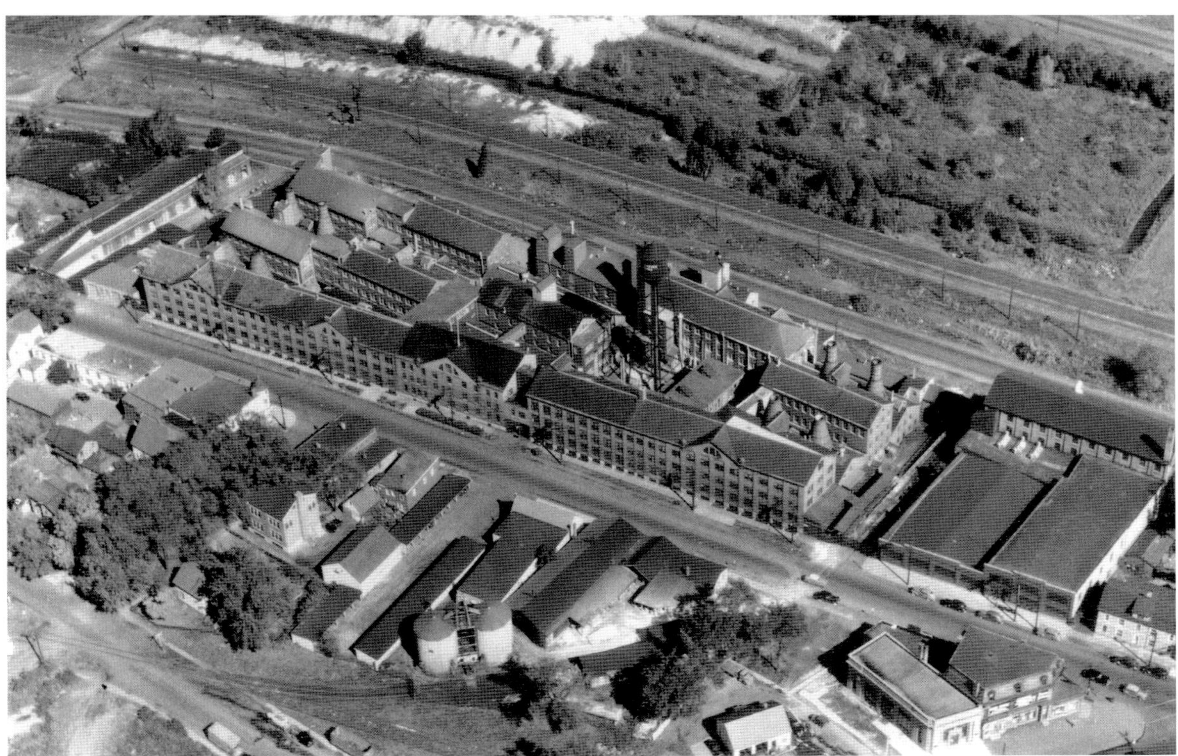

6.6. Aerial view of Fayette Street plant, 1948.

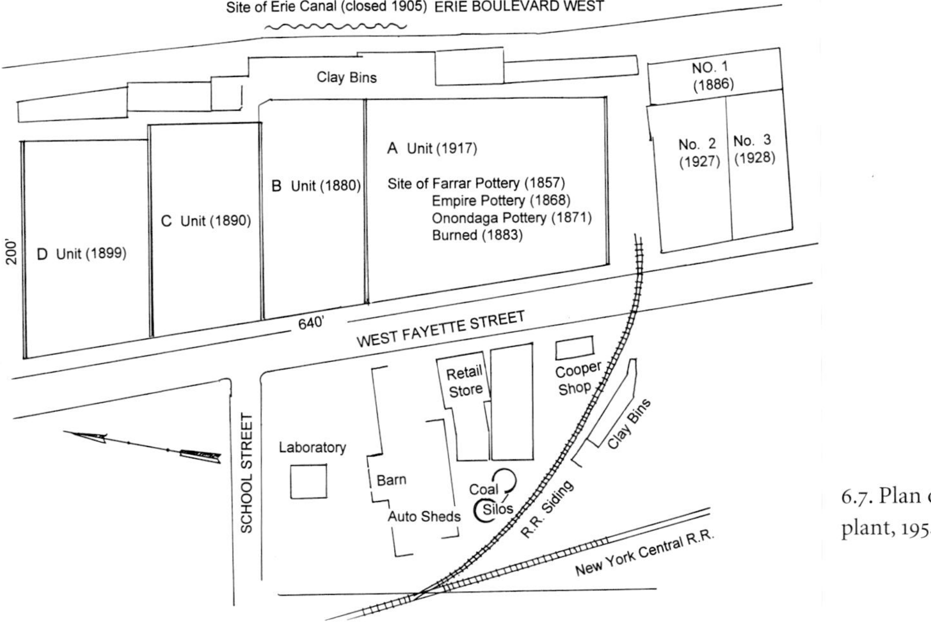

6.7. Plan of the Fayette Street plant, 1954.

further expansion. Its location on the Erie Canal had ceased to be an advantage because the canal by 1906 had been supplanted by the wider New York State Barge Canal routed north of the city. The pottery's rail siding transported most of the ware made at Fayette Street.

The fastest-growing and most profitable side of the pottery's business was hotel ware; it needed more and better space. After the war, the pottery's directors instituted a search for a large site on which to build a new plant for hotel ware manufacture. It needed to be close to rail transportation and accessible to good roads for the trucks that had begun to transport ware. The company located an ideal site in a three hundred-acre farm just north of the city on Court Street, in the town of Salina. On 17 July 1921, three days before the pottery's fiftieth anniversary celebration, Salisbury, in the presence of the pottery's employees, turned the first spade of earth to break ground for the new plant.

In August 1921, with construction of the Court Street plant underway, *Syracuse China News* reported that the pottery had 850 employees at Fayette Street, not counting officers, in 44 separate skilled operations. Of these, 250 worked in the decorating department, 200 were associated with the clay room, 240 could be found in the various branches of kiln work (placing, firing, cleaning, selecting), 40 in lithography, 50 in ware room operations, packing, and shipping, and 60 in repair, maintenance, and grounds.

The opening of the Court Street plant on 7 June 1922 enlivened the pages of the *News* for months. The June issue reported the dedication ceremonies for the new factory. Accompanied by a brass band, Richard Pass and his bride, Ruth, led the parade of hundreds of employees around the new building. Bert Salisbury cut the ribbon to open the facility. After tours of the plant, a festive day of refreshments, dancing, and games followed.

The design of the one-story Court Street complex incorporated the latest thinking in mechanical and industrial engineering (fig. 6.8). From a main artery running the full length of the building, a distance of about one-fifth of a mile, the production departments extended at right angles, each contiguous to the succeeding operation, allowing a straight-line flow of materials and ware. The plan also permitted future expansion, which began almost immediately and continued periodically for the next fifty years. On a single level, and without changing direction, one could follow a piece of china through the plant from its origins in the mixing room of the clay shop to its departure into the world from the shipping docks. The "streamlined" linear arrangement of production departments reduced conveyance problems to a minimum.

Once operative, the building promised major improvements in working conditions. The building's single-story plan provided better light and ventilation. Cheap and plentiful electric power provided a system of exhaust fans, ducts, piping, and vacuum outlets that reduced dust and waste heat to levels dramatically below those that had prevailed at the antiquated Fayette Street units. The new plant manufactured hotel china exclusively. Thin household china production and the pottery's business offices remained at Fayette Street.

The first trials of the two new kilns had occurred in April; the first trial ware, a ***Round Edge*** saucer, had emerged in May.[7] It took some months after the June dedication to get Court Street staffed and in full operation. In September, John Wigley opened the plant's decorating shop,

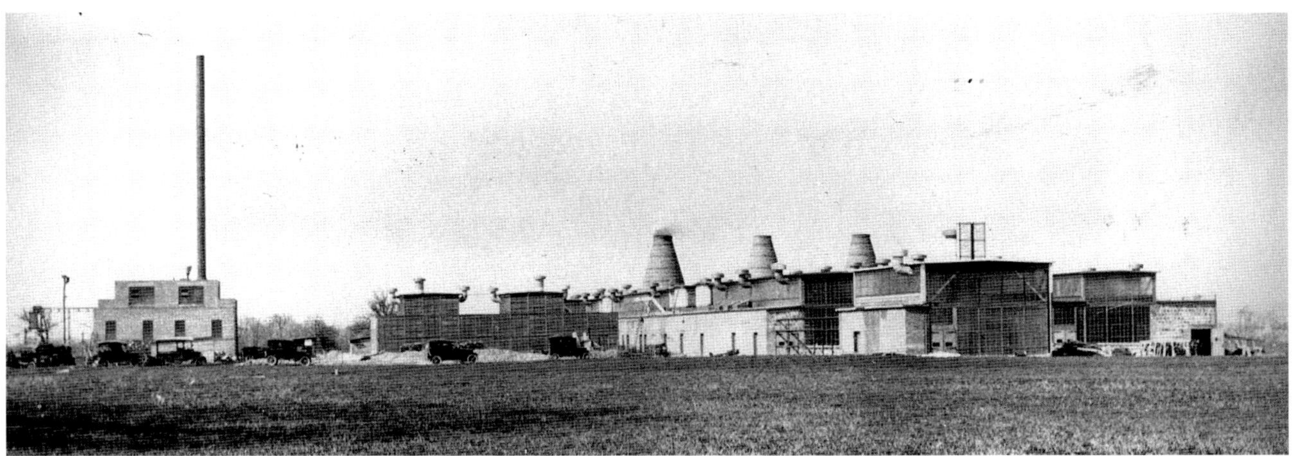

6.8. The Court Street plant, under construction in 1922. The bottle kilns, supplanted by tunnel kilns beginning in the 1930s, were not taken down until the 1940s.

organizing the system and training the new decorators.[8] By the end of the year, Court Street was ready to ship large volumes of hotel ware. Early in 1923 the company's overall annual production rate had increased 25 percent over the previous year.

In 1920, the pottery had begun a new system of date-coding hotel ware. It seemed unnecessary, however, to put dates on fine dinnerware for purposes of quality control. The pottery continued to use the O.P.Co. Syracuse China mark, initiated in 1897, but dropped the imprinted date code at the end of its cycle in favor of a rubber-stamped code that would be printed as an integral part of the backstamp. This included a capital letter for the year, and a number for the month, separated by a hyphen (A-1). After 1923, to distinguish between factories, the letter came first for ware made at Fayette Street plant (D-1) and the number first for ware made at Court Street (1-D).

In spring 1923, the company introduced another new shape from the prolific Mark Haley, **Governor Winthrop** (see plate 20). Although its simple elegance was in keeping with the widespread interest in things colonial during the 1920s, the shape did not survive the decade.

The first significant plant expansion at Court Street occurred in 1923, with the construction of a state-of-the-art lithography building next to the factory. In the original lithography facility at Fayette Street, a lack of climate control had impeded production. In planning the Court Street shop, Salisbury called on Willis H. Carrier for advice. In 1921, Carrier had devised the first successful industrial-scale centrifugal air-conditioner in the world. Built in Germany, it had been installed in the Carrier Company offices in Newark, New Jersey. Carrier sold that machine to Onondaga Pottery in 1925 to provide the constant temperature and humidity required in a lithographic plant. The Carrier staff, advised by Augustus Koch, consulted on the structure of

O.P.CO.
SYRACUSE
- CHINA - -
B-2

O.P.Co./Syracuse/China
(stamped date code)
(1920–46)

the building. Built with one inch of cork as wall and roof insulation, it had double-glazed windows facing north to prevent direct sunlight from falling into the working space. The inside temperature did not vary more than two degrees, and the humidity not normally more than 2 percent. Thirty-five years later the air-conditioning unit still operated.[9] In April 1924, four new kilns went into operation at Court Street, bringing the total there to six.

In 1923, Pass took on larger responsibilities and was made superintendent of all manufacturing operations. Working from offices at Fayette Street, he oversaw both facilities. One of his first innovations was the institution of a factory order system to standardize and synchronize production at the two plants. The system ensured the quality and consistency of the pottery's china. It recorded every step in the development of a new item and every change made to it thereafter. At every point in its production, the manufactured item needed to match the specifications of its factory order precisely. The subtlest design alteration in the curve of a handle or in the flare of a rim called for a factory order change.

The factory order (F.O.) system consisted of a series of memoranda, generated initially by the sales department. They circulated through established channels for endorsement or amendment by every department involved in the manufacture of an item. The factory orders specified the clay body, dimensions, weight, and all else needed by the pottery to produce, even after the passage of decades, an object identical to its approved prototype or "standard." The factory order recorded an item's history, beginning with its creation by the pottery's designers and ending, years or decades later, as a D.I.F.O. (discontinued item factory order). For nearly half a century, no change in design, specifications, or production occurred at the pottery without a factory order. The system extended also to changes in backstamp designs and prices. Preserved in the Syracuse China archives, the factory orders constitute for the decades covered an exceptionally thorough historical record of individual item manufacturing in the American ceramic tableware industry.

In addition to this paper record, the pottery preserved a specimen of every item in every shape as a factory standard. (This was common practice in the industry.) Each was a biscuit-fired piece of ware on which were inscribed, with dates, the item's factory order number, measurements, weight, and other data. For each item, the pottery also preserved a master mould and a profile (used to make the tool that formed the item). The decorating department kept its own written standards for decal placement, hand painting, and other treatments, as well as an approved sample of finished ware for each custom order. The lithographic department maintained stones and progressive sets for each decoration; the printing department kept the copper printing plates for each transfer design.

The company foresaw that the opening of the Court Street plant would create a need for more skilled pottery workers. In fall 1921, it initiated an apprenticeship clay shop training course, the first of its kind in Syracuse. All boys between fourteen and seventeen enrolled in Continuation School work could receive credit for apprentice training through the the New York State Department of Education.[10] Thirty boys were taught the fundamentals of jiggering, casting, and pressing under the supervision of skilled pottery workers. Boys over seventeen could receive regular wages for time spent in the course, as could Continuation School students when that school was not in session. In addition, boys were eligible for weekly bonuses. They were granted a diploma signed by an officer of Onondaga Pottery on successfully completing the twenty-four-week session. In 1923, the program extended to elementary apprentice training and included such fundamental courses as mathematics, English, civics, and hygiene, as well as clay shop training in lining throwing, cup jiggering (fig. 6.9), and cup turning. In October 1923, an advanced program was added to train the apprentices in speed, skill, and workmanship. It readied them to take their places as journeymen. Besides special classes in pressing, casting, and cup making, the boys studied the history of pottery and related mathematics, and attended special shop lectures to acquaint them with the operations of every department in the plant. This stage required fifteen months of training and guaranteed a job if satisfactorily completed. Just how many new pottery workers completed this program, and how many remained with the company, is unknown, but the program was always spoken of with pride by those who ran it and those who took it.

Changes in taste constantly influenced the decoration of china. W. L. Huber reported that in the early days of the pottery, hotel proprietors and dining-room stewards had selected the patterns, but by the 1920s they usually deferred to architects and interior decorators. The most prestigious hotels, resorts, and clubs sometimes shied away from decalcomania decoration as

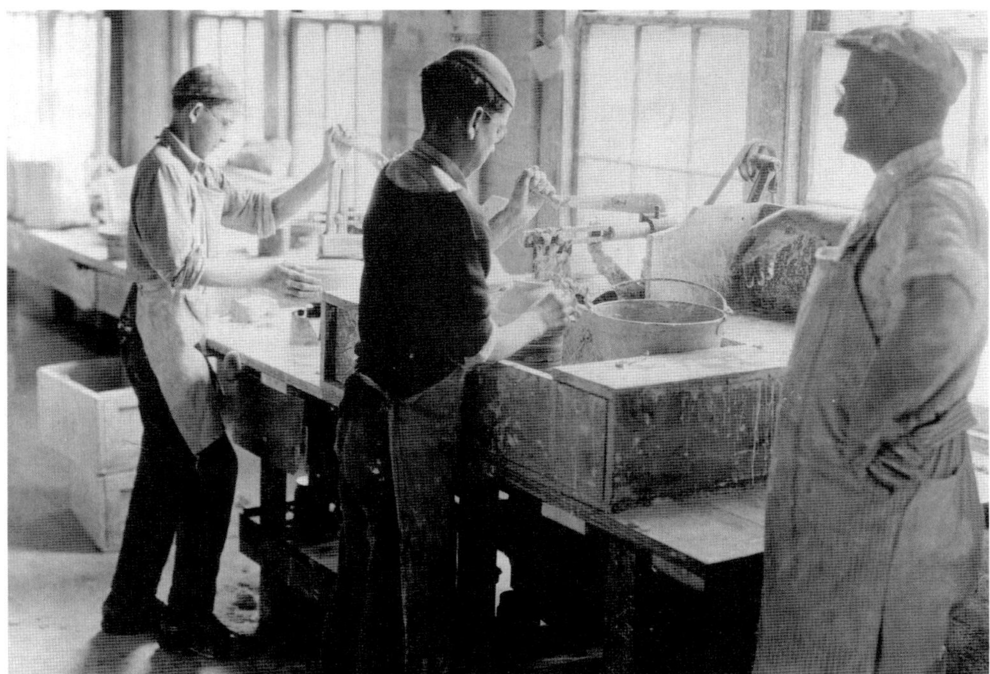

6.9. Continuation School apprentices learning to jigger, 1923.

"mechanical" in appearance and "lacking a human touch." They often called for hand-painted china. Hand-decorated ware from Italy enjoyed special popularity at the time.

In response, Harry Aitken developed a new approach to hand-filled decoration. The first evidence of it, an intentionally primitive floral pattern called *Peasantry* (or P-1), appeared as an illustration in 1924 in the pottery's first major catalogue since 1914. These "underglaze hand-painted treatments," as they were soon renamed, embraced myriad styles. To make them, charcoal was pounced on a stencil pricked with tiny holes to delineate a design on biscuit ware. Decorators would then apply color to the ware using the pounce marks as guides. By 1927, Aitken had designed 26 "P-series" decorations, and by 1933, 116 (plates 20 and 21). He sometimes invited the pottery's hand decorators to submit designs.[11] The series reached 188 designs in 1941. By then the vogue for primitive hand painted patterns had run its course.

Besides Aitken, others who designed patterns, both in the decorating and lithography departments, included the now very senior Elmer Walter, John Wigley, A. J. Koch, and some newcomers, Hanley Hennock, Charles McKaig, E. M. Otis, and Douglas Bourne (the son of Hamlet Bourne, who had designed **Imperial Geddo** in the late 1880s). Since 1903, when the pottery obtained its first patent, to the end of 1931, the pottery's designers had patented nearly 160 patterns for applied decoration. These designs represented only a small part of their work, however. They had created over two hundred regularly available patterns for the decoration of household dinnerware, several hundred more stock and custom designs for commercial (hotel) ware, and many special commemoratives. Such a variety of patterns applied to a few good shapes allowed customers nearly endless possibilities in matters of style, from chaste Victorian florals to the latest modern designs.

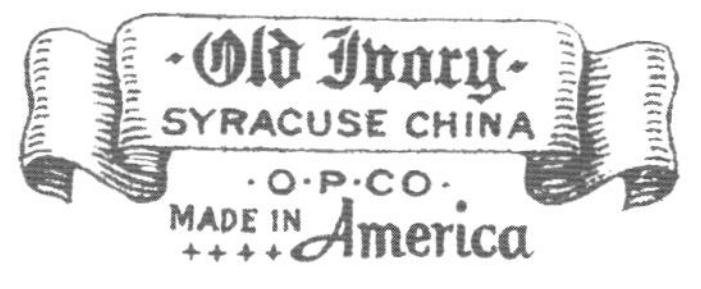

Old Ivory (1927–60)

To keep pace with changing tastes and consumer demands for new and colorful dinnerware, "Doc" Schramm began to develop colored china bodies. In 1927, the company introduced a pale, cream-colored clay body and named it *Old Ivory*. Ware made from the new body was marked with a distinctive backstamp, a furled ribbon with gracefully folded ends, carrying the name "Old Ivory" in Old English script and "Syracuse China," under which appeared "O.P.Co." Sometimes the backstamp also included the slogan "Made in America." Schramm had hoped to color the clay bodies with natural materials but in the end, the *Old Ivory* body was essentially a precisely controlled formulation of the pottery's standard clay formula with the addition of a body tint. To get the new ivory body on the market rapidly, the decorating department created four new patterns, "Berkshire," "Chiquita," "Tapestry," and "Meridale" exclusively for the **Mayflower** shape.

Bertram Watkin, an English pottery designer who had trained at Wedgwood and had come to the United States in 1920, succeeded Haley in 1926. Watkin went to work immediately on **Winchester** (fig. 6.10), a new dinnerware shape that would be exclusive to *Old Ivory*. In 1927, the company introduced the shape to the public. Like **Mayflower,** it was not patented, and like that older shape, it proved to be one of the firm's more successful products. The ivory body was used mostly for household dinnerware, but occasionally also for hotel ware. It redefined the aesthetic relationship of the colors of food, pattern, and glaze. It was now possible to eliminate

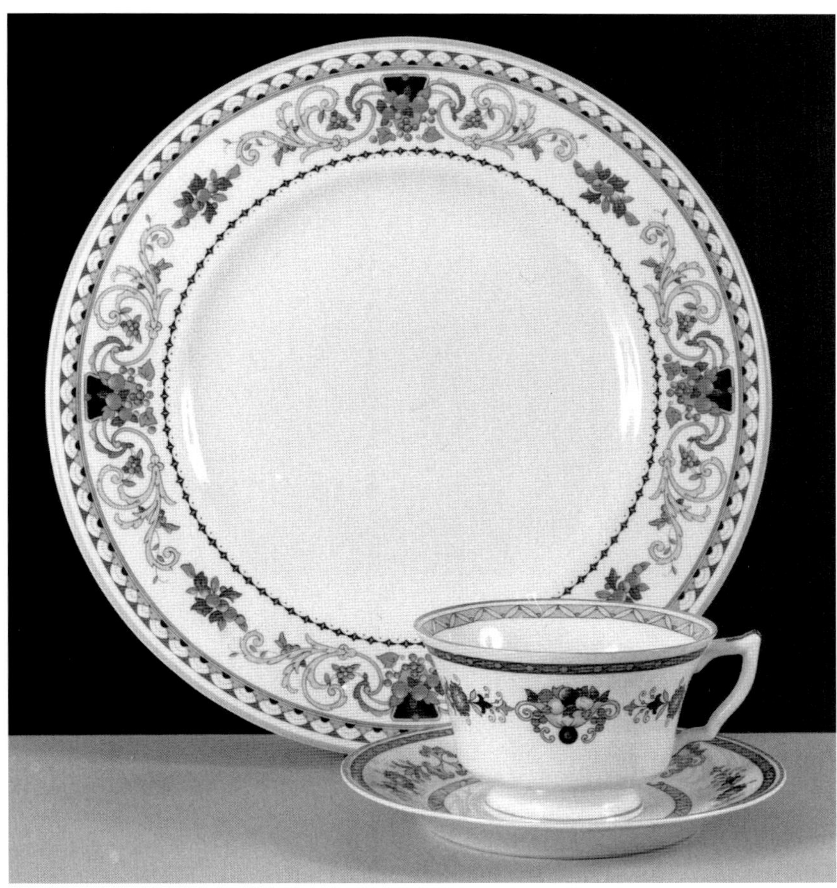

6.10. *Winchester* shape with "Pembroke" pattern on plate and "Midlothian" pattern on cup and saucer, 1927.

white as a component of an elegantly set, color-coordinated table. Putting his special skills as a portrait sculptor to work, Watkin also created "Toby" mugs of Al Smith and Herbert Hoover in a limited edition for Patriotic Products Associates of Philadelphia, to commemorate the 1928 presidential elections (fig. 6.11).

In its 1928 Christmas issue, *Syracuse China News* reported on the pottery's campaign to sell *Old Ivory* in hotel and restaurant publications. With a nod to the public's steadily increasing awareness of style and decoration, and the "present so-called wave of color," the pottery prepared a campaign of two double-page, four-color advertisements, pointing out that "the modern hotel man should pay just as much attention to the design of his china as he does to draperies and other equipment of his hotel," and that, "*Old Ivory* Syracuse is a new, beautiful china that is not only practical, not only economical, but entirely in keeping with the new ideas in hotel decorations." The article noted that the company had three hundred dealers, including hotel supply houses, retail china stores with hotel china and equipment departments, and department stores that catered to the hotel business as well as household customers.

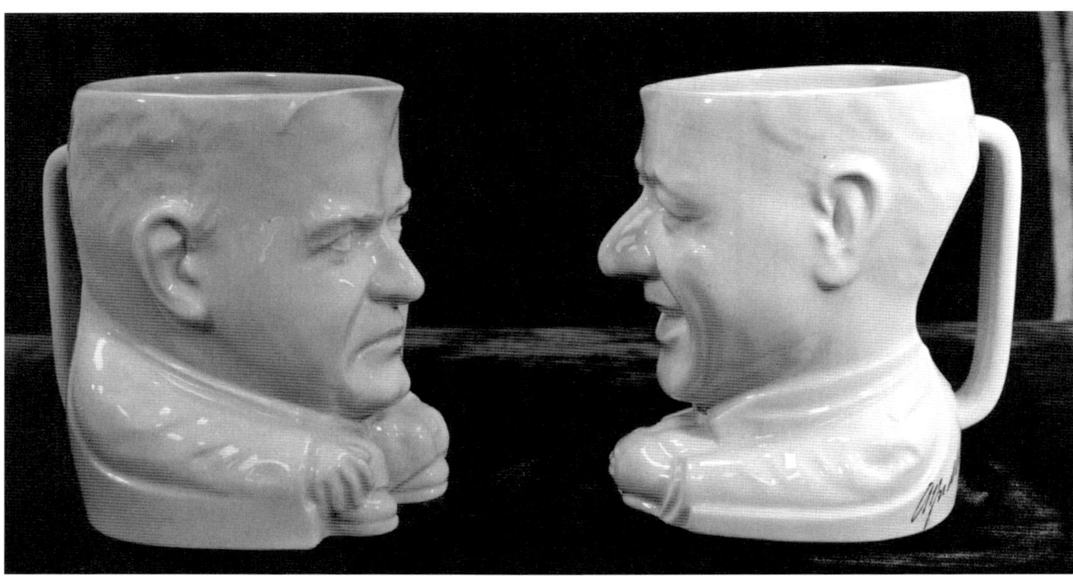

6.11. Bertram Watkin's Toby mugs depicting Herbert Hoover and Alfred Smith, candidates for president of the United States in 1928, made for Patriotic Products, Philadelphia, Pa.

In a few large cities, the company had more than one distributor, but in most, Syracuse China was an exclusive proposition. The company sold only through its distributors, making no direct sales to the consumer. This meant that a comparatively small sales force could cover the entire country, each salesman traveling with six or seven good-sized sample trunks. The dealers, in turn, kept in contact with hotels and restaurants, railroads, clubs, tea rooms, lodges, churches, and hospitals. Max Palmer, the legendary salesman who became manager of the Hinrichs's distributorship in Chicago, affirmed that many of his customers looked forward to the chance to "upgrade" to Syracuse China as soon as they could afford it.[12] With Lenox, Lamberton, Haviland, and a very few others, Syracuse China ranked as a prestigious product.

Business in hotel ware ranged from a one-hundred-dollar order for a tea room to a twenty-thousand-dollar order for a large hotel chain such as Ellsworth Statler's. Statler was a pioneer hotel builder who, when he opened his first restaurant in Buffalo in 1901, used Syracuse China. He came close to bankruptcy early in his career and never forgot those suppliers, including the pottery's distributor, who stood by him in hard times. He would not betray old loyalties, and when he built his hotel empire he always used Syracuse China. He was a generous benefactor to his profession. The American Hotel Association, established in 1910, opened, in 1922, the first school of hotel management in the United States at Cornell University. It faltered, but Statler bailed it out in 1923. The school's students held an annual banquet for members of the AHA at a facility they called the "Hotel Ezra Cornell". Every year from its beginning until the present (1997) the pottery has designed and manufactured a commemorative service plate for the event (plate 29).[13]

For large orders, the pottery's staff of artists often made many original sketches before a pattern or crest was found acceptable. For smaller orders, stock pattern usually filled the bill. With a new product such as *Old Ivory,* the pottery sent reprints of its advertisements directly to selected hotels along with sample plates. This campaign paid off.

In 1929, Aitken designed his "Poppy Pattern" plates, among his most distinctive achievements as a decorator (plates 8,9, and 10). A skillful watercolorist, Aitken painted the poppies in his own garden for a series of twelve service plates, and described his project for *Syracuse China News:* "Never in my experience as a decorator has there been such demand for color in decoration. The decoration of the poppy starts with a freehand engraving on steel, from which we print. It is then ground laid [with solid colors], painted by hand and lined, all of these processes being under the glaze, after which we snap up a brilliant highlight in [overglaze] enamel color, which is the finishing touch."[14] This began Aitken's hand-painted *Nature Studies* series, a group of forty designs depicting flowers and birds (plates 7, 8, 9, and 10).

Although the *News* had diminished somewhat in size and frequency of publication, it continued to report on the pottery's latest achievements and employee activities. In early November 1929, the company exhibited in New York at the annual National Hotel Exposition. It had gratifying responses to its display of several hundred sample plates and seventy-five examples of different patterns labeled with the name of the hotel using the pattern. Aitken had painted a colossal poppy plate as a centerpiece for the exhibit (fig. 6.12), which had made its debut at the New York State Fair at the end of the summer. The company had produced a four-reel, sixteen-hundred-foot motion picture showing the various processes involved in the making and decorating of Syracuse China. Its purpose was to take the factory to its distributors' salesmen, to show how the product was made. The *News* also reported a large order for the Chalfonte Haddon Hotel in Atlantic City. The pottery had won the competition for that job from over "hundreds of designs submitted by the leading potteries in this country and abroad." In its January–September 1930 issue, which was its last until after World War II, there was no mention of the crash of the stock market in October 1929.[15]

The pottery made many of its outstanding custom orders for railroads, serving, in all, nearly two-thirds of the railroad custom over the years beginning around the turn of the century. For the Missouri Pacific Railroad in 1929, it created a handsome service plate featuring a steam locomotive in its center, a reproduction of William Harnden Foster's painting "The Sunshine Special," with a border of panels depicting the state flowers of the eleven states through which the Missouri Pacific's trains passed (plate 18).[16] This service continued until 1949, when it was replaced with a similar plate, "The Flight of the Eagle" (plate 19), this time with a diesel locomotive and a border depicting the states' capitols. When, in 1930, the Union Pacific Railway System began a new rail service between Chicago and Denver, it named the train the *Columbine,* after the state flower of Colorado. The flower motif expressed throughout the decoration of the train appeared even on the menus. Aitken created an original treatment of it in hand-filled transfer print for the train's china, much in the spirit of his *Nature Studies.* The patent for "Portland Rose" for the Union Pacific's Chicago to Portland, Oregon, service bears Aitken's name. Honoring this "City of Roses," it bore a distinctive backstamp that included the name of the rose variety, "Madame Testout," and the words, "For you a rose in Portland grows." The "Columbine" and "Portland Rose" patterns were used on Syracuse China *Old Ivory*.

These years of innovation and buoyant success ended as the Great Depression deepened. By 1932, the company felt the depression's full force and faced its most difficult struggles since the curtailments of the 1890s. Orders fell from a record peak of over a million dozen items in 1929 to a low of four hundred thousand dozen in 1932. To keep as many employees as possible working, even for one day a week, the company began promoting "Syracuse Specials," offering decorated china at substantially reduced prices. It advised its distributors that "current conditions in our industry are such that cost of production is not a major factor in the determination of sales prices." It was clear that there were major challenges ahead.

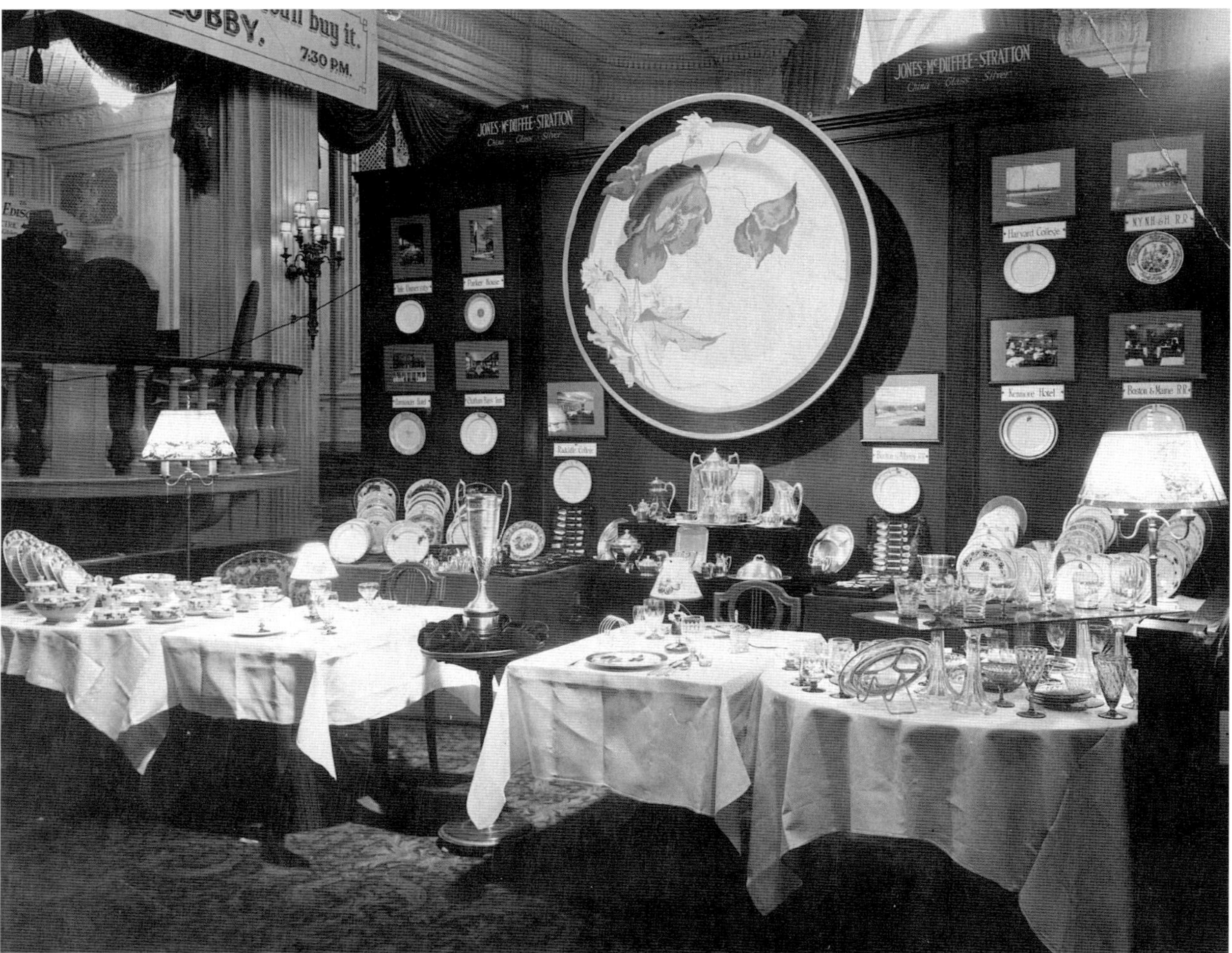

6.12. Harry Aitken's colossal poppy plate in the Jones-McDuffee-Stratton's booth display of Syracuse China at the National Hotel and Restaurant show, New York, 1929.

Seven

Econo-Rim and the Great Depression
1932–1941

In moving to Syracuse, I felt that I was isolated from the art world
that I had enjoyed [but] during these five years it has been possible to
bring the center of the ceramic art world to this city.

—R. Guy Cowan, 1937

DESPITE the debilitating effect of the Great Depression on the American tableware industry, the 1930s proved to be a time of remarkable vitality and creativity at Onondaga Pottery. More innovative shapes and decorating techniques arrived in that decade than in any
previous one.

Having broken away from the near-universal whiteness of fine china when it introduced its
Old Ivory clay body in 1926, the pottery now strove steadily to develop new colors. It was an exacting challenge for the laboratory, for any colorant needed to be compatible with the other
components of the clay body and perfectly uniform as a color in the finished product. In 1931,
Edward Schramm and his colleagues arrived at two promising colored bodies, one blue and one
tan, the latter named *Adobe*. Bertram Watkin designed a new dinnerware shape exclusively for
Adobe, **Clinton** (see plate 20), decorated with charming hand-painted floral decorations, but it
remained in production only until 1933. The blue body attracted a few customers, but it was
discontinued after several months, perhaps because of production problems. The pottery also
offered the **Governor Winthrop** shape (see plate 20)in *Adobe* between 1932 and 1935, when it,
too, was discontinued. **Governor Winthrop** was basically a white shape.

Watkin's only hotel ware shape, **Pilgrim** (fig. A.1.), designed in 1931, a few months before
his retirement, was a great success. It was in a way a retrospective shape echoing the design of
the original **Round Edge** in the configuration of its underleaf. **Pilgrim** struck a compromise between the heavy, revised **Round Edge** and the more refined **Empire** shapes. All three shapes provided a flat-leafed rim as a palette for decoration. **Pilgrim** sold well in all body colors and still
remains in production. *Adobe* and Watkin's **Pilgrim** had done much to see the pottery through
the very difficult first years of the Depression.

Watkin's successor, R. Guy Cowan, rapidly became a central figure in the further revitaliza-

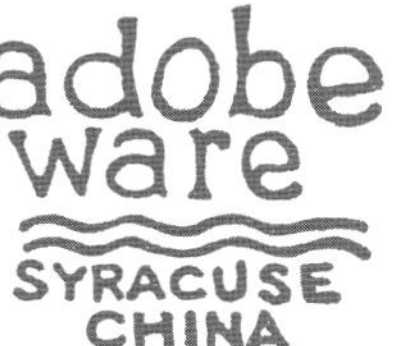

Adobe Ware (1932–72)

tion of the company's products during the 1930s. When he joined the pottery in 1932, at first on a part-time basis, Cowan was already an illustrious figure in American ceramics.[1] Born in 1884 in East Liverpool, Cowan had come to Syracuse in 1893 when his father, Louis Cowan, joined Onondaga Pottery as head of its decorating department. Like James Pass, he grew up in the pottery business. His father taught him decorating skills, and the younger Cowan began to work part-time at Onondaga Pottery in the summer of 1898, at age fourteen. When classes resumed, he lined and gilded ware after school. His interest in decorating soon gave way to one in ceramic engineering, however, and in 1902 he enrolled in the recently established New York State School of Clay Working at Alfred University.[2]

When lack of funds threatened to interrupt Cowan's freshman year, Charles Binns, the school's director, wrote to Pass on his behalf, evidently unaware that Louis Cowan might need to leave the pottery. On 23 February 1903, Binns wrote:

> I want to consult you about [Guy] Cowan. He is in trouble about money and says he fears his father cannot keep him here. . . . The boy is a good student . . . and it would be a thousand pities to stop him now. He has no tuition to pay but a small amount for laboratory fees and his room and board. . . . [He] does not know I am writing nor does his father. In fact I do not want them to know. . . . Do you think it would be possible for him to receive a small advance from your firm, to be worked off during the summer. . . . I am most anxious to retain the boy if it can be done as I feel sure he will be a credit to the school. At present his burden is greater than he ought to bear.[3]

Pass's reply explained that with his problems with Louis he found it inappropriate to advance funds. His letter prompted this response from Binns: "I appreciate your confidence and sympathize with your position. I have said no word about you to the boy and have arranged with President Davis to give him time until he shall be able to earn a trifle."[4] Even with help from Alfred, Cowan was obliged in time to interrupt his studies.[5] He wrote often to Binns; his letters to his teacher and Pass's letters survive in the archives at Alfred.[6]

In 1907, Guy Cowan received the first degree in ceramic engineering conferred by Alfred University. He worked in the trade, then went to Cleveland to teach ceramics at East Technical High School. In 1912, he set up the Cleveland Pottery and Tile Company. In 1917, he established Cowan Pottery and in 1920 moved it to the Cleveland suburb of Rocky River. There he achieved his goal of producing art pottery of fine workmanship in a distinctive modern style.

Cleveland was then the single most important American center for "the modern tendency" in ceramic art. Art Deco, Art Moderne, and Vienna Secession styles all found homes there, and intermingled. Cowan drew successfully on the distinctive talents of a number of younger ceramic artists (some of them his students), all imbued with this modern spirit. These included such well-known ceramic figures of the era as Viktor Schreckengost, Waylande Gregory, Thelma Frazier Winter, Don Winter, Paul Manship, and Arthur Baggs. Although some were associated with Cowan only periodically, they constituted as important a group of allied ceramic artists as then existed in the United States. After 1920, Cowan himself did little designing at his pottery, giving his attention instead to clay chemistry, including the development of new glazes, and to the management of his firm. In 1928, he became a faculty member of the Cleveland School of Art, where he taught its technical courses as a "Master in Ceramics." This was the background Cowan brought with him when he joined Onondaga Pottery, and it had a great influence on the character of the shapes and patterns that came from the pottery in the 1930s and 1940s.

By the mid-1920s, Cowan Pottery's national dealer organization included some twelve hundred outlets. Its popular, large-volume production ceramic figures kept the pottery financially sound, while its small edition art pieces, selling at much higher prices, earned it an illustrious reputation. The crash of the stock market in 1929 and the Great Depression that followed presented insurmountable obstacles to survival, forcing Cowan to close his pottery in December 1931.

In 1932, Cowan accepted dual appointments in Syracuse, sharing his time between Onondaga Pottery and Pass & Seymour as a part-time consultant.[7] In the summer of 1933, he succeeded Watkin as chief of product design at the pottery. Cowan brought with him not only outstanding experience in the aesthetical, technical, and managerial aspects of ceramic production, but also a commitment to modern design.[8] His combination of strengths, enhanced by his national reputation, was unique within the industry.

Almost as soon as he arrived in Syracuse, Cowan assumed a central role in the establishment of the annual Ceramic National exhibitions, begun in 1932 at the Syracuse Museum of Fine Arts. He worked with Anna Wetherill Olmstead, the museum's director, in organizing and producing the series. The first exhibition honored Adelaide Alsop Robineau, who had died in 1929, and also included work from many able ceramic artists, including several from Cleveland. As the years passed, the Ceramic National provided American ceramists with the first continuing national theater for their talents. Some of the exhibitions traveled to New York City, and also to Europe. Purchase prizes funded by Onondaga Pottery and others allowed the museum to begin to assemble what soon became the premier collection of twentieth-century American ceramic art.[9] Many of Cowan's Cleveland colleagues and former students exhibited in the annuals and some served as judges. The Ceramic Nationals kept Cowan closely involved with modern ceramic art and undoubtedly contributed to his success in designing tableware shapes for the 1930s. Together, Onondaga Pottery's fine china and the museum's steadily growing collection of ceramic art gave Syracuse an important reputation as a center for American ceramics.

Cowan arrived at the pottery during a period of important social and artistic change. He helped usher in an age of design for industrially produced ceramics influenced by the Art Deco style. That style had derived its name from the Exposition Internationale des Arts Decoratifs et Industriels Modernes held in Paris in 1925. Although Art Nouveau and the Arts and Crafts movement had left their marks on surface decoration, neither had inspired much change in tableware shapes. But the Art Deco style did. As Bevis Hillier has observed, it was an

> assertively modern style, [and] drew inspiration from various sources, including the more austere side of Art Nouveau, Cubism, the Russian Ballet, American Indian Art and the Bauhaus; it was a classical style in that, like neo-classicism but unlike Rococo or Art Nouveau, it ran to sym-

metry rather than asymmetry, and to the rectilinear rather than the curvilinear; it responded to concrete and vita-glass, and its ultimate aim was to end the old conflict between art and industry, the old snobbish distinction between artist and artisan, partly by making artists adept at crafts, but still more by adapting design to the requirements of mass-production.[10]

Econo-Rim (1933–67)

Perhaps more than any other designer in America, Cowan and his associates at Onondaga Pottery, under the direction of Richard Pass, changed the face of hotel ware in the 1930s. This was a remarkable achievement. A plate's function limits the possibilities for alteration in its form. It must rest evenly on a flat surface and hold food without run-off. Tradition-bound restaurants and hotel dining rooms had always been wary of any change in tableware shapes. Cowan's **Econo-Rim** hotel ware of 1933 probably amounted to as great a departure for the time as could successfully be made in industrially produced tableware, and it was an immediate success. In this shape, he greatly reduced the width of the rim on plates and other flatware items. This increased the capacity of the well without changing the plate's outside diameter. (Within the factory the top of the rim was called a "leaf".) Customers who always ordered by outside measurement now had a shape of significantly different proportions. Cowan decorated the narrow rims with a moulded, well-defined, relief design that transformed the classical egg-and-dart motif into a distinctly Art Deco form. **Econo-Rim** (see Fig. A.2)was perhaps the first mass-manufactured tableware shape to succeed as modernist design. It was particularly appealing in *Adobe* ware decorated with hand paints, Native American and Western motifs, and rustic patterns. (The **Econo-Rim** backstamp always took precedence over the *Adobe* one.)

As hotel ware, **Econo-Rim** was economical to make and use. Cowan conceived it in part for use in public eating places where space was at a premium, since the larger well allowed the use of smaller diameter plates. This was important in railroad and steamship dining rooms and for hospital trays (fig. 7.1). But **Econo-Rim**'s acceptance was probably due at least as much to its looks as to its practicality. In March 1947, fourteen years after its introduction, *Ceramic Industry* reported that the shape's immediate and sustained success in the hotel trade was "sweet to the soul of the sales department, and doubtless to the designer, but [it also] carried other Onondaga hotel lines along with it to better sales records."

In the 1950s, Huber recalled his involvement in the origin of **Econo-Rim.** As was so often the case, his contact with the marketplace had put him in a unique position to know what clients wanted. He said that the idea for a narrow-rim plate came first to him as a special request from Frank Lewis, dining car superintendent for the Union Pacific Railroad. Lewis apparently told him that in the near future dining cars would be, "narrower to get the speed. . . . But the trouble is that [because] those cars are narrower . . . the tables are smaller. We haven't really got room on the tables to serve well."[11]

Some hotels and restaurants found that **Econo-Rim**'s Art Deco relief designs did not always emerge cleanly from dishwashing machines. Statler suggested that if the shape were made with plain rims, he would use it in his coffee shops. A factory order in December 1935 for the backstamp for this "plain **Econo-Rim**" specified that it be marked with the **Econo-Rim** backstamp, but with the words "patent-pending" removed. This alteration of the shape was then called

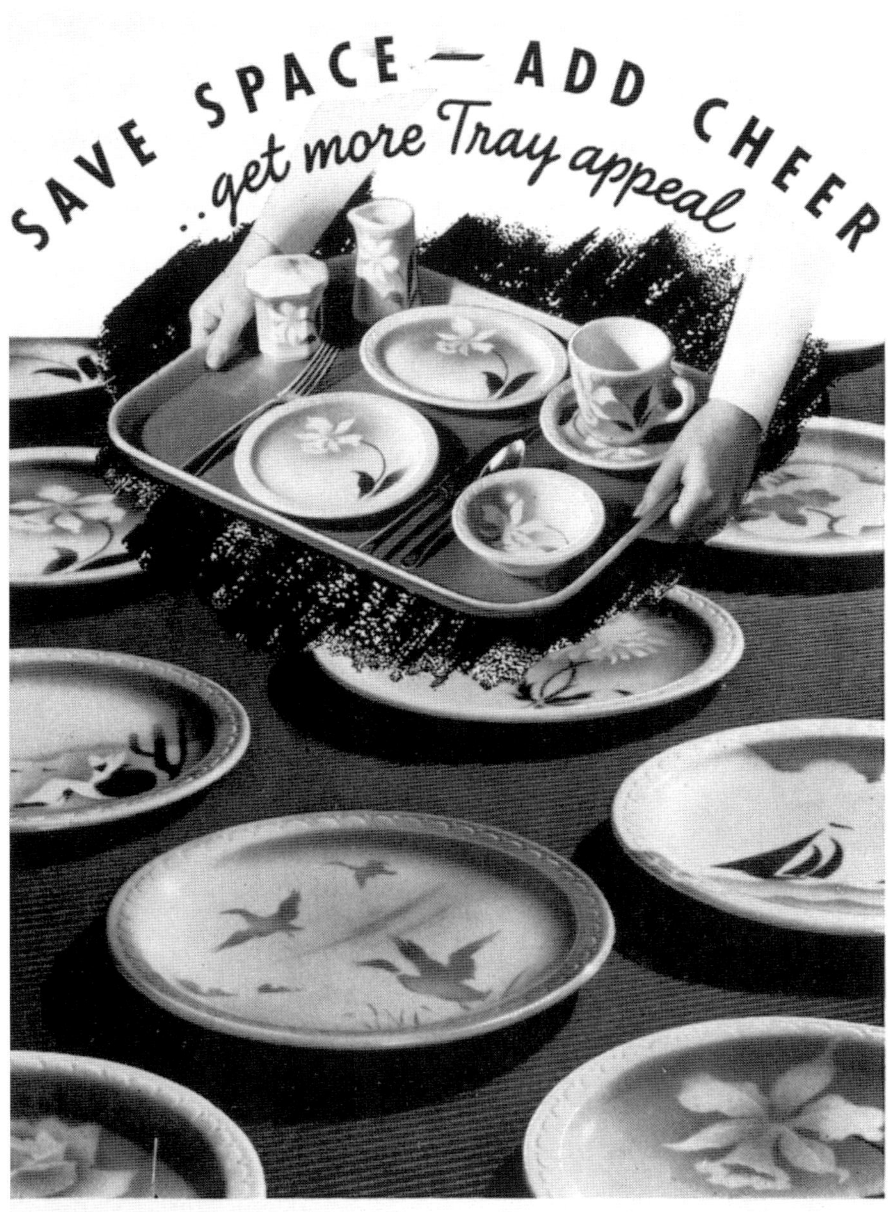

7.1. An advertisement for *Econo-Rim* in *Modern Hospital*, February 1939.

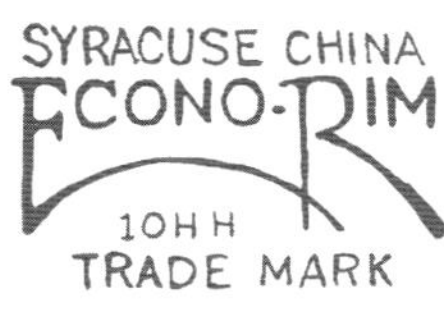

Morwel (1936–67)

Statler (though it was by no means exclusive to that chain) until 1939, when Edward Otis of the design department, who had named **Econo-Rim,** also named this variant shape: "Because there's more well and less rim: *Morwel*" (fig. A.3).[12] (Both shapes remain in production in 1997.)

It was considered an achievement that **Econo-Rim** required only two coffee mugs and four teacup shapes.[13] Although household dinnerware shapes usually required only one cup form in several sizes, hotel ware spawned myriad cup shapes. No item lent itself so well to a restaurant's or hotel's need to establish a distinctive identity. Unlike plates, cups are held and brought repeatedly to the face, making them, in the course of a meal, the most clearly visible item of tableware in a place setting. Every change in form, capacity, or even the set of a handle, could signal a name change. The pottery had made more than seventy different cup shapes to accompany its earlier hotel ware lines. But this quantity was impractical to maintain in an age of streamlining and cost-consciousness. In his memoir, Huber told how one such cup shape came to bear his name.

> We began to make a heavy cup with a heavy handle and of course they sent one sample to Euniac [the company's sales representative in New York] and one to me. On the tag on the bottom of my cup, they had written the name "Euniac." I knew that there had been a mistake when they had packed it up. They had sent Euniac's cup to me while he got my cup with "Huber" on the bottom. Euniac apparently didn't realize that. He began to send orders in for Huber cups and . . . that's where the name . . . came from.

In 1937, the company's marketing department hired Helen Graley to head its internal advertising department. In October, she produced and copyrighted a shiny new product catalogue, the first since 1924. With graphic design in the Art Deco style, an iridescent green cover, and bold bright-blue and black printing, the catalogue in its design, production, and content expressed striking optimism in nationally troubled times. It began with Longfellow's poem "Keramos,"[14] and a statement titled "Achievement" that reviewed the company's history. It said that the pottery had been the first to manufacture commercially an American china, had produced more fine chinaware than any other pottery in the United States, had been the first in America to produce a rolled edge shape, the first to make its own ceramic decalcomania, and the first to succeed with underglaze decalcomania. The company was a pioneer in the production of colored bodies when, in 1931, it introduced its tan *Adobe* (and the short-lived blue body), to augment its *Old Ivory.* Although the pottery's primacy in each case may not have been absolute, it had been in the forefront of each development.

The 1937 catalogue offered a large selection of items in the **Mayflower** and **Winchester** thin dinnerware shapes and in the **Empire** and **Round Edge** hotel ware shapes. But in special layouts it emphasized its newer products. It promoted **Econo-Rim** ware in both white and adobe. The catalogue also introduced two more of Cowan's shapes: **Doric** and **Shelledge.**

The company patented the **Doric** shape, an accessory line of mostly cast, once-fired ware (Fig. 7.2). It consisted of tea and coffee pots, creamers, sugars, and ashtrays. The pottery at first offered it only in the *Adobe* body. In this shape, Cowan had again transformed classical motifs

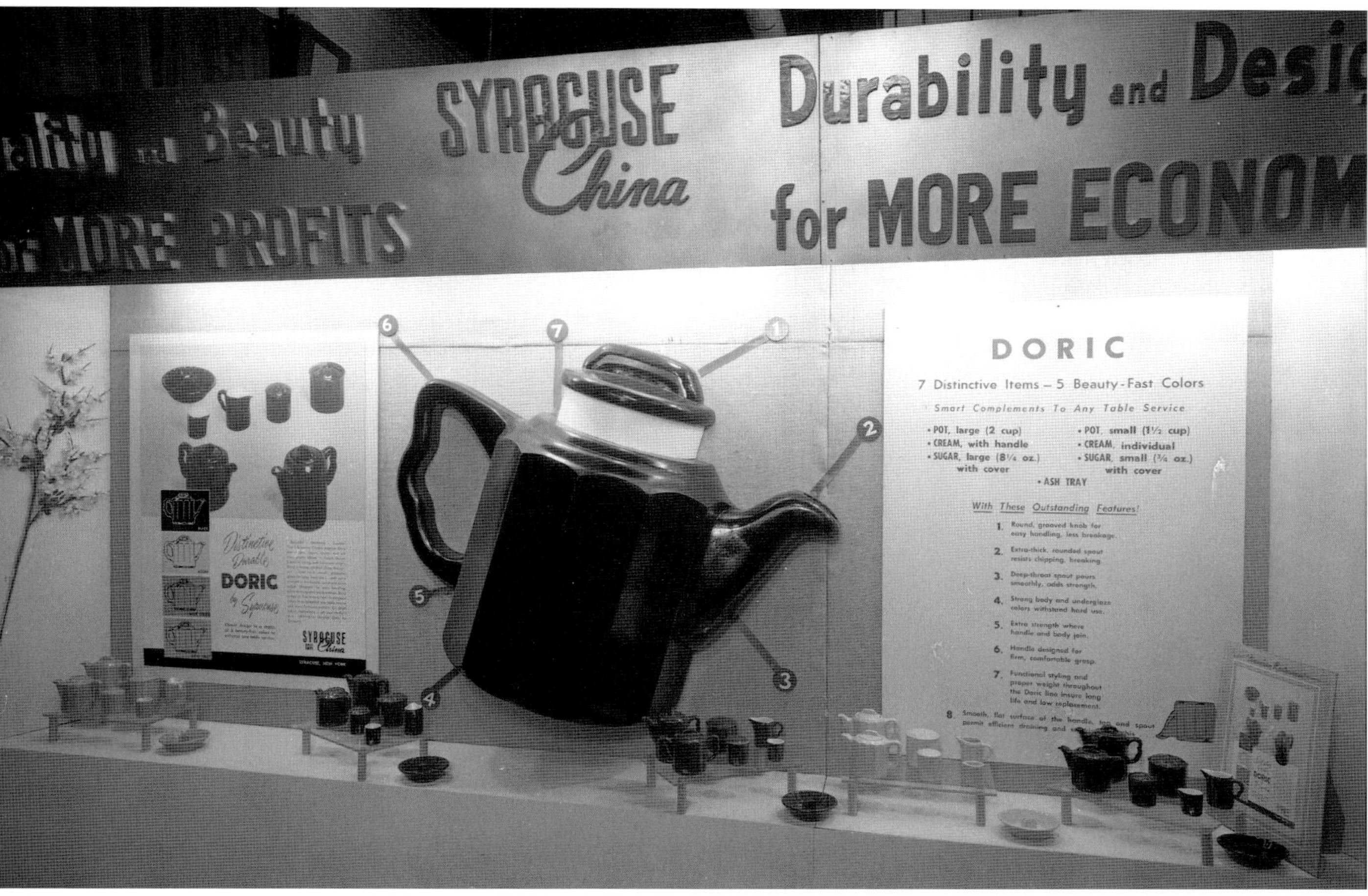

7.2. **Doric** display for a national hotel ware convention, c.1937.

by touching them with Art Deco feeling. The catalogue stated that **Doric** was "in fine classic form combining simplicity with beauty—designed for hard service. A necessary requisite to complete any decorated service." The shape derived its name from its fluted forms and was meant to complement such shapes as **Econo-Rim** and **Morwel,** which did not offer the variety of accessory items that earlier hotel ware shapes had. In time, the pottery offered **Doric** not only in *Adobe* but in the solid underglaze *Vitritone* colors added to the line (sage green, dove gray, forest green, and black).[15] This is the closest the company came to making the popular bright-colored utilitarian pottery called "refrigerator ware" offered by Hall China and others in the 1930s. Most items of **Doric** ware were cast in moulds, each imprinted on its base with the company's initials: O.P.Co.

The company called its process for decorating with solid color *Vitritone.* This method of producing solid colors was achieved by dipping or spraying ware with a pigmented slip after it was mostly cast and dried, but still in the "green" unfired state (masking off the parts that did not receive color), not yet glazed. Theoretically, it was cost-effective to once-fire, but because many problems arose from glazing in a biscuit kiln, much of the ware needed to be fired twice. This process was already in use elsewhere in the industry. The pottery gave names to two other color processes it originated in the 1930s. *Artint* was a method developed in 1931 and used exclusively for spraying and rubbing underglaze colors on the relief decoration of **Econo-Rim.** In 1933, *Syratone,* another solid-color process, involved spraying color on biscuit rather than green ware.

The 1937 catalogue contained three pages of excellent quality color reproductions featuring a selection of underglaze decorations in a variety of treatments on wide-leafed and **Econo-Rim** plates, twelve to a page (plates 21 and 22). It illustrated examples of various decorating techniques: prints, decals, hand-paints, crests, and lines. The catalogue also contained a colorful centerfold showing three "service plates" in full-scale, state-of-the-art reproductions, with three types of elaborate decorations. They were printed in color and gilt as facsimiles of the plates. The reproductions included embossing to imitate acid-etched gold, a solid-black *Syratone* plate overlaid with a gold print pattern, and an ivory plate with an acid-etched Nile green *Syratone* rim.

Hotel ware service plates, named **Elton, Oxford,** and **Rugby** according to their diameter and thickness (and difficult to tell apart), were larger than dinner plates, usually between ten and eleven inches in diameter. Often custom-decorated to echo and intensify a theme or color scheme in a club, hotel, or restaurant dining room, these plates greeted diners at preset tables, offering a treat to the eye. Removed when the meal was about to be served, the handsome plates played their part in company with a table's linens, crystal, silver, and floral centerpiece to establish a tone of quiet splendor. Less common in domestic settings, they nonetheless were acquired by those who desired a "complete" set of china.

The third of Cowan's shapes introduced in the 1937 catalogue was **Shelledge.** This was a fine china shape for household dining. The catalogue described it as "a ceramic accomplishment in pure white, unusually thin dinnerware suitable for many service combinations, for festive occasions or everyday use" (fig. 7.3). This modern line consisted of plates in four sizes, whose wells were either plain or decorated with intaglio (carved) stylized relief images of fish, fruit, or flowers. Its narrow rims echoed **Econo-Rim** styling. The shape's bowls and cups had gracefully flared rims. Their finely scalloped and crimped edges and delicate thinness suggested the name **Shelledge.** Handles looped over the tops of the teapot, creamer, and sugar bowl. The company patented the shape in 1936 and generated a new backstamp for it.

Before this, Cowan had sent prototypes of the shape to Mary Ryan, a decorator who operated a celebrated showroom in New York. She had been his own pottery's New York sales representative from 1927 to 1930. Ryan was a trendsetter who knew what buyers from Macy's and other large-volume department stores wanted. She seemed to have the final say on all of the early **Shelledge** designs.[16] Perhaps on the strength of her approval, Cowan entered his **Shelledge** intaglio plates in the 1935 Ceramic National. Between 1939 and 1941, Cowan remodeled some **Shelledge** items. Manufacture of **Shelledge** was interrupted during World War II. Production resumed in 1947, when the pottery reissued it with additional items to make it a full dinnerware shape.

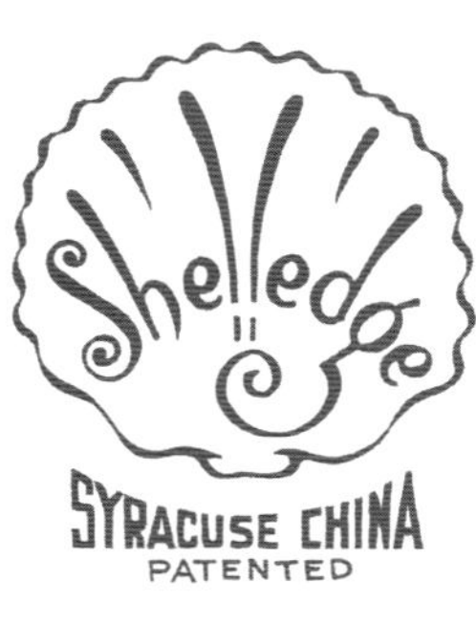

Shelledge (1936–49)

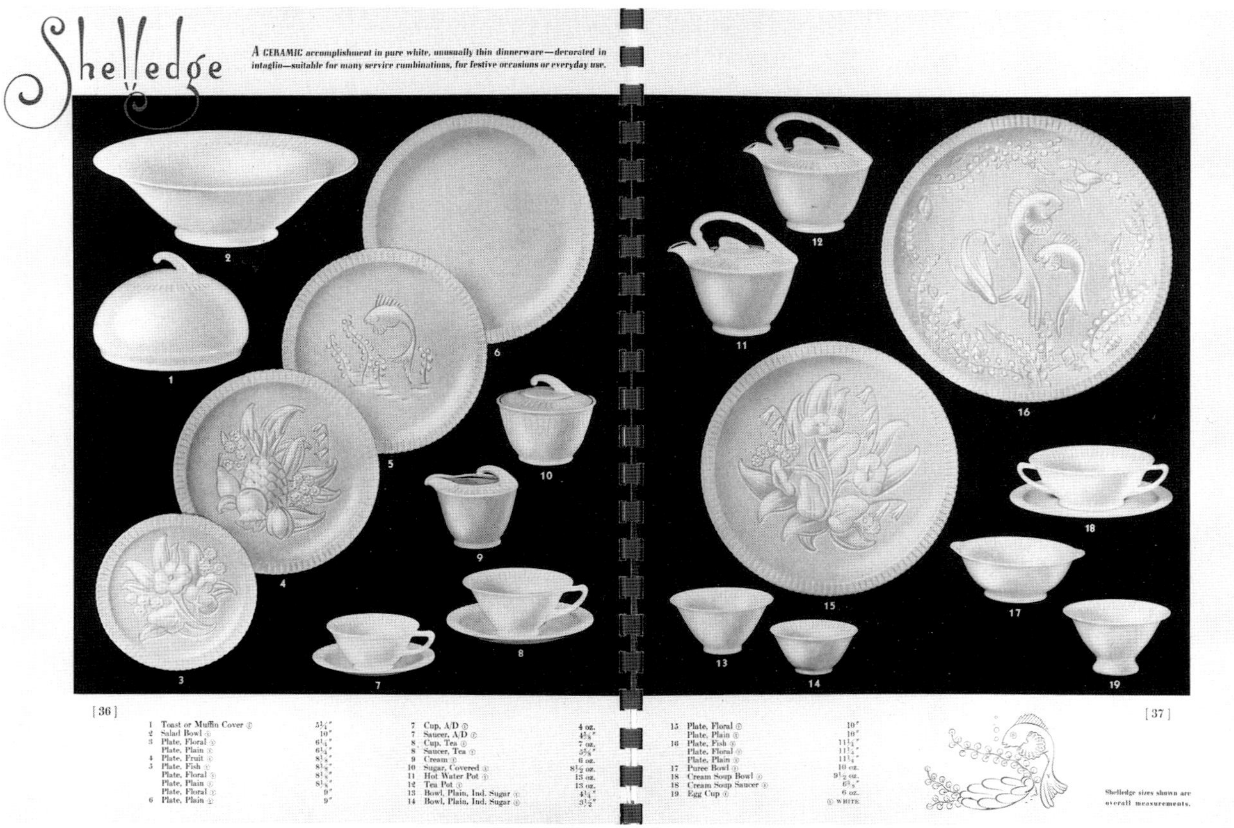

7.3. Introduction of new **Shelledge** shape in 1937 catalogue.

The 1937 catalogue also introduced the Colonial Revival **Virginia** shape. In a two-page spread, it was shown with a blue floral pattern, "Lady Mary" (fig. 7.4). **Virginia** always carried the Old Ivory backstamp. It was a variation of the **Winchester** shape with rounded handles instead of pointed ones; it used the same flatware and turned items (made on the same moulds) as the **Winchester** shape. By creating different appurtenances and new serving pieces, Cowan had created a new shape line. When **Winchester** was discontinued in 1950, some patterns that had originated on it were transferred to **Virginia**.

In 1938 Cowan designed another thin dinnerware shape in the spirit of the Colonial Revival and named it **Federal** (fig. 7.5). With delicately fluted and scalloped moulded decorations, and curved handles and knobs, this shape contrasted sharply with the Art Deco styling of **Shelledge**. **Federal** was offered in *Old Ivory* with delicate floral patterns. It was clear that when setting a table, many American housewives still chose to preserve traditional values in decor rather than make a modern statement. **Federal** carried its own backstamp. Beginning around 1940, all dinnerware backstamps also carried the name of the pattern.

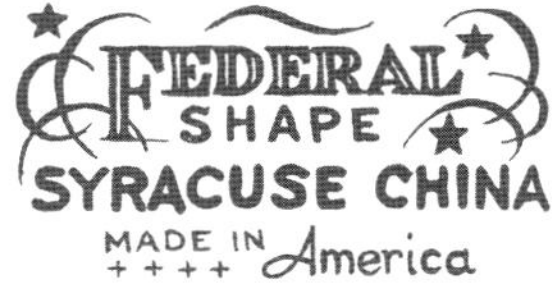

Federal (1938–69)

Lady Mary

A NEW pattern on the new Virginia dinnerware shape—modern, striking and effective.

No.	Item	Size	No.	Item	Size
1	No. 8 Plate	$9\frac{3}{4}$″		No. 8 Plate	$9\frac{3}{4}$″
2	A. D. Cup	3 oz.		No. 10 Plate	$10\frac{1}{4}$″
2	A. D. Saucer	$4\frac{3}{8}$″		No. 7 Plate, Rim Deep	$8\frac{7}{8}$″
3	Tea Cup	$7\frac{1}{2}$ oz.	9	Cream Soup Bowl	10 oz.
3	Tea Saucer	$5\frac{3}{4}$″	9	Cream Soup Saucer	$6\frac{1}{2}$″
4	Teapot	27 oz.	10	Covered Sugar	12 oz.
5	Covered Casserole	42 oz.	11	Cream	9 oz.
6	No. 10 Platter	$12\frac{1}{8}$″	12	Bouillon Cup	$7\frac{1}{2}$ oz.
	No. 12 Platter	$14\frac{1}{4}$″	12	Bouillon Saucer	$5\frac{3}{4}$″
	No. 14 Platter	16″		No. $4\frac{1}{2}$ Fruit	$5\frac{1}{8}$″
7	Handled Boat	20 oz.		Double Egg Cup	$4\frac{1}{2}$ oz.
7	Sauceboat Stand or Relish	$8\frac{1}{4}$″		No. 8 Baker	$10\frac{1}{8}$″
8	Rim Oatmeal Bowl	$6\frac{1}{4}$″		No. 2 Small Deep Bowl	5″
8	No. 4 Plate	$6\frac{1}{4}$″		Cake Plate	$10\frac{1}{4}$″
	No. 5 Plate	$7\frac{1}{8}$″		Chop Dish	$12\frac{1}{2}$″
	No. 6 Plate	8″		No. 2 Comport	$8\frac{7}{8}$″
	No. 7 Plate	$8\frac{7}{8}$″		No. 3 Comport	$7\frac{7}{8}$″

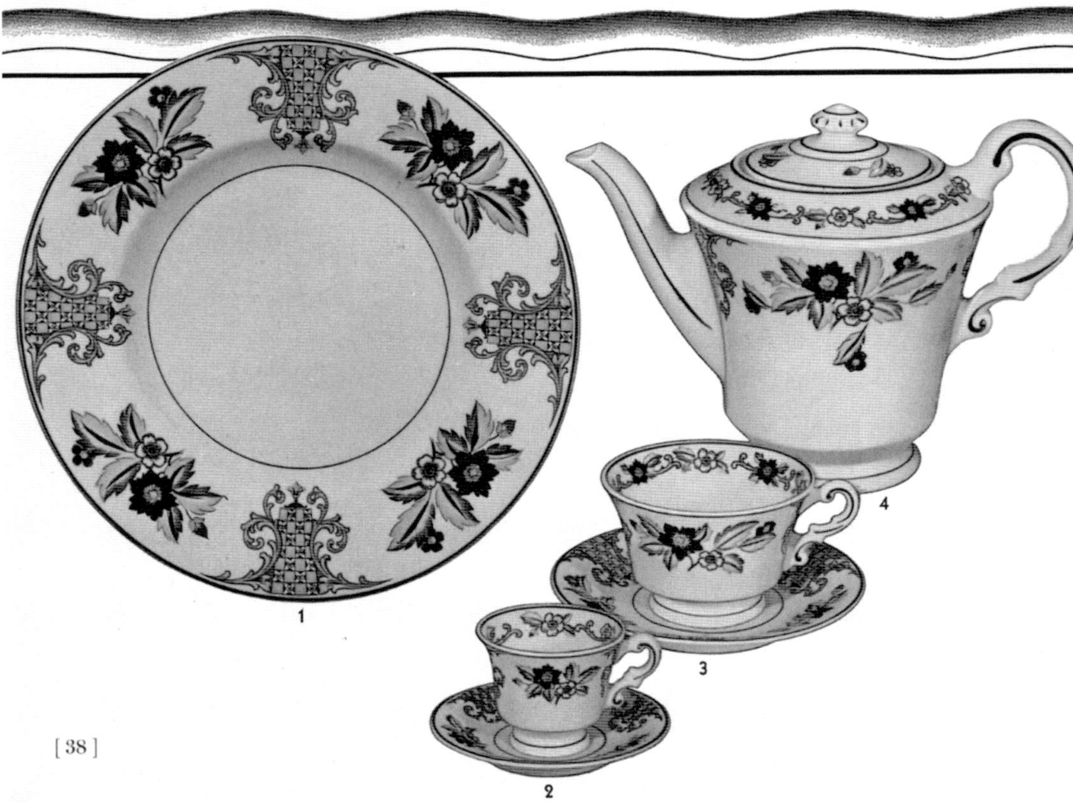

[38]

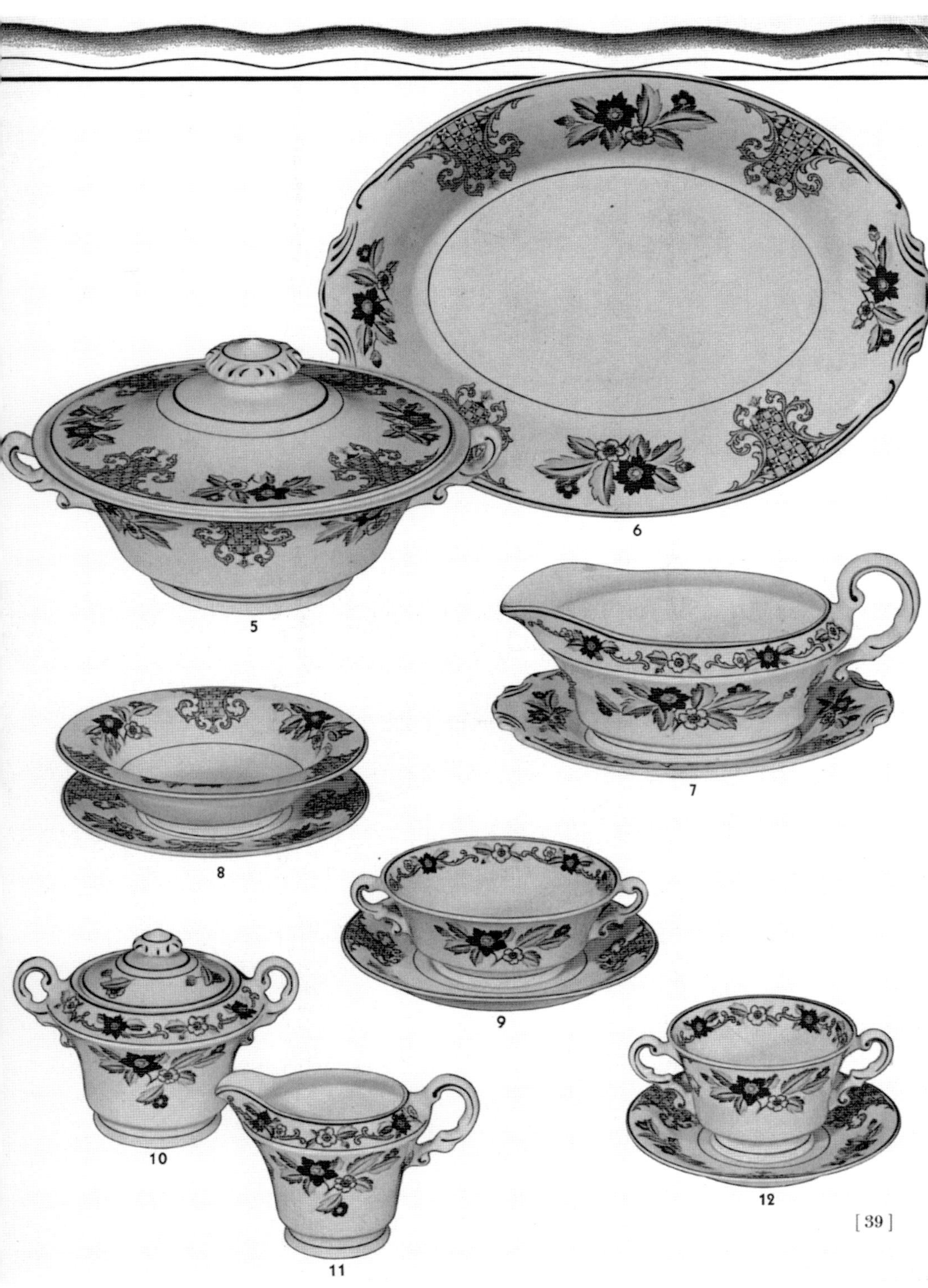

7.4 "Lady Mary" pattern on new *Virginia* shape, introduced in the 1937 catalogue.

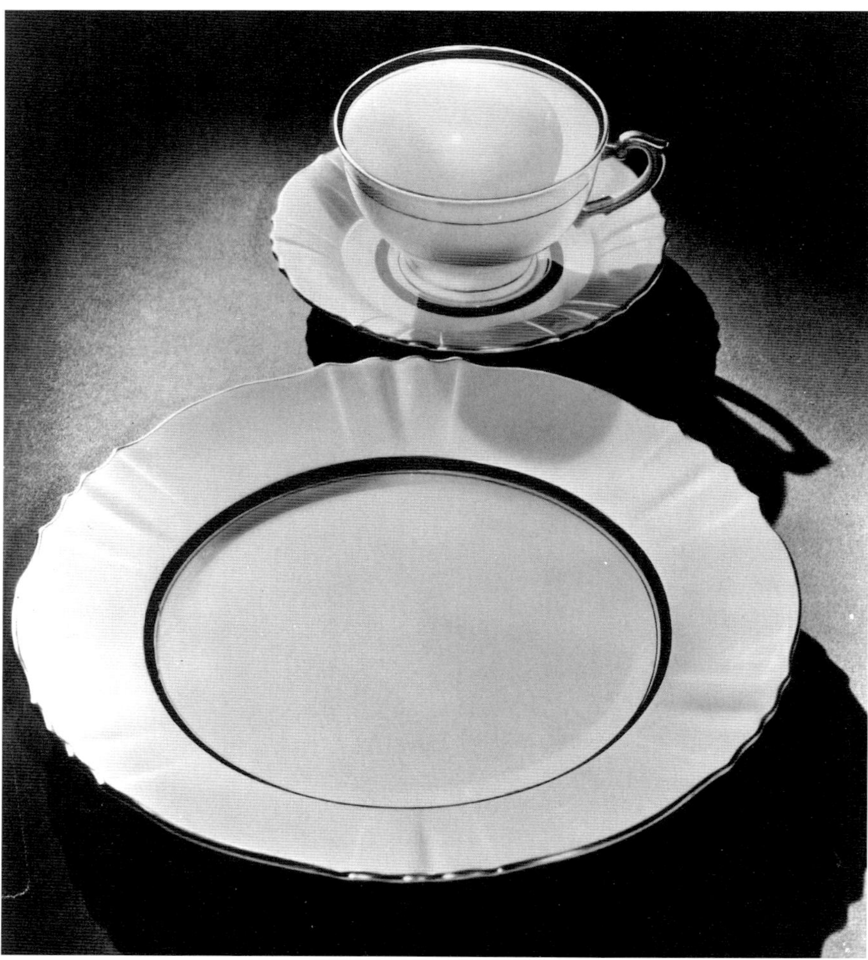

7.5. *Federal* shape, "Cornwall" pattern, photographed for earliest product brochure, c.1938.

The initial success of Cowan's shapes owes much to the company's skillful promotion through advertising and to its sales force's sensitivity to developments in specialized parts of the market. That the shapes continued to sell well for decades, and remain much admired, testifies to their intrinsically good design, to Cowan's remarkable ability to imbue traditional functional forms with contemporary feeling, and to the excellence of the designers, modelers, and pattern artists who worked with him. In creating shapes, he did not break with the past so much as select elements of good design from ancient times to the recent past and transform them to serve distinctly modern products. Some of his thinking about his approach to design is reflected in an unsigned rough draft of an apparently unpublished promotional piece written for the pottery, probably in the late 1930s.

We frequently hear the descriptive term "classical modern" applied to new creative work. By which is meant that a new creative design has in it certain desirable characteristics which can be recognized as having been inspired by the work of the classical periods or by their revival in the eighteenth century. The Shelledge shape in Syracuse china is one of the best examples of such a point of view. In design it is unquestionably a modern creation. It reminds one of some of the beautiful wares of the past, yet one cannot find its counterpart in the past. . . . The intaglio carving is a process never before used in manufacture. No cup has before been made with a carved decoration on the inside. The plates with their narrow rims are both beautiful in form and the utmost in efficiency in the serving. One of the largest and most luxurious of the new airlines, where every detail is ultra modern, uses a service of Shelledge. Here is recognition of a product as being in the spirit of the times and yet on the other hand this service could be used in the most correct of Georgian interiors and it would fit its surroundings. . . .

The Onondaga Pottery during its whole history has proven itself to be "design conscious." During the past few years it has given many instances of its ability to . . . make the product most suited to a new field. Thus it is that most of the new streamlined trains of America use services of Syracuse China. One of the latest of these services was developed for the series of Super-Chief streamlined trains of the Santa Fe railroad. . . . Here was an instance where some of the local color of the Southwest was called for. The conditions also called for a decoration which would be in keeping with a decidedly modern interior. The motifs for the decoration were finally chosen from an extinct tribe of Southwestern Indians, known as Mimbres, who had evidently risen to a very high level in the arts, whose work had great variety, delightful humor, and, most surprisingly, fitted in beautifully with the modern spirit. To adapt the ancient Indian decorations to modern china, and to a modern setting, required imagination and decorative judgment. That the commission was carried to a successful conclusion was due to the talent of Miss M. E. J. Coulter . . . [of] the Harvey System and our own R. Guy Cowan.[17]

The Mimbreno Indian decoration stands as one of the most distinctive patterns made by the pottery for railroad service. The Harvey restaurant chain, which served major rail terminals in the West, sent its veteran designer M. E. J. (Mary Elizabeth) Coulter to the pottery to work with Cowan on the project. She based her designs on the black-and-white painted pottery of the Mimbres Indians, an agricultural people who had lived in Arizona and New Mexico during the Pueblo III period (1100–1200 A.D.).[18] The distinctive underglaze pattern featured stylized animals and borders in brick red and blackish-brown on **Round Edge** and other shapes in the ivory body (fig. 7.6). Coulter tried to keep the decorations close to the primitive style of the Native Americans.[19]

Already in process, but not in time to reach the 1937 catalogue, was Harry Aitken's *Shadowtone* technique of decoration. Aitken, who by this time had headed the decorating department for over thirty years, had always worked closely with the design department coordinating patterns with shapes. In 1938, he completed development for use in mass production of *Shadowtone*, a process of airbrushing (spraying) colors through stencils (see fig. 7.1). Airbrushing was not a new idea; it had been popularized in the nineteenth century by the Rookwood Pottery of Cincinnati as a means of decorating art pottery with a mouth atomizer.[20] Aitken modified this process and fitted it to his system, introduced in 1924, of cutting pounce masks as guides for hand-painted decorations. Complicated, often many-layered, kinetic stencils were hand cut

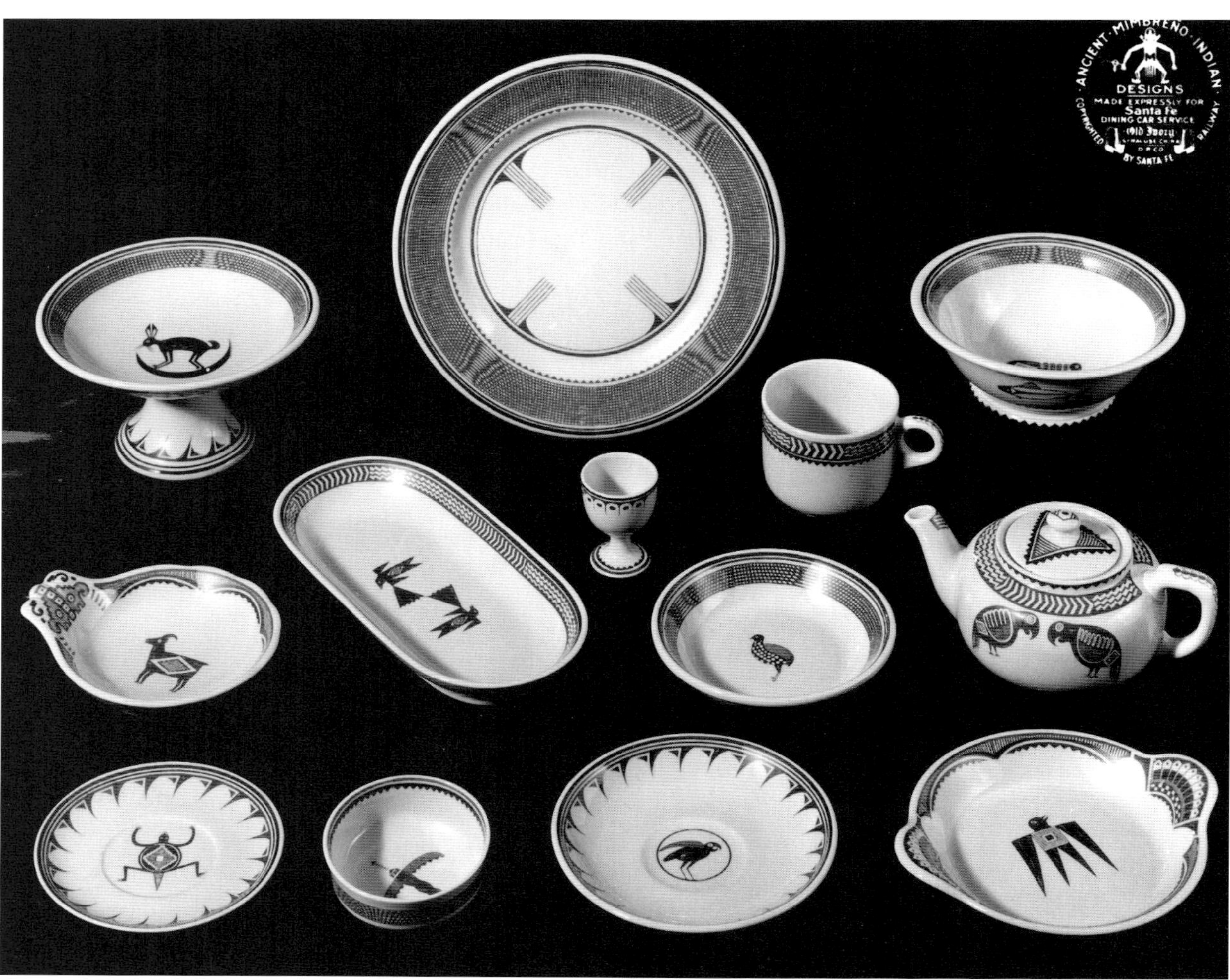

7.6. M. E. L. Coulter's "Ancient Mimbreno" patterns, based on Indian designs, for the
Santa Fe Railroad, 1937.

and soldered into shapes that fit each item's contours. Spraying the decorations required great
skill of manipulation. For "Rose" (SH-1) alone, a two-color decoration used on the forty-six
items available in **Econo-Rim** in 1938, nearly one hundred masks were needed. In all, Aitken
created nearly three hundred *Shadowtone* decorations.[21]

Cowan had mixed feelings about his situation in Syracuse. He found it satisfying despite
his frustration over the limits on individual expression inherent in working in an industrial
setting. After five years at the pottery, however, he saw that he had certain advantages (beyond
the good salary he received). He expressed this in a letter he wrote on 2 November 1937 to the
head of George Rumrill Pottery in Little Rock, Arkansas. He was replying to a proposition that
he relocate:

My position here has many compensations in spite of the fact that there is not very much creative art work to it. However, I have been able to do about the only new things which have been done in the hotel china business, and at the same time been able to make money for our firm. The chief advantage here is that I am working for the most prosperous and successful firm in this field, and can count on an income above which you consider high, for as long as I live. . . . In moving to Syracuse, I felt that I was isolated from the art world that I enjoyed. During these five years it has been possible to bring the center of the ceramic art world to this city. For five years we have had here the national Ceramic Art Exhibitions. From here [they] have toured the country, being shown in the major European museums . . . and in New York, at the Whitney Museum. We have had two of the last three conventions of the Art Division of the American Ceramic Society in Syracuse, and just last week opened a new and beautiful art museum, of which I am a trustee and secretary, and which is specializing in ceramic art development. . . . In carrying on these activities there is considerable prestige to a firm which uses my name. That is the reason why my present firm is interested in having me carry on. There is also a great deal of stimulation and the possibility of getting many new ideas while working with the leading artists and designers of the country. . . . Perhaps from a strictly commercial standpoint, I put too high a value on my name, but I do have considerable pride in what the name Cowan has meant in the past, and I believe still means. That pride would never allow me to have my name on a product which was below the standard I have set for myself.[22]

The Great Depression took most American potteries down with it. Tableware production in Trenton, which had begun to decline early in the century, survived the Depression only in the case of fine Lenox China. Tableware production in East Liverpool survived chiefly in the case of the Homer Laughlin Company, which had built its reputation on "premiums" and other cheap ware, and had been the major supplier of inexpensive tableware to the Woolworth chain.[23] Nationwide, the production of fine china tableware had been reduced to a very few companies, including such pioneers as Onondaga Pottery, Mayer, Scammel, Shenango, Walker, and Lenox. Many earthenware potteries had closed; others moved into the production of vitreous hotel ware; Buffalo, Jackson, Sterling, Shenango, and Iroquois (which began early in the century in Syracuse) were among the best known of the latter. It was often stated that ***Econo-Rim*** had saved the company during the Great Depression and this was perhaps the case. Except for ***Round Edge, Econo-Rim*** remains the longest-lived shape the company has produced.

By 1937, the economy showed some signs of recovery, if only fitfully. Years later, a former employee recalled in an interview that in that year he obtained a job in the pottery's biscuit warehouse after trying unsuccessfully to find work in nearly every other factory in town. Impressed with the quantity of stored ware, he was surprised that production was continuing. Indeed, even during the lowest point of the Great Depression, the pottery made ware partly to keep its people working. In the grinding department, work was down to three days a week. In the biscuit warehouse, men worked full time for a while, and then half time before things began to pick up. There was no time-and-a-half for overtime and no paid vacation, but no one got laid off. Keeping workers on during hard times, despite a lower wage scale, paid off for the pottery. It encouraged loyalty and good will. By 1940, workers had returned to a forty-hour (or more) week.

The sales force now sold nearly 850 items of dinner and hotel ware in three body colors. In November 1940, the company published two new price lists, one for household dinnerware and the other for hotel ware. It offered nearly 250 items in its dinnerware shapes: ***Mayflower, Winchester, Shelledge, Virginia,*** and ***Federal.*** For hotel and restaurant use, it offered nearly 580 items in six shapes: ***Round Edge, Empire, Pilgrim, Econo-Rim, Doric,*** and ***Morwel,*** as well as scores of items such as teapots, jugs, pitchers, cups, saucers, and mustards in other specialty shapes.[24] To all this the pottery offered over a thousand pattern decorations in myriad styles.

Onondaga Pottery worked hard to sustain its position within its competition. When the company sent an exhibit of china to the 1939 World's Fair in New York, Helen Graley initiated a large national advertising program for household dinnerware. As had her predecessors, she emphasized that Syracuse China was American-made, translucent, and vitreous, but she now devised a campaign to help consumers to distinguish "true china" from cheaper grades of tableware. The pottery made and boxed a small china bell that she distributed to dealers and other customers, with a motto, "True to its tone," which the company copyrighted in 1940 (fig. 7.7). "Hold it to the light. Hear it ring!" one advertisement insisted. Onondaga Pottery sight and sound tests became part of retail sales presentations. The bell was offered as a mail order premium. Advertisements in leading women's popular journals such as *Bride's Magazine* (winter 1938) proclaimed that Syracuse China "Rings with the clear music of wedding bells!" and another in 1938, "'Romance' for that enchanting first dinner in your own home." In *Better Homes & Gardens* (May 1939), an advertisement said, "It is true china—a term that means to china what sterling means to silver."

In another promotion, the marketing department issued a set of six small (4⅞"), square plates, each with its own printed scene from American history set within colored *Vitritone* rims, for two dollars a set (plate 23). It published a brochure with a mail-in order form; an order brought both the set and a large broadside. Graley's ads went into not only women's magazines but also trade journals such as *Crockery and Glass Journal* and *China, Glass, and Lamps.*[25] One advertisement pictured a table set with Syracuse China in front of a large picture window with a view of the World's Fair in New York.

Graley continued her campaign until October 1942: "An Echo of America's Golden Age," "Home Atmosphere at Johns Hopkins" (for hospital promotions), "American Beauty All the Way Through," "Medieval Legend Inspired This Modern Beauty." Many of these ads focused, as the pottery's ads had since 1913, on dinnerware patterns: "Sherwood: As American as Plymouth Rock." One promoted ***Shelledge*** as: "So lovely it was shown at the Metropolitan Museum."

The decade of the thirties had proved to be one of the company's most successful in design, employee relations, reputation, and at certain points, despite the Depression, even in sales. In a brief, unpublished history of the firm, Frank Adler observed: "As a result of new products and design innovations, orders once again rose to over a million dozens in 1940. The following year, as the nation's entry into World War II loomed imminent, a record entry of over 1.8 million dozens of china on order prompted the company to advise its customers that it would begin

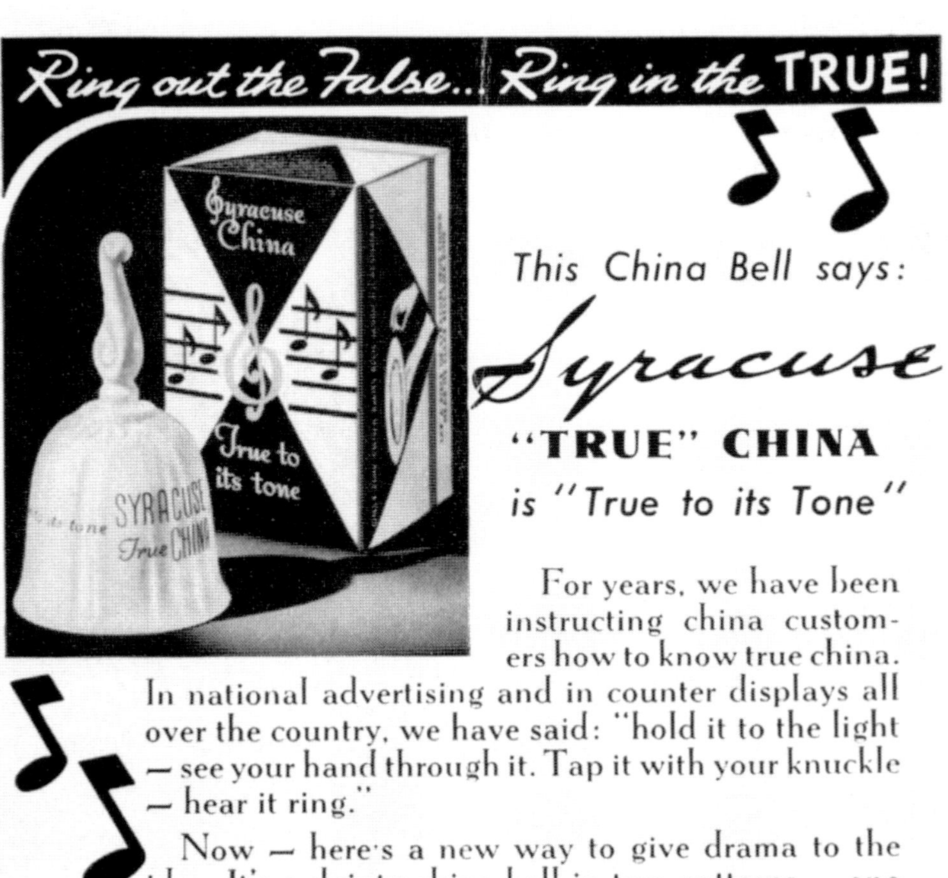

7.7. "True to its Tone." Onondaga Pottery's "China Bell" advertisement in 1940 trade journals. The bell was a promotion piece for the **Federal** shape.

shipping china on a priority basis of essential types of food service operations. A week later, the United States entered the war."[26]

The outbreak of hostilities in Europe in 1939 put an end to the importing of European ware. It also cut off sources of English china clay. Not until the entry of the United States into the world conflict, however, did the full impact of the war begin to be felt at the pottery. In August 1940, Bert Salisbury had announced that Schramm, after several years of experimental work, had developed a completely satisfactory substitute for English clay as the essential ingredient in the manufacture of Syracuse China. He explained that with the outbreak of war in Europe, the company had stockpiled a year's supply of English clay and English-made transfer paper to see it through until the clay room could make the adjustments required for the substitute china clay.

By this time, the company employed 1,150 people at its two plants. Though Salisbury was optimistic about the future, he worried that a new federal wage-and-hour law, under which the forty-hour week would go into effect in October 1941, would injure the company by instituting overtime rates. Echoing the conservative viewpoint of his day, he expressed his concerns to the *Syracuse Post-Standard* on 14 August 1940:

> Our industry requires a large proportion of skilled workers. A person without experience has to have three or four years of training to become skilled. Therefore we cannot easily nor quickly make a very large increase in numbers of employees, even if business warrants such an increase. Continuous overtime work, at overtime rates, for present skilled employees, would mean an automatic raise in production costs and selling prices of our products, and higher prices probably would impair demand.

One and one-half years later, these concerns were irrelevant as the nation entered a period of war in which all the old assumptions about production would need to be revised. The world of ceramic tableware production would never be the same. By that time, Salisbury had retired to become chairman of the board of directors, ending his twenty-eight-year-long tenure as president.

The decade prior to the war had been in every sense remarkable, however. The intrinsic high quality of the pottery's ware, the shapes and decorations developed by Cowan and Aitken and their able staffs, imaginative advertising, and a superbly effective sales force had made a success of a dismal decade. Through the Syracuse museum's annual Ceramic National exhibitions, a quiet dialogue opened between production ceramics and studio ceramics. Fine vitreous dinnerware remained a hallmark of an American home—countless brides-to-be registered their dinnerware patterns with their china retailers. Although the Depression had closed many potteries, those that remained open had every reason to be optimistic about the future, at least until war became a certainty.

Eight

Richard Pass and Industrial Humanics
1942–1958

Enter to be and find a friend.

—motto over entrance to Court Street plant

O N 30 September 1941, a month after the pottery's seventieth birthday, and with Europe at war, Richard Pass was elected president of Onondaga Pottery. In his role as vice president of Research and Production for almost twenty years, he had been largely responsible for sustaining the company's progressive product development and high morale during the most trying of times.

He called his enlightened approach to employee-centered management "industrial humanics." This approach had developed from the company's Statement of Policy, which had been written in his father's time. It held that the business must benefit not only its customers and its stockholders but also the people who worked for the organization at all levels. Pass believed fervently that the interest of all three groups was in the end the same, and was best served by mutual respect, cooperation, and good will. He devoted his life to these ideals.[1]

In March 1947, *Ceramic Industry* offered a summary of the events that marked the beginning of this policy:

> From the old days of cold dipping rooms and tallow candles to modern conveniences, the way was slow but, as one old-timer put it, always upward. James Pass knew what he wanted and how he wanted it done. He asked cooperation and he worked alongside [his employees] with a . . . sense of humor, no matter how rough the going might at times become. This laid the foundation for Onondaga's program of what Richard Pass later called "industrial humanics."[2]

Now that he was in charge, Richard intended to put more of his ideas into practice. In less than two months, however, America had entered the war. Men and women alike left for military service. The work force soon consisted mostly of women, who, now trained in most pottery skills, contributed to the production of almost five million dozens of china during the four years of war in spite of manufacturing restrictions. After the war, many women stayed on. From that time, they continuously constituted over half the work force.

Buoyed at first by the wartime economy, orders came in thick and fast, and production rose to its highest peak in years. Moving to get expenses down and to increase efficiency with new ways and new ideas, Pass put Harold "Bud" Allen in charge of the Court Street plant; John Wigley continued to supervise the decorating department there. Clifford Parmelee took charge of a newly created department of machine design to speed up the mechanization process. (In the 1930s, Parmelee had perfected a walking-beam-construction tunnel kiln.) Pass had both factories converted to natural gas power. He put more rigid specifications in place, improving the brilliance of the glaze and the straightness of the ware. He developed a suggestion system in which he awarded many employees with cash for successful ideas.

To involve more people in the decision-making process, Pass created three new committees: the Management Council (with representatives from various departments); a Quality Club at each plant (made up of foremen) to improve the already high production standards; and the Art Council.

The Art Council, headed by Pass himself, included artists, designers, and representatives from production, sales, and management. Beginning in 1942, the council met to develop product shapes and patterns for postwar markets. Guy Cowan and Harry Aitken were members, as were the artists Douglas Bourne, Charles McKaig, Edward Otis, Adam Schylinski, and John Wigley; Alfred Hoffmann, head of lithography; and Edwin Hinrichs and MacFarland Wetmore from sales.

It was difficult for nonessential industries to obtain enough raw materials and fuel to stay in operation, but "war work" guaranteed supplies of these things. Orders for army nappies and navy mugs helped to sustain the pottery (fig. 8.1). A more important (and secretive) contribution to the war effort began when the pottery set out to work with army ordnance on a mine development project. Pass teamed specialists from the two companies he headed, including the pottery's research director, "Doc" Schramm; Flemmon P. Hall, director of research at from Pass & Seymour; and five other men. The army needed a nonmetallic land mine that could not be located by electronic mine detectors. Specifications called for it to function buried in any soil and even under water. It needed to withstand the pressure of running soldiers but explode under the pressure of moving vehicles. It had to function at temperatures ranging from 40 degrees below zero to 170 degrees Fahrenheit. Nonmetallic land mines had been produced by the Germans, but none that met these specifications.

The pottery's research team designed and developed the casing for the mine by adapting the Syracuse China clay body for use as the basic component. The challenge of making a nondetectable fuse was more difficult. After seven months of research at Pass & Seymour, the team created one that would detonate in any weather. The electrical workers at Pass & Seymour made nearly four million fuses; Onondaga Pottery workers manufactured and assembled the mine casings (fig. 8.2).

The public did not learn about the mine project until after the war, when the *Syracuse Herald Journal* reported it on 4 November 1945, but, at the close of the war, the company received a citation for outstanding service to the country. That summer, the pottery began its transition to

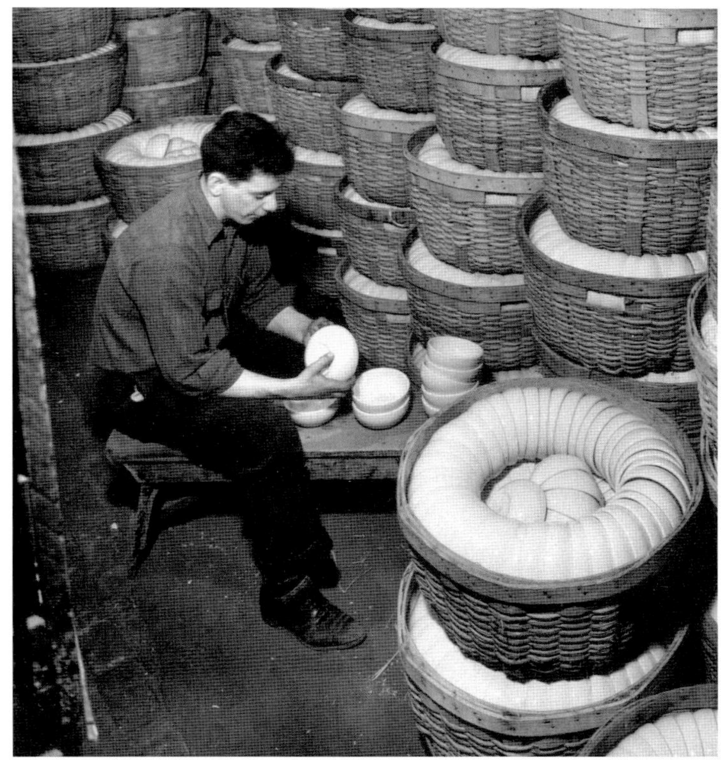

8.1. Wartime production for the U.S. Army: baskets of nappies arrive for selection.

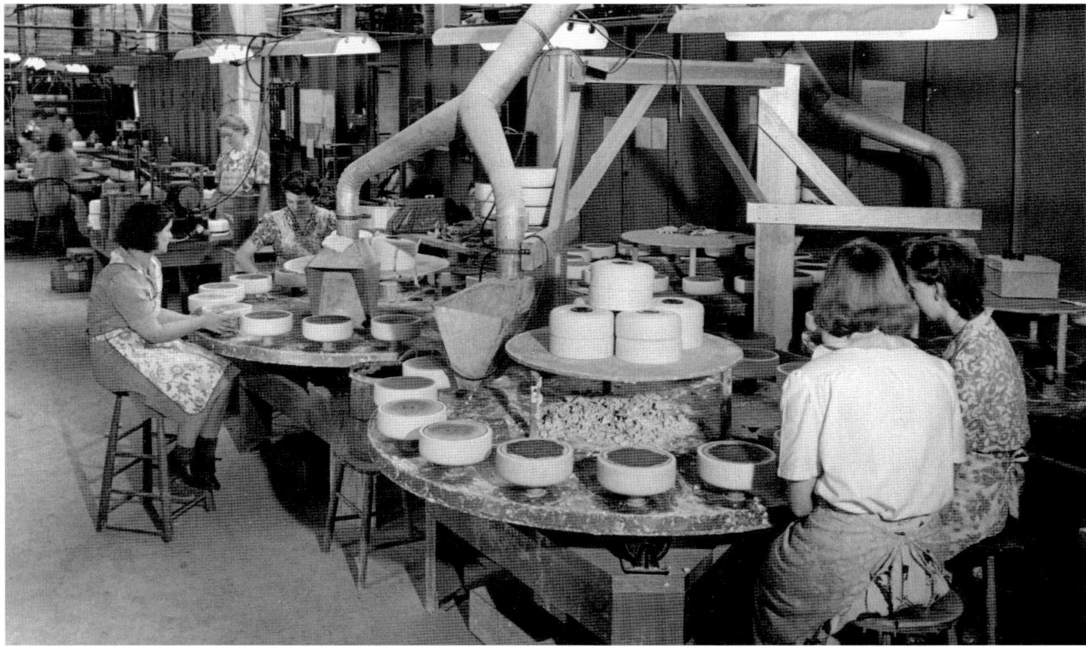

8.2. Wartime production, at the Court Street plant, of the ceramic casings for land mines.

peacetime. As Pass and his colleagues knew, this would not be a return to the prewar world. The American society that took shape in the years after 1945 would present the pottery with exceptional opportunities and daunting challenges.[3]

The end of World War II ushered in an age of optimism for the company as well as for the nation. With the war behind him, Pass was finally able to put his own impress on the pottery. The Art Council continued into peacetime and was renamed the Syracuse China Creative Design Studio. During the war it had developed so many new product ideas that it took the company more than a decade to put them all into production. Pass and his associates believed that a new generation, freed from the constraints of depression and war, would be open to fresh, modern ideas. The American public was ready to resume living, but in a "modern" way. American potteries, which had been more fully automated at the beginning of the war than were European potteries, and which had stayed in production during the war, were ready to produce a new product for the new age.

While American forces had fought to liberate the free world, designers who had fled from Europe in the 1930s helped liberate American design from its ties to tradition. They represented not reform but rather revolution. They had brought with them the International Style in architecture, Bauhaus-inspired graphics, and other approaches to design in all the visual arts that broke cleanly with older styles. This influence extended to ceramic tableware. American tableware designers synthesized this new modernism with their own penchant for the older Art Deco and Art Moderne to produce a kind of democratized modernism that suited the uncluttered, increasingly casual style of living that developed in the postwar years. Consumers now demonstrated greater interest in modernism, both in shapes and decorations, and less interest in older conventional designs than ever before.

Postwar Americans were more outdoor, casual, and informal than were Europeans. Michael Farr, in his analysis of the recent history of modern design in American ceramics for *Design* magazine in December 1952, wrote:

> Its immediate origin can be found in the designs developed by Californian potters, mainly since the war, although elements in the style can be attributed to Swedish and Danish work in the 1930s. After several years of being rather precious and obscure the style has made a progressive conquest of the markets in the US from west to east coast. A large percentage of American manufacturers have since taken it up because it accords convincingly with the current American practice of "informal living," a practice which implies that meals, preferably of the buffet type, can be taken in the garden, off the kitchen table or on the living room floor.[4]

This spirit of casualness and informality would persist and in time make fine, translucent dinnerware, and the formal dining it represented, a less important part of middle-class life. In the late 1940s, the alliance of optimism, modernity, and informality could also be found in American interiors, furnishings, textiles, other decorative arts. In 1947, *Ceramic Industry* had predicted this development in an issue celebrating the seventy-fifth anniversary of Onondaga

Pottery. It said that American dinnerware would become lighter without loss of strength, freer in its shapes, and less traditional in its decoration.

The next year, Pass affirmed the crucial importance of design when he said that the history of America's progress in tableware manufacturing could be written in terms of its styling, and that since the 1920s, style had evolved from a minor to a major consideration. He saw five factors responsible for this growth in the importance of design: (1) an elevation in the public's level of education, and with it a rise of interest in artistic and other cultural values; (2) a new national self-confidence that had replaced an earlier perception of American decorative objects as inferior to European; (3) improvements in the overall standard of living that allowed more people to afford things once considered luxuries; (4) the intrinsic aesthetic appeal of newer ceramic products; and (5) the influence of Syracuse's Ceramic National exhibitions in bringing about a new understanding that ceramic art could be fine art.[5]

It seemed all but certain that the next generation of ceramic tableware would be more self-consciously modern, but with an American flavor. Edwin Hinrichs, who had come from the Chicago dealership in 1943 to become sales manager, pointed out that Onondaga Pottery's determination to style its products for a confident new America paralleled that of the earlier successful decision of the American automobile industry to stop copying European designs and make cars frankly styled for American taste.[6]

The first major test of one of the Design Studio's new ideas came with *Airlite* china for American Airlines (fig. 8.3). Cowan had began work on this lightweight shape in 1940. Interrupted by the war, it went into production late in 1945. The pottery claimed *Airlite* to be the lightest weight commercial chinaware ever mass-produced. It was meant to replace the plastic tray service then in use by airlines. Cowan designed thirteen items, but only three went into production. Use of a coupe plate and a dessert dish began on the airline's fleet of DC-4 planes in February 1946. A cup was added for use on the DC-6 service inaugurated in April 1947. With them, the airline expected to serve ten thousand meals a day. A sample menu from the DC-6 day plane and sleeper included fruit cocktail, celery heart, radish rose, fried chicken, buttered rice with peach half, chef salad with cucumber wedges and French dressing, Parker House roll with butter, peppermint stick candy ice cream, and a choice of three beverages, all compactly served in coupes and a cup on a small tray. With this drastic reduction in items of dinnerware and dining space, air travel offered the antithesis of the Victorian table.

The pottery found itself unable to manufacture *Airlite* at a profit. The shape's extreme thinness—less than five gauges ($\frac{5}{64}$") thick—led to excessive breakage in manufacturing, a problem American Airlines also experienced. Its production ceased in 1949.[7] This was a great disappointment for Cowan, but it probably did little to dampen the optimistic spirit that had propelled the pottery into the postwar world of modern living. Pass forged ahead.

Modern living can have unexpected ties to the past. Some practices at the pottery survived long after one might suppose that they had been supplanted by newer ways. A case in point was the shipping of china. The revived *Syracuse China News* in October 1947 reported the death of one of the company's last coopers, Hervey Austin. George Hartman, who had headed the

Airlite (1946–49)

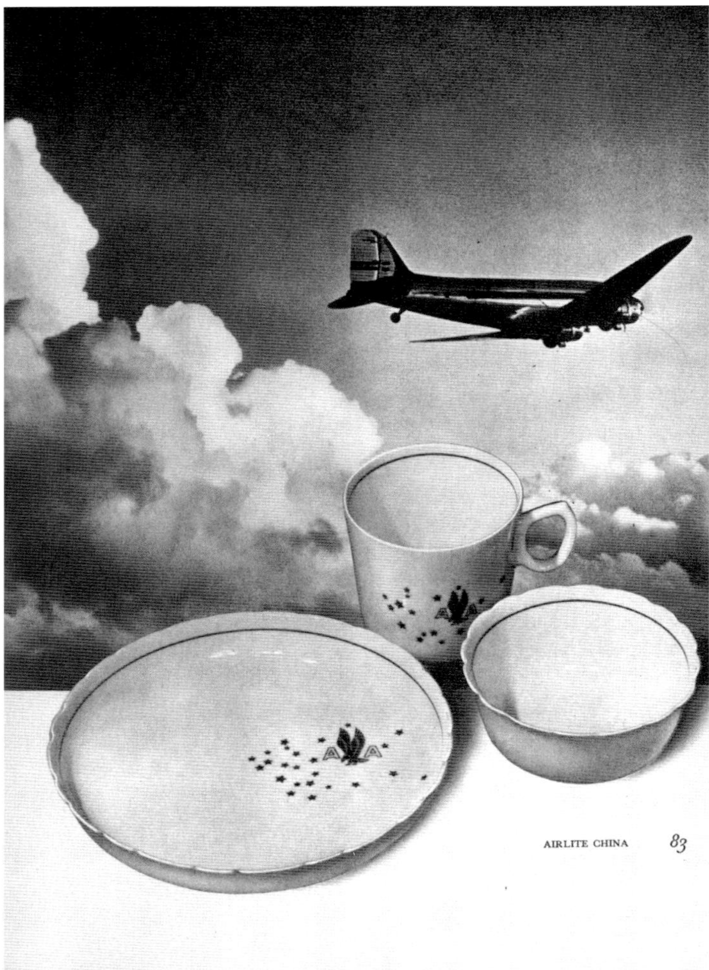

8.3. *Airlite* illustrated in the seventy-fifth anniversary issue of *Syracuse China News*, 1946.

cooper's shop of four men since 1907, had retired in 1946 (fig. 8.4). Although the pottery no longer made wooden barrels to ship its ware, it continued to use them until the last one was packed on 7 June 1957. In view of the technological advances in nearly all other operations, it seems surprising that the practice of packing china for shipment in barrels endured as long as it did in the industry, but expertly packed barrels, in which the ware was cushioned with straw, were in fact the safest mode of transport for china. Packing a large barrel took skill, strength, and time. Loaded with ware, the barrel could have weighed up to three hundred pounds. By the 1950s, however, dealers and distributors found cardboard cartons easier to handle and more efficient for storage.

The October 1947 issue of the *News* noted that 10 current employees had worked for the company for more than fifty years, 34 for more than forty, 66 for more than thirty, 233 for more then twenty, and 222 for more than ten.[8] Despite the loss of employees who perished in

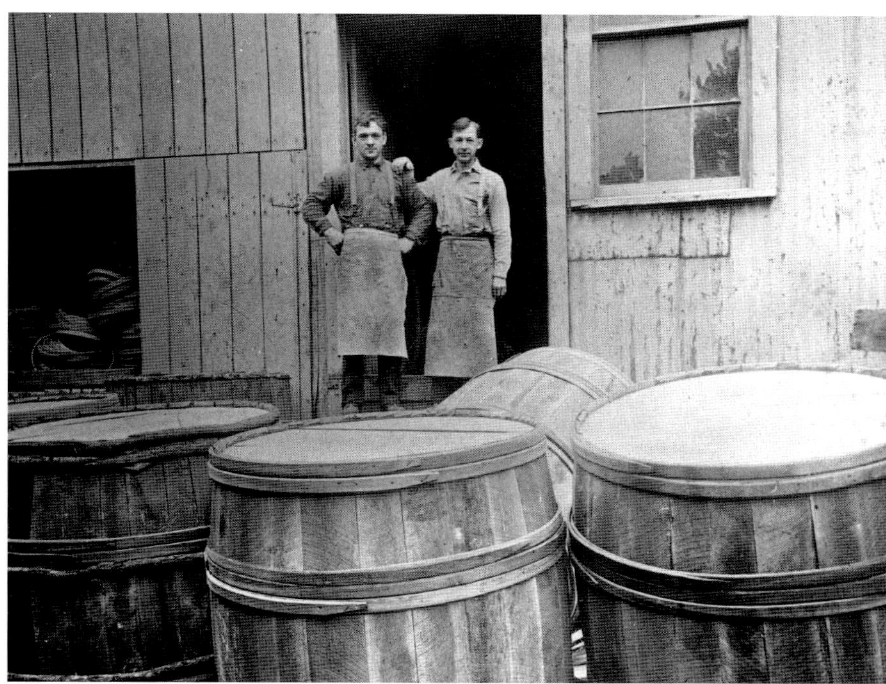

8.4. Coopers George Hartman and Hervey Austin at the Fayette Street cooper's shop, c. 1920s.

the war, and growing opportunities for employment in other Syracuse industries, the pottery's workforce remained remarkably intact.

In 1947, the *News* began a series of features about some of the company's best known and most prestigious customers. One was Harvey's Hotel and Restaurant System, which decades earlier had spread from Ohio to the Pacific coast. It reported that, "Other frontiersmen tamed the west with six-shooters and swinging fists. A sparse, punctilious Englishman, Fred Harvey, helped civilize it by a method uniquely his own—he sent thousands of attractive 'Harvey Girls' to work in his restaurants along the Santa Fe Railroad and their effect on the badmen of the range was devastating."[9] This romantic account, already popularized by an MGM film starring Judy Garland, *The Harvey Girls,* might have said more of Fred Harvey's discerning use of American china. In addition to the "Mimbreno" and "Santa Fe Poppy" patterns custom-ordered from the pottery, the Harvey company commissioned Syracuse China patterns for its Crossroads Room in Chicago's Dearborn Station (fig. 8.5), his restaurants in the Los Angeles Union Station, and the El Tovar Hotel on the rim of the Grand Canyon, as well as for its railroad dining car enterprises.

There was little that was distinctly modern about Harvey's patterns, however, and Onondaga Pottery did not find modern design to be universally popular in spite of the forecasts. In 1948, it introduced a gold and gray "Governor Clinton" pattern on its **Virginia** shape in *Old Ivory.* Named for Governor DeWitt Clinton, the great proponent of the Erie Canal, this traditional pattern made its national debut in advertisements in *Better Homes & Gardens, House*

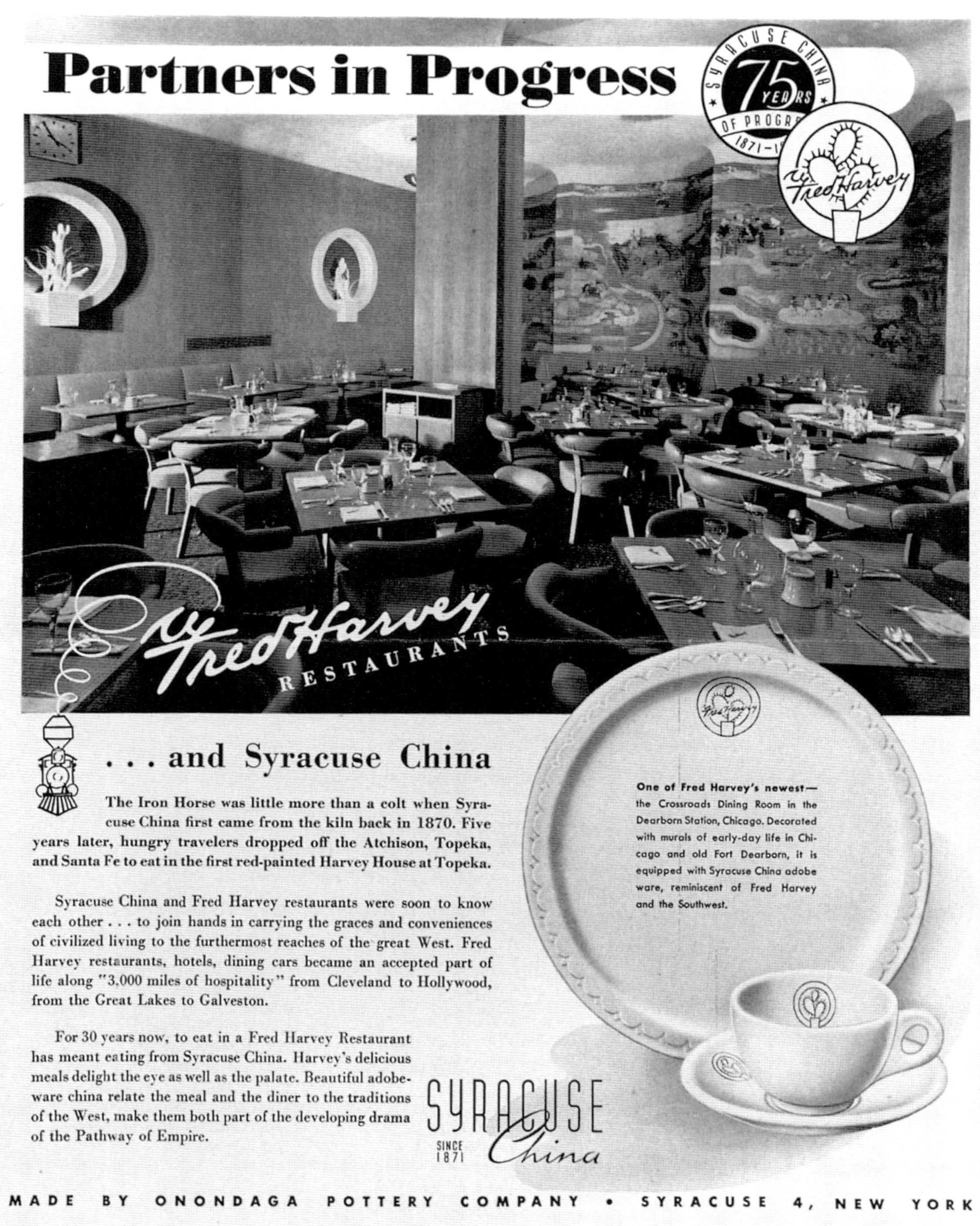

8.5. A Syracuse China/Fred Harvey advertisement in *Fortune* magazine, March 1946.

and Garden, and *House Beautiful* magazines.[10] Selling well, it did little to support the predictions made in 1946 and 1947 that distinctly twentieth-century, modernist design would be the wave of the future. The demand for traditional and historic shapes and patterns remained strong.

Construction of forty-nine thousand square feet of additions to the Court Street plant, completed in 1949, included a new biscuit kiln, a new clay shop, and new space to house the *Shadowtone* and *Syratone* departments in the decorating shop. By that time, most of the older hand-work processes (fig. 8.6) disappeared replaced by entirely mechanized operations (fig. 8.7).

Although expansion of the plant reflected the company's optimism about its future, there were also reasons for concern. At that year's annual shop meeting, an occasion when the president summarized the company's annual operating statement to the entire workforce, Pass pointed out that although earnings had doubled since 1940, factory wages and benefits had

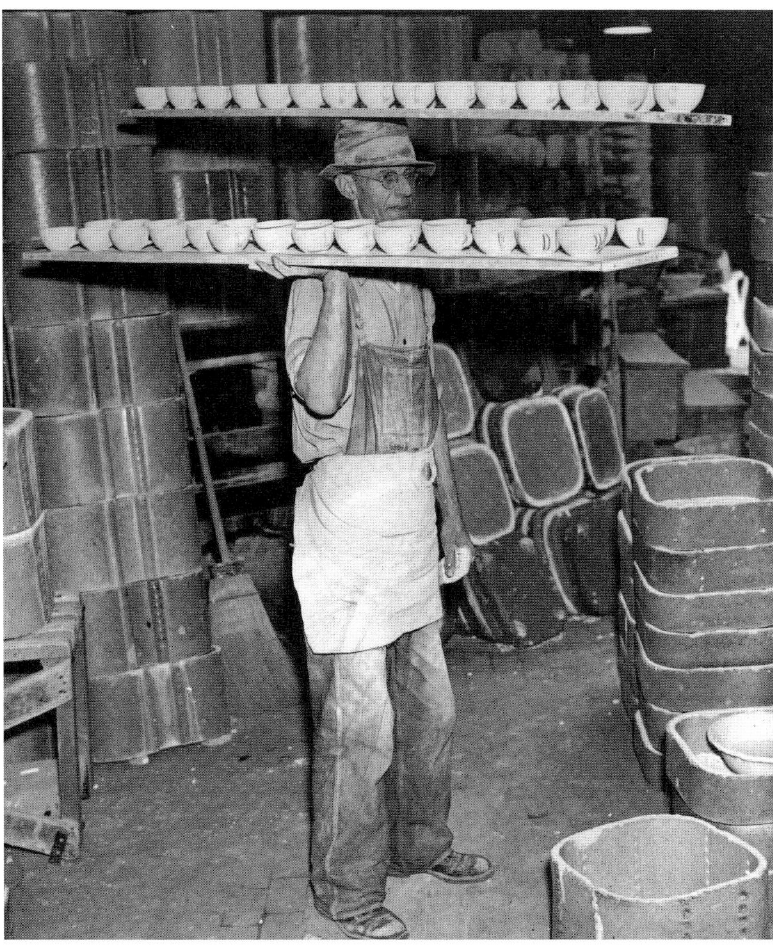

8.6. A pottery worker carrying boards of cups on his head and shoulder at the Fayette Street plant amidst piles of saggers.

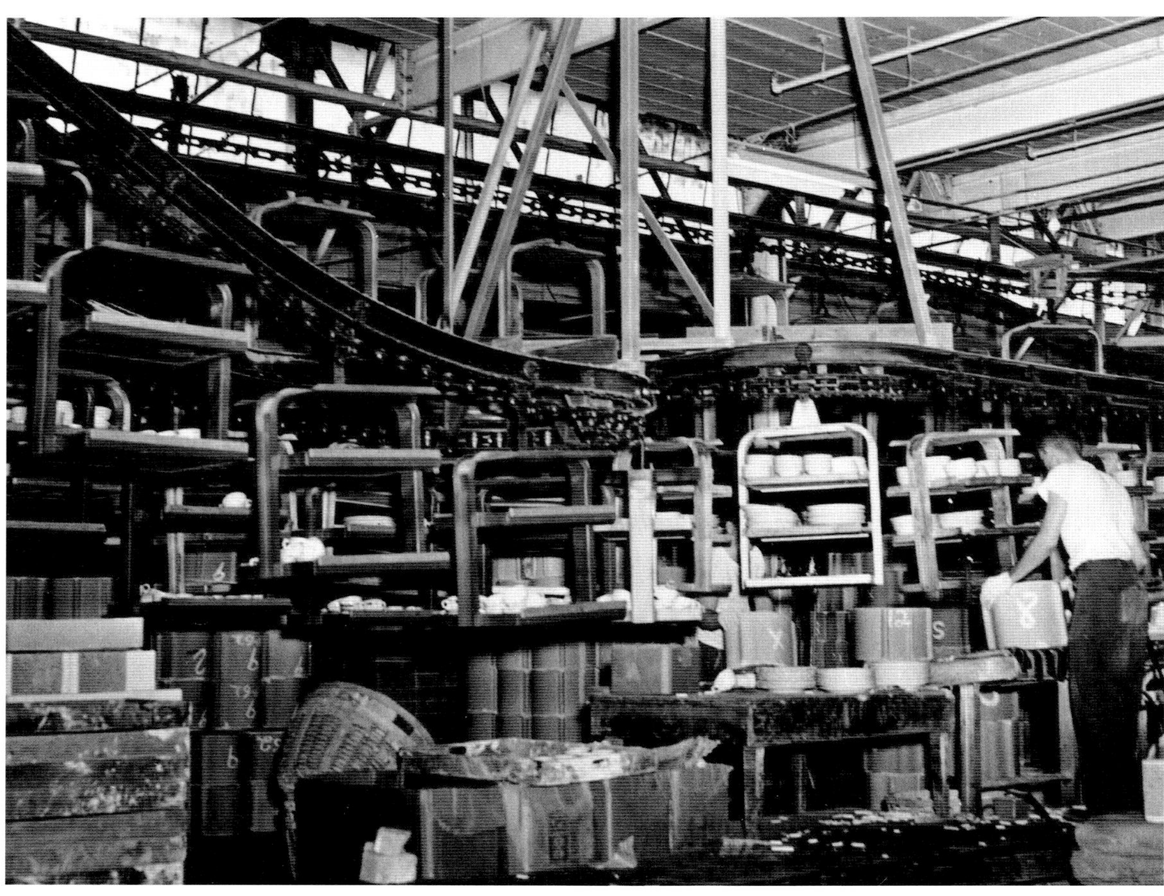

8.7. Employees unload saggers to conveyors at a glost tunnel kiln, Court Street plant, c. 1963.

risen to represent half of the company's expenses by 1949, emphasizing the high cost of labor.

He reported a worrisome drop in the demand for china over the last year. Less optimistic than two years earlier, he attributed this drop to a decline in "eating out" and to growing competition from plastic and glass tableware manufacturers. This increased competition, Pass stressed, made it necessary to keep the selling prices of the ware down. Selling prices had risen only 45 percent since 1941, while hourly wages had risen more than 100 percent. In the same 1949 report, he described the company's problems as "serious" and urged everyone to "work together in the spirit of mutual respect, confidence and good will which has built this organization."

In the meantime, the company continued impressive displays at the annual trade shows. Syracuse China remained a prestigious product. Dorothy Draper, the celebrated interior decorator, designed a pattern to suit her revitalization of the grand Greenbrier Hotel at White Sulphur Springs in West Virginia. The Pierre Hotel's Cotillion Room in New York City and the

Drake Hotel's Camellia Room in Chicago (another Draper project) were both redecorated with exclusive Syracuse China dinnerware patterns to suit. The company's china could also be found in such less-elegant places as Miami Beach's "Pickin' Chicken," where a *Shadowtone* pattern portrayed a chicken on every plate.

At the National Restaurant Show in Chicago in November 1949, the company introduced its **Winthrop** shape, hotel ware that featured scalloped wide-rimmed plates (see fig. A.4). To achieve the scallop it was necessary to replace the rolled edge with a welt running under the rim. It also introduced its *Hospitality Group,* a line of fourteen patterns on **Morwel** and **Rolled Edge** hotelware aimed at "current market demands for better quality, fresh new decorations and fast delivery." The *Hospitality Group* patterns were stocked for immediate shipment. A national advertising promotion issued fifty thousand flyers and a new price list for the company's dealers. *China News* reassured employees in March 1950 that "those who sell our product have been given the information necessary to promote this new line and now they have another opportunity to 'talk' Syracuse China and convert the 'talk' into dozens for factory production." This began a national campaign for the **Winthrop** shape.

The nature of the increasingly severe competition faced by the company with **Winthrop** and its other shapes was summed up by an incident that happened in 1950. Jules Gulden, a war refugee who once had headed a ceramics factory in Hungary and had become director of product development and advertising for the household tableware division at Onondaga Pottery, wrote to New York's Senator Herbert Lehman that he had been stunned to learn that the State Department had ordered foreign-made ware for use in its embassies abroad. The State Department had used Syracuse China dinnerware since the late 1930s. He argued that an American embassy using fine quality American products served as a showcase for American industrial achievement. "I know about the endeavors of the American offical policy to help European countries to get back on their own feet. But, I believe that even if all American embassies were equipped with European products, this would scarcely count for anything in this great economic [effort]!"[11]

In time, Syracuse China and other American producers of fine china would again be asked to fill orders for embassy china, but the problem that Gulden had cited, that of the United States government's policy to support the economic revitalization of foreign potteries even to the detriment of its own industry, remained.[12] It steadily undermined the success of American potteries. In time, competition from overseas would nearly eradicate the American fine china dinnerware industry. Spurred by the government's free trade policy, the war-ravaged economies of Europe and Japan, with much lower labor costs, steadily and drastically cut into Onondaga Pottery's markets.

Pass reiterated the problem in *Syracuse China News* in October 1950. He said that cheap imports from low wage countries continued to plague the industry and that he had mounted a letter-writing campaign urging Washington to reinstate a protective tariff.

> One of my earliest recollections is of a time when the pottery was closed for lack of orders, the only time it has been closed since I have been alive. That was in the 1890s when the free traders

were last in control of our government. . . . When my father was in charge of this company, . . . he fought for adequate tariff protection; so did Bert Salisbury; . . . so has our vice president, Mr. Torbert, who has been with this pottery for fifty years; and so am I fighting to protect this pottery. . . . No doubt this battle will go on indefinitely.

In 1950, the company published an attractive new hotel ware catalogue, streamlined in its presentation.[13] It now offered 309 items in seven shapes: **Rolled Edge,** "the first practical hotel shape"; **Pilgrim,** "a refined Rolled Edge"; **Empire,** "hotel ware version of fine home dinnerware weight," with a plain edge in place of roll, "well adapted for banquet, club or tea room services"; **Econo-Rim,** "distinct embossing on edge, unusual space-saving dimensions plus engineered durability"; **Morwel,** "natural result of **Econo-Rim** pioneering, plain edge allows on-rim decorations; **Doric,** "a classic form combining simplicity with beauty, designed for hard service, auxiliary serving pieces provide a colorful complement to your Syracuse China service"; and **Winthrop,** the "newest hotelware shape, *actually dramatizes food!* Combines bridge-type strength with lovely shape for beauty." The directness and insistent tone of the catalogue's text probably reflected the company's competitive mood.

Attractive double-page spreads in the 1950 catalogue illustrated the standard pattern options in decalcomania, print, Syratone, Shadowtone, line, hand paint (now numbering over 200), and crest techniques, as well as items and stock decorations available for each shape. Its "Hospitality Group" offered the fourteen stock patterns in prepackaged three-dozen multiples of **Rolled Edge** and **Morwel,** a promotion that proved highly successful. The catalogue continued to offer ware in three colors: white, adobe, and ivory.

In 1950, late in his career, Guy Cowan had designed and patented hollowware items for another household dinnerware shape, his elegant, thin **Berkeley** (fig. 8.8). Distinguished by spiraling swirls, the wide-rimmed plate of this shape was of a type made by other potteries (and that remained popular throughout the world), but here it received unusually fine treatment. The company offered several new patterns for **Berkeley,** among them "Apple Blossom," "China Spring," "Gardenia," "Gray Baroque," "Jewel Tree," "Lilac Rose," (plate 25) and "Temple Bells." The company issued separate illustrated price lists as small flyers that fit in pocket-sized ring binders for all household dinnerware patterns.

In 1952, B. Altman & Company of New York commissioned the American artist Adolph Dehn to create a limited edition set of twelve decorative service plates reproducing, in overglaze decalcomania, a group of his original watercolor paintings called "The American Scene" (fig 8.9 and plate 14). In a colorful brochure, Dehn described his choice of subjects:

To see the soft, intimate hills of New England and then to be surrounded by rugged and monumental piles of rock and snow known as the Rocky Mountains, to pass from the fertile green plains of the Middle West, to the dazzling sun and sand of the Mojave Desert are experiences that every American should know. Another aspect of our landscape . . . is our man-made landscape . . . the great cities, the factories, bridges and mines, unique and particularly American. . . . These twelve subjects show the character of the whole country.

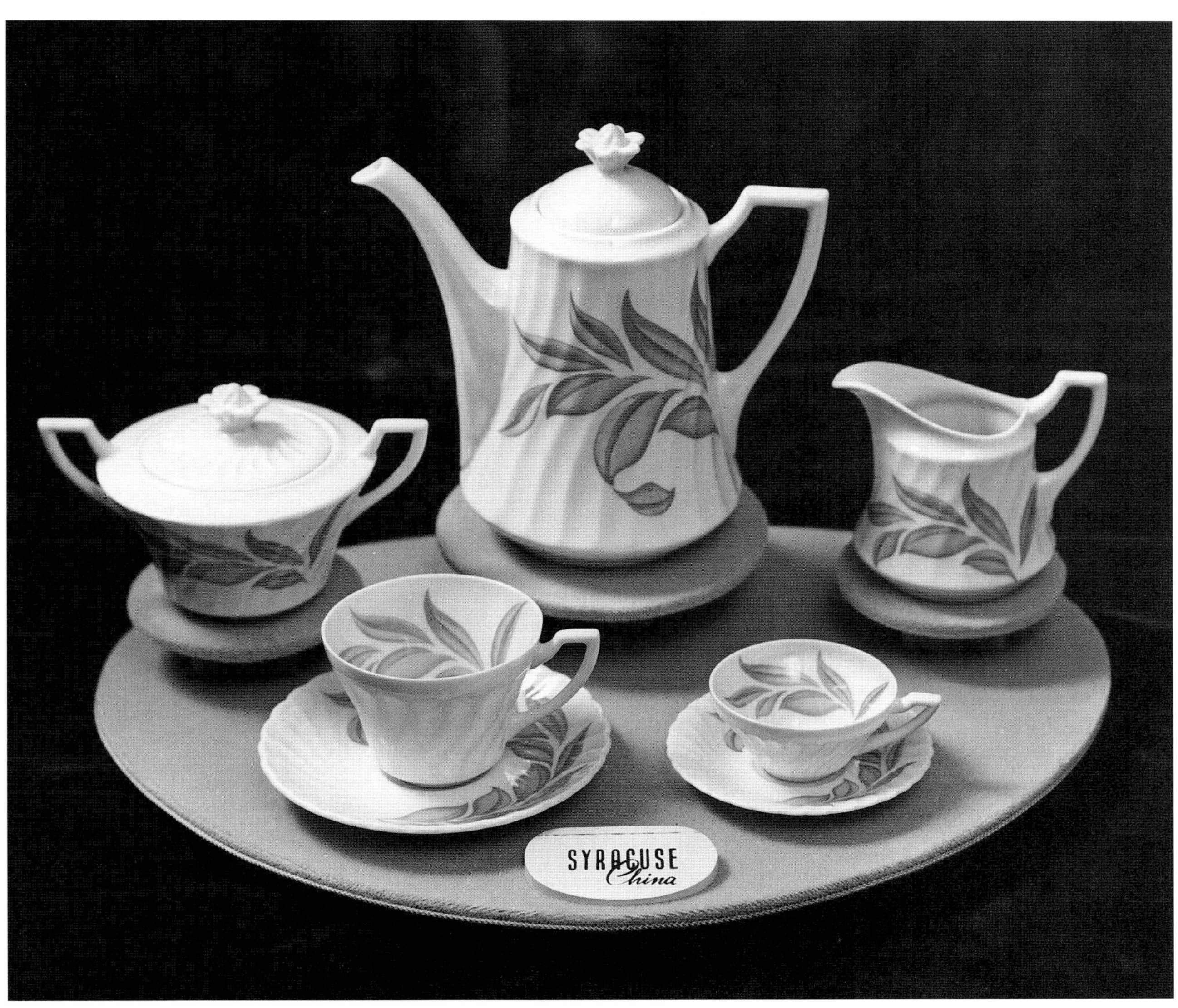

8.8. Display of R. Guy Cowan's new **Berkeley** line with "Dawn" pattern, 1950.

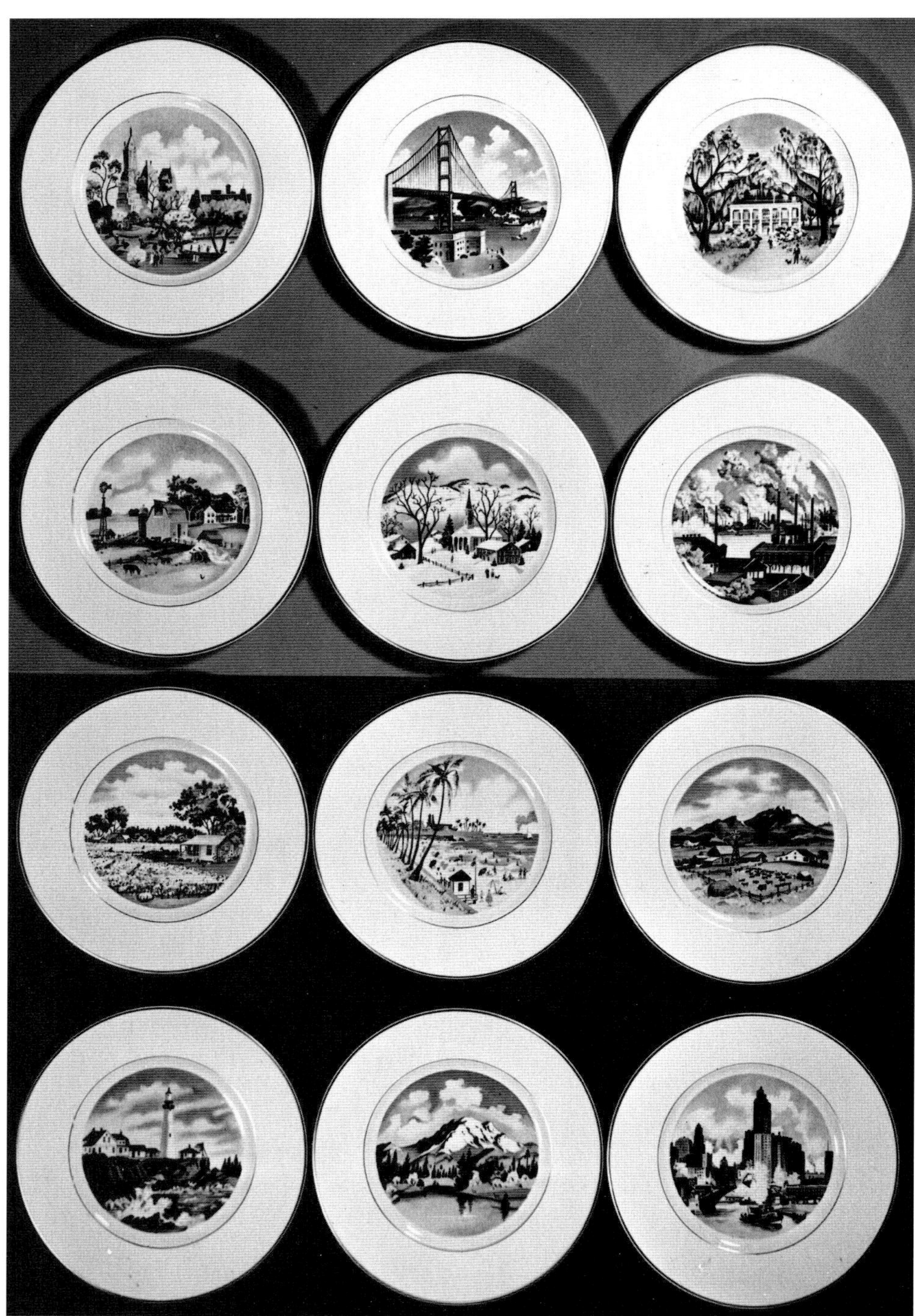

8.9 Twelve plates designed by Adolph Dehn depicting "American Scenes" made for B. Altman, & Co., New York, 1962. From top, left to right: "Central Park," "Golden Gate Bridge," "Southern Mansion," "Middle West Farm," "New England Winter," "Pittsburg," "Southern Cotton Field," "Florida," "Western Ranch," "Light House in Maine," "Mount Ranier," "Chicago."

Another distinctive limited edition consisted of eight "American Song Birds plates." Based on original paintings by Atlanta's Athos Menaboni, the series was issued in 1955 for the Cerebral Palsy School-Clinic of Atlanta, to be sold to benefit its programs (plate 27). The project was supported by Mills B. Lane, Jr., president of the Citizens & Southern National Bank of Atlanta, who was a personal friend of the artist and a patron of the school. Although Onondaga Pottery made nearly all of its own decals, in the case of this exceptional job, which required twenty colors, the Commercial Decal Company of Mount Vernon, New York, produced the bird prints.[14]

The 1950s brought in a flood of new shapes in both commercial ware (as hotel ware was now called) and household dinnerware. In 1951, Cowan created his last shape for the company. A dinnerware shape, **Paul Revere** (fig. 8.10), consisted of ovoid hollowware forms to accompany **Federal** shape flatware. Richard Garvin succeeded Cowan when he retired in 1952 as head of design. The artists in his department (which now combined decorating and design), notably Michael Szymanski, Edward Otis, Marge McMahon, and Charles McKaig, created patterns for the new shapes, many of them available on all three of the pottery's clay body colors. In 1953 the design staff created **Essex,** a rimless (coupe) shape with a scalloped edge (a modification of **Winthrop**), and **Copa** (see fig. A.6), a plain-edged coupe shape, not part of the **Winthrop** line.

In 1955, the company introduced **Trend** (see fig. A.7), a commercial ware shape with squared contours and "econo-rims," promoted especially for use as hospital tray service. Edward Lund of the sales force saw a need for a shape that would economize space on a tray. Following the example of the then-new square milk bottles, Don Folley modeled **Trend.** In 1959, the company

Trend (1955–78)

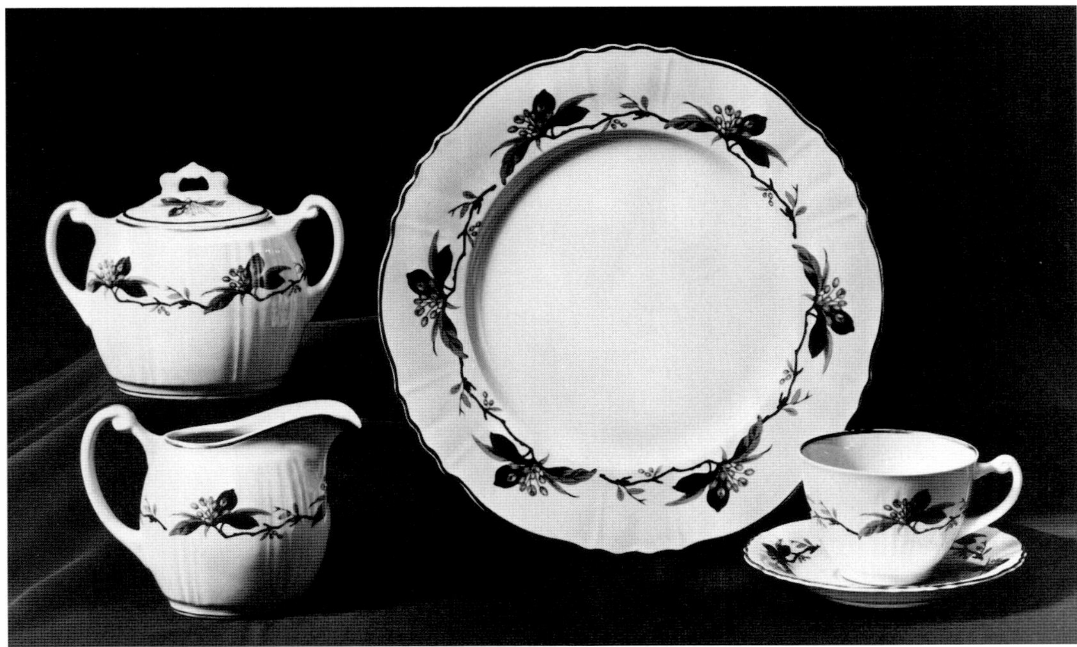

8.10. **Paul Revere** shape with "Westvale" pattern, 1951.

introduced **Kent** (see fig. A.6), a narrow rim variation of the **Winthrop** shape. All of these commercial shapes of the 1950s were touched one way or another with postwar modern feeling, but no true revolution in design had occurred at Onondaga Pottery or elsewhere in the industry.

After a lapse of twenty-two years, the company returned in 1951 to the Pottery and Glass Exhibit, a trade show that had met in Pittsburgh every January since 1880. This was the trade's premier gathering. *Syracuse China News* reported that the company's display was the "hit" of the show. Where competitors offered two or three new patterns, Onondaga Pottery introduced more than twenty for the **Berkeley, Virginia, Federal,** and **Shelledge** shapes. Before he retired, Cowan had remodeled **Shelledge** as a full dinnerware shape line. He designed a new cup for it by removing the flared rim and moving the cup's fluting from the inside to the outside, making it a more conventional and practical shape for everyday use, as well as easier to produce. At the Pittsburgh show, **Shelledge** was presented as a dinnerware shape to be decorated with patterns. In place of intaglio carving, it had a series of eight new underglaze patterns called the "Suburban Line." Billed as a modern, casual line, it was shown with pastel *Shadowtone* decorations, making this the first time a household dinnerware shape had used mostly stencil work and mostly underglaze decorations.

Household dinnerware shapes continued to roll out of the Design Studio. The *News* reported in April 1953 that in January of that year, at the highly competitive Pittsburgh National Dinnerware trade show, the company presented its new **Carolina** shape (fig. 8.11). It staged the display

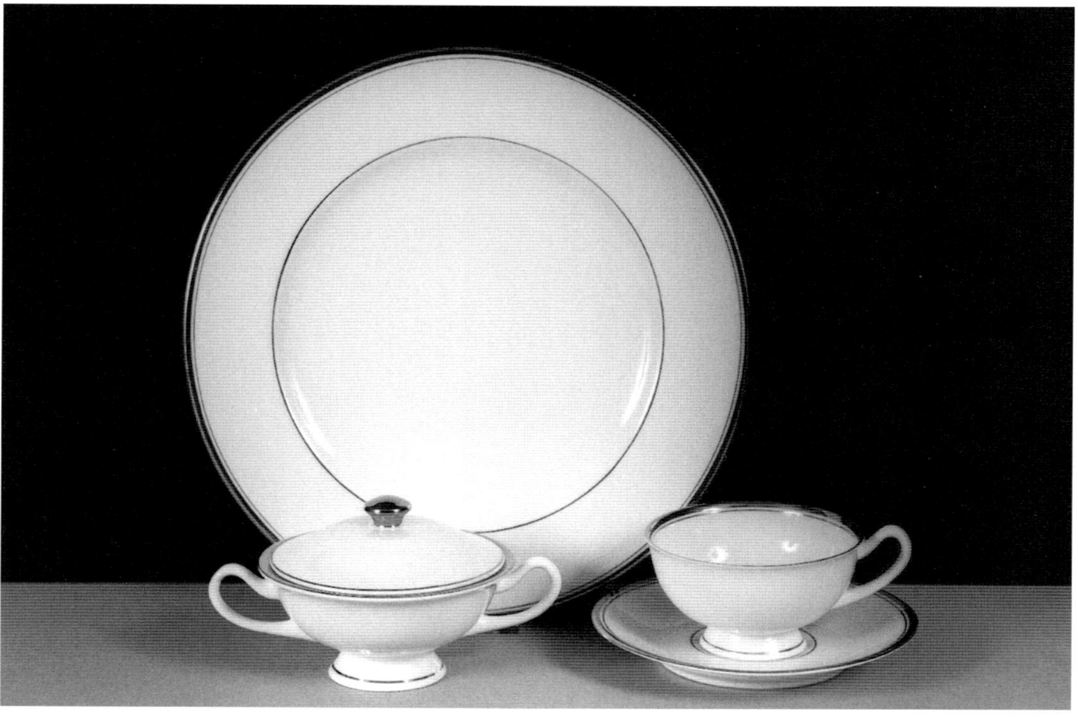

8.11. **Carolina** shape with "Countess" pattern, 1953.

as a wedding reception, complete with wedding cake, wrapped gifts, and flowers. An advertising campaign followed in four leading bridal magazines. For its flatware, *Carolina* borrowed from the *Winchester* shape, with its items made lighter in a soft white body. The delicate modern hollowware forms carried flared rims and stood on flared pedestals. Five new patterns, collectively named "Bridal Chorus," introduced the line. At the same time the company introduced a new pattern for *Berkeley* and exhibited fifty-four other representative patterns.

Because the Pittsburgh dinnerware show was limited to American manufacturers, foreign competition set up a rival exhibit. Pass visited it. He examined the ware of eighty makers of all types of ceramic ware from England, France, Germany, Italy, and Portugal, as well as Japan. He found the Japanese exhibits the most numerous and most interesting, but he observed that all the foreign makers were copying American ideas and even hiring American designers. Even though a Japanese copy of Onondaga Pottery's "Gardenia" pattern sold at a fourth of the price of its Syracuse China original, he found nothing to dampen his belief that American china still led the world in "quality and beauty."

In spite of improved productivity and product innovation during the early 1950s, the household china division lost business steadily to foreign competition. Its problems intensified as American dining habits increasingly called not for fine china but for a less formal kind of dinnerware and sometimes only for paper or styrofoam. The commercial division fared better. At the National Restaurant Show in Chicago in May 1954, the display included eighty-seven new services installed in the last nine months. It highlighted the three markets in which Syracuse China was now most widely used: transportation, hospitals, and hotels and restaurants.

The pottery introduced yet another household dinnerware shape at the 1954 Pittsburgh show: *California* (fig. 8.12). Its flatware had wide feet, curved edges to prevent spilling, well-balanced creamers, full oval platters, and easy-to-hold handles. *California* was a well-styled, straightforward coupe shape, modern and stylish. It was introduced with eight minimal decorative treatments of small patterns or platinum lines. Meanwhile, in an effort to understand the fast-changing household dinnerware market better, the company invited groups of women from throughout the country to evaluate new designs. Then a training program, the Fine China Sales Guild, came into being to train a new generation of retail salespersons in the fine points of Syracuse China. The company advertised in thirteen of the leading women's magazines to tell "Mrs. America Housewife where she will be able to buy her Syracuse China in her nearest marketing area."[15]

The 1957 Pittsburgh show launched the pottery's newest shape on the high seas of dinnerware competition. Designed for wear, *Carefree* was heavier-bodied than standard fine china, more durable, practical for casual living, and guaranteed one year against breaking, chipping, or cracking. *Carefree* offered a compromise between commercial ware and dinnerware. Its patterns were underglaze, making the ware ovenproof and safe for home dishwashers. The sleekly styled *Carefree* at first featured gleaming copper covers for its sugar bowl, serving dishes, and salt and pepper shakers (fig. 8.13), but these were soon dropped in favor of china covers. Aimed at increasingly informal lifestyles, *Carefree*'s advertisements claimed it to be equally at home on

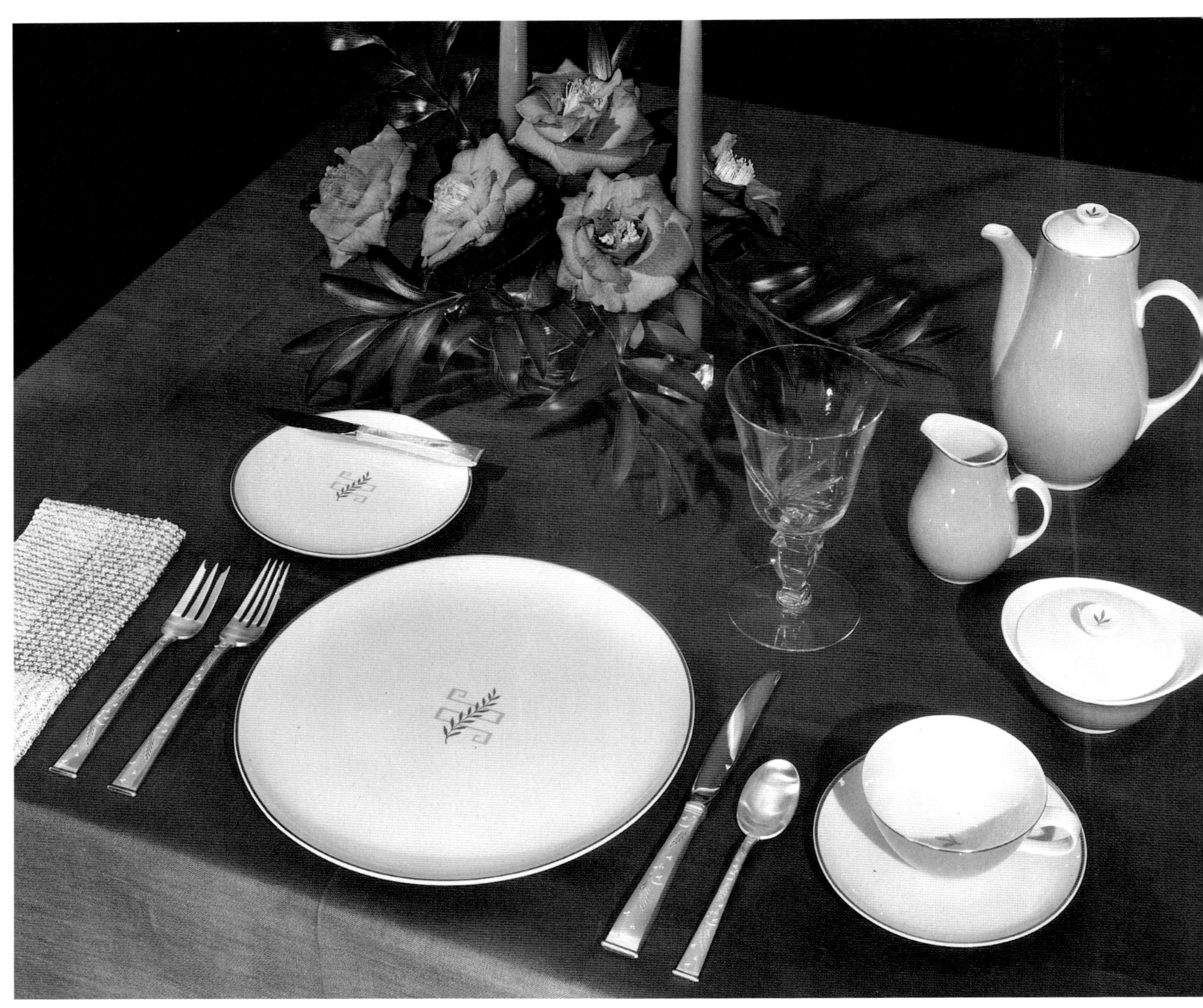

8.12. *California* shape with "Dorian" pattern, 1954.

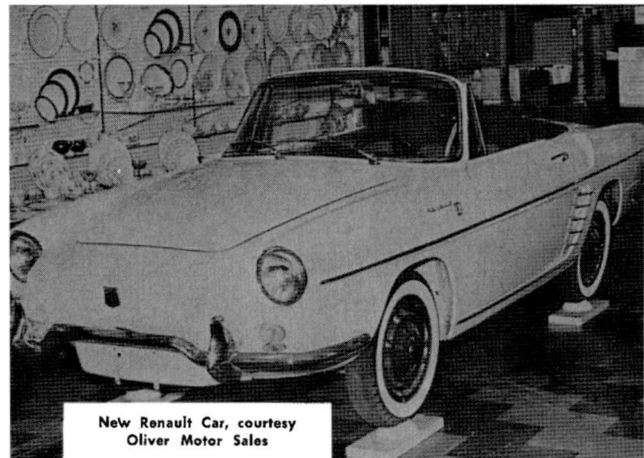

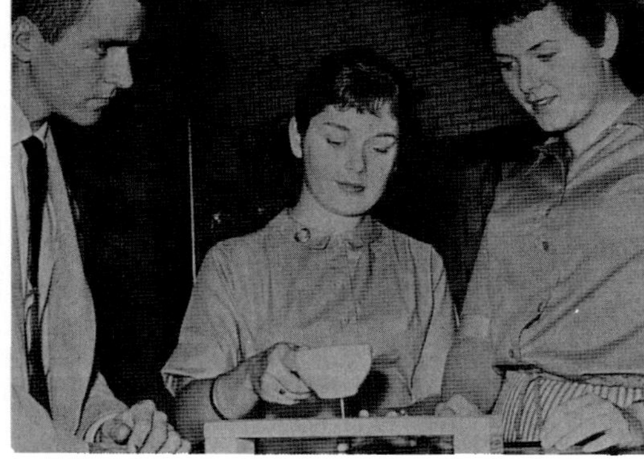

CAN YOU PARK A CAR ON YOUR CHINA?
POUND A NAIL THROUGH A BOARD WITH IT!
You can with Syracuse Carefree China—Here's Proof!

Stop in now and see this dramatic demonstration — make your own durability test. Drive a nail through a board with a piece of Carefree China as our Miss Donaldson is doing in the picture taken above in our store. See the 1650 lb. Renault car setting on four Carefree Soup Plates without any damage.

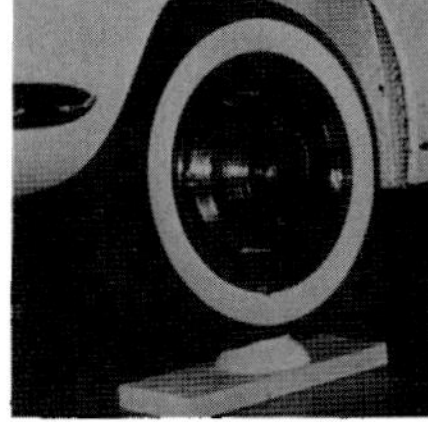

Carefree is beautiful veritied China . . . not plastic . . . not earthenware! It's venproof and **guaranteed 1 year** against breaking, chipping or cracking. So elegant for formal entertaining . . . so durable for rough - and - tumble family use!

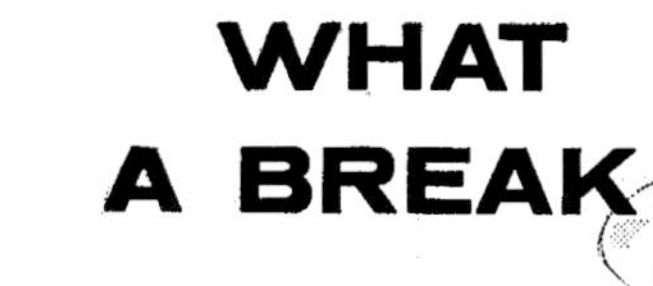

WHAT A BREAK!

NOW ON SALE!

AT SAVINGS UP TO **40% OFF!**

SALE ENDS OCTOBER 22ND

4-PIECE PLACE SETTINGS
AS LOW AS **$3.99**

16 - PIECE STARTER SETS
SAVE 20%!
SERVICE FOR 4 from **$15.95**

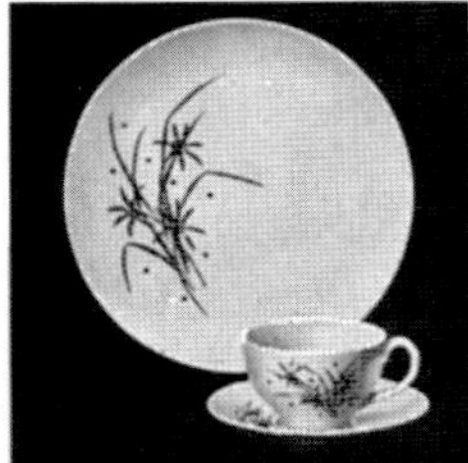

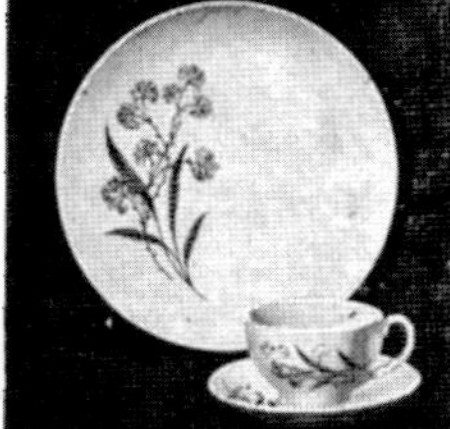

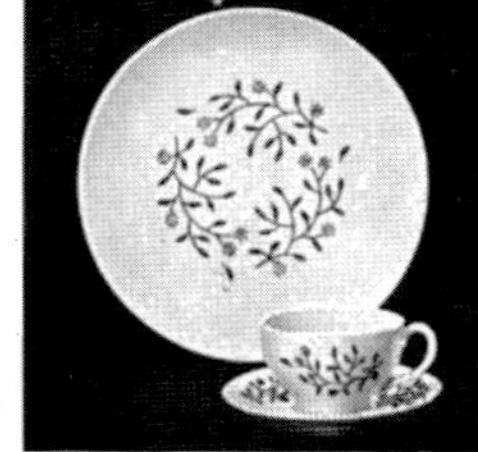

See the enclosed folders for pattern selections and complete price list. Also enclosed is a free sample of a new and wonderful silver polish. Take note of our Introductory Specials on a new and beautiful silver pattern by 1847 Rogers Bros. . . . Lelani . . . a perfect companion for your new china!

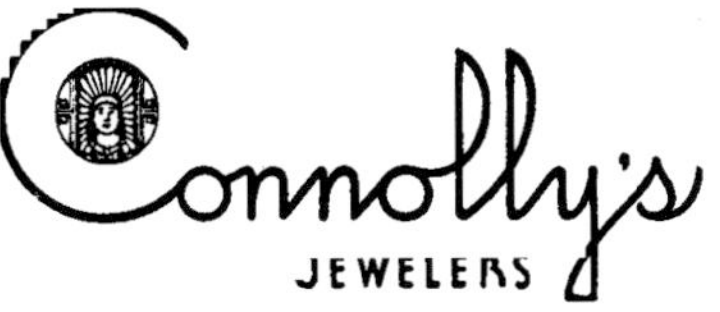

Buy On Convenient Budget Payments — Only 10% per Month. Small Deposit will Lay-Away Until Christmas!

SPECIAL INTRODUCTORY OFFER!
LELANI By 1847 Rogers Bros.

BUFFET SERVER
Useful with all pastries, frozen desserts, canapes, molded aspics. Beautifully gift boxed!
REG. PRICE WILL BE $3 **$1**

COFFEE SPOONS
After-dinner coffee, demi tasse or as cocktail muddlers. For serving any condiments! Set of 4 gift boxed.
REG. PRICE WILL BE $4 **$2**

8.13. Local promotion by jewelry store touting benefits of new *Carefree* casual household china, 1957.

the patio, in the kitchen, or as the central element of a formal dinner setting. Its low-key, modern decorations reflected these choices. Yet, despite this effort, dinnerware production remained a barely profitable enterprise.

The company's struggle against foreign competition took several forms. In 1959, it acquired the controlling interest in the former Vandesca Pottery Ltd., founded in Joliette, Quebec, in 1947. (Pass brushed up his French in advance of visiting the new plant, so that he could welcomed his new employees in their native language.) Vandesca-Syracuse Ltd. was the only Canadian company manufacturing vitreous china for hotels and restaurants, but English potteries had begun to duplicate American commercial ware and export it to Canada tariff-free to sell at prices below those for American china. Some of this British ware found its way to the United States, underselling Syracuse China despite the 45 percent tariff. In an attempt to gain a stronger competitive position, Onondaga Pottery shipped biscuit ware to Vandesca in Canada, paying the tariff rate for unfinished goods, a rate much lower than for finished ware. The company decorated and glazed the ware in Canada. For sales within Canada, this put its goods on an even footing with British imports. The company continued this practice until 1994. By 1966, Vandesca employed one hundred workers and made four products: Syracuse China of Canada, Vandesca China, Vandesca Fine China, and Vandesca Fireproof Cooking Ware.

The pottery continued to encourage a family spirit among employees. In issue after issue, the

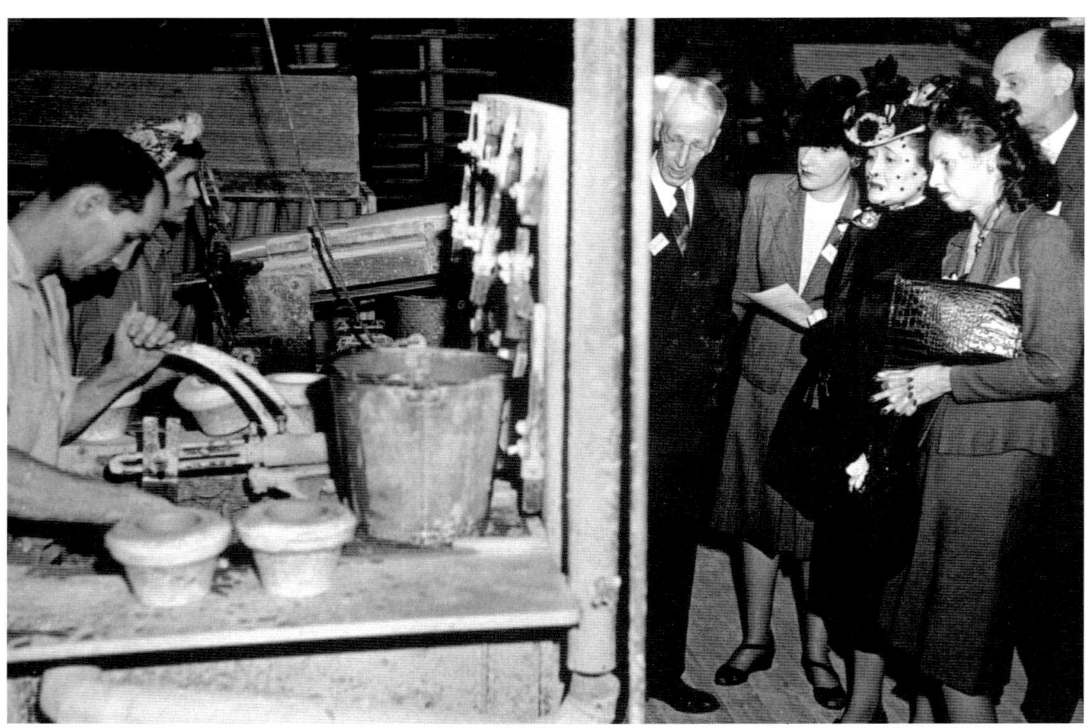

8.14. Bill Rybak jiggers cups for a plant tour led by Guy Cowan in mid-1940s.

News filled its pages with reports on the annual company picnic, the spring dance, golfing and baseball awards, bowling team banquets, basketball and horseshoe tournaments, fishing and hunting exploits, young men in the armed forces, weddings, new babies, hirings and retirements, and deaths. On the practical side, it explained new tax and insurance laws, described benefit packages, reported on the latest product news, and gave pep talks on such issues as the buying power of bonuses earned through increased production. Plant tours were easily arranged and encouraged; Guy Cowan sometimes led them for important visitors (fig. 8.14). An open house in May 1955 attracted thirty-six hundred people: "potters," families, and friends. The company opened its thousand-acre Star Sapphire Forest Preserve in the Adirondacks to employees. Pass took justifiable pride in what he and the pottery had accomplished through industrial humanics.

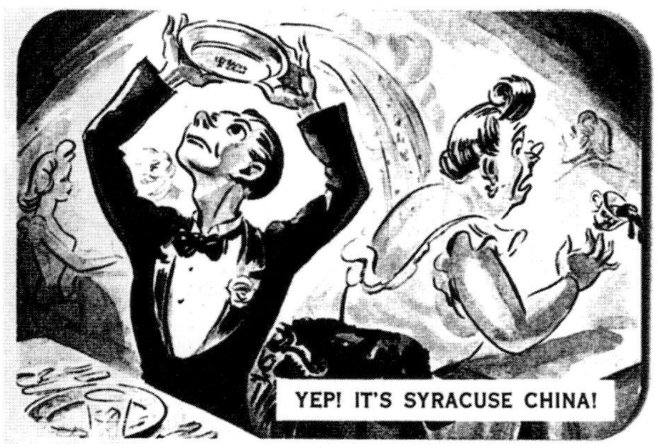

8.15. Richard Pass initiated the "Turn-over Club" card in 1946, a popular, amusing, and long-lived promotion. Design by Charles McHaig.

The man himself had contributed vitally to the company's success in times of economic crisis and war, and to its continued survival in the postwar years. He initiated highly successful programs to promote his company's ware (fig. 8.15). In his struggle against foreign competition, Pass had "charged again and again" the halls of Congress, testifying at committee hearings, battling for fair trade rather than free trade. At the same time he took on the stepped-up challenges posed by the National Labor Relations Board.[16] With a work force and production space each more than double the size of that of his father's day, with a requirement to meet constantly increasing government regulations, his job by the 1950s was very different from the "hands-on" post filled by James Pass. The pottery employed twenty-seven hundred people in its two plants. Richard felt deeply responsible for the welfare of his employees. He found it increasingly difficult to maintain the close relationships he fostered in "industrial humanics" or to convince others that they should work as hard for his goal as he did. In 1958, he retired after forty-three years with the company and left its management to "younger men."

Nine

The Final Years of the Old Company
1959–1971

IN 1958, Foster Rhodes succeeded Richard Pass as president. Rhodes had come to the company in 1928 from Cornell University with a degree in mechanical engineering. Trained by Pass, he became general foreman of the Fayette kilns in 1931, then moved to the clay shop and dipping section, and in 1933 became assistant superintendent of the Fayette plant. He had experimented with the use of gas to fire periodic kilns and helped to install the first biscuit tunnel kiln in 1935. In the 1940s he was, successively, the Fayette plant superintendent and production manager of Court Street. He became a director of the company in 1949 and executive vice president in 1956. Now as president, he had big shoes to fill, but he had been well prepared.

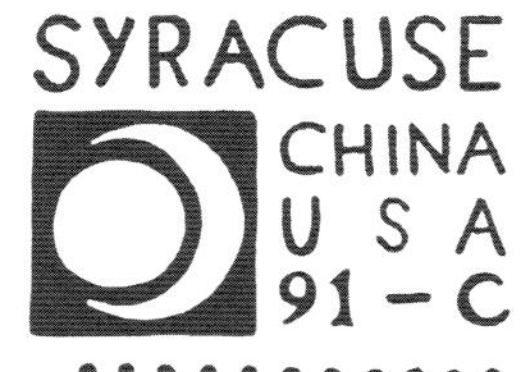

Peter Piening
(April–August 1962)

Rhodes was optimistic about the future of the pottery's hotel ware and he meant to reassert its distinctive place in that field. He called for a new trademark to reflect the company's stability, progress, and confidence in the future. To achieve this, in 1961 the company commissioned the internationally distinguished graphic designer Peter Piening to design a trade logo for its commercial ware, packaging, letterhead, forms, and commercial product catalogue. After a career in graphic design, Piening had joined the faculty at Syracuse University. He was widely known for his many designs of interlocking circles, including his Ballantine Ale logo. The logo he designed for Syracuse china also incorporated a circular motif. It consisted of a stylized representation of a small plate (an orange circle) placed on a large one (a white disk), both on a black square suggesting a table or placemat. His graphic style integrated the company's program of advertising and promotion. Despite the excellence of Piening's designs, they had a brief life. The logo was perhaps too abstract for its times. After five months as a backstamp and a little over a year as a logo, Piening's designs were replaced by a variation of the standard backstamp.

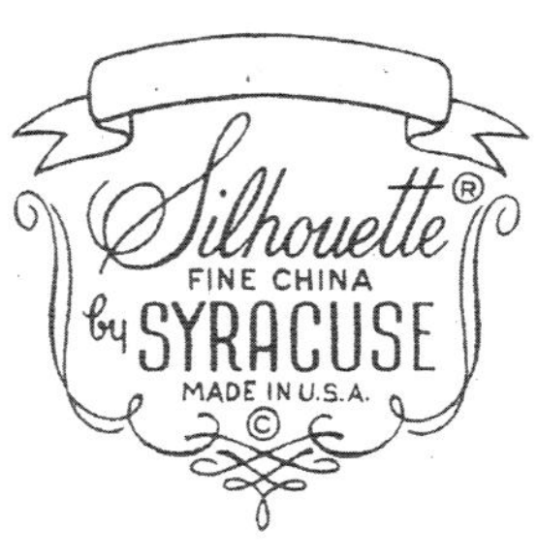

Silhouette (1960–70)

The company introduced its fine china **Silhouette** shape in 1961 at the Atlantic City China and Glass Show. **Silhouette** combined baroque fluting and a modern coupe shape (fig. 9.1). The company claimed it to be the "whitest china" with the "thinnest and most delicate cup produced in the entire field of American china." The lab created a special whiter and more translucent flint body, which was used only for this shape. Retaining the structural strength inherent in all types of Syracuse China, **Silhouette** became one of the pottery's more popular shapes. Still,

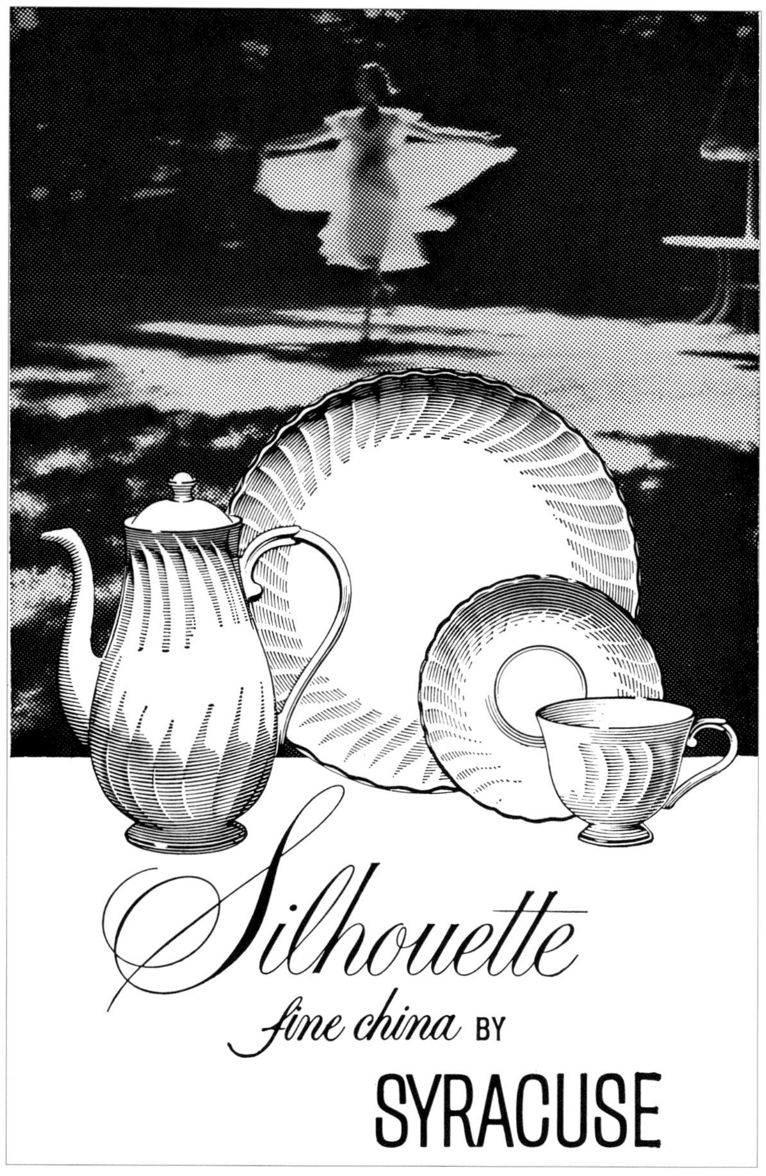

9.1. Advertising material for the new *Silhouette* shape, 1961.

many customers preferred the flat-rimmed **Berkeley** plate to the new **Silhouette** coupe; good designs, traditional or modern, can coexist happily.

In the same year, the pottery created for Colonial Williamsburg a faithful reproduction of an eighteenth-century English Bristol Delft "Yellow Rooster" pattern for use in the newly restored Chowning's Tavern. It also recreated the colonial "Blue Canton" pattern for other Williamsburg sites, as well as plates for Christiana Campbell's Tavern that duplicated the decorations of the original "William and Mary" plates in the Colonial Williamsburg Museum collection.

In January 1957, the pottery unveiled its **Gallery Collection** at the Atlantic City show
(fig. 9.2). This consisted of twenty-two attractively boxed gift items in four distinct styles.
The first two revived and modified some of the company's historical shapes, a **Marmora** fancy
dish and two **Puritan** salad bowls.[1] The second group complemented the **Silhouette** line with
a centerpiece bowl, compotes, vases, a cigarette holder and ashtray, and a candleholder. The
third group, a modern cigarette cup and ashtray, bore the name of its creator, the New York
industrial design firm Donald Desky Associates. The company tapped the talents of another
Syracuse University design professor, Arthur Pulos, for the last shape in the line, "Facet Ware,"
a thin modern shape in matte yellow, brown, orange, and undecorated white.

Under Rhodes, for the fiscal year 1962 the company reported the greatest volume of orders
and the largest dollar shipments in its ninety-one-year history. Employees received a six-cents-
an-hour pay raise. That year, Rhodes became chairman of the board, following in his mentor's
footsteps. Richard Pass himself had served as senior advisor during the Rhodes years. Pass died
in May 1964. He had served his company and his profession with great distinction, carrying on
the Pass family tradition of innovation and integrity in its field. He had served as president of
the U.S. Potters Association and had sat as a member of the Board of Councilors of Alfred University's College of Ceramic Engineering. Alfred University recognized his contributions to the
advancement of ceramic art with an honorary degree in 1954. As a tribute to his program of
industrial humanics, and to carry on his good work, at the time of his death friends and fellow
workers established in his name an annual memorial scholarship to be awarded to deserving
children of Onondaga Pottery and Pass & Seymour employees.[2] The untimely death of his
protégé Foster Rhodes in 1967 at age fifty-eight effectively ended the Pass years.

In 1962, William R. Salisbury, great-grandson of Mills Pharis and son of Bert Salisbury, became the ninth president of the Onondaga Pottery Company. He had joined the company in
1934 after graduating from Williams College. Continuing the family tradition of beginning at
the bottom and working up, he started in the clay shop and soon became foreman. He took
leave to earn a degree in ceramic engineering at the Massachusetts Institute of Technology. In
1942, he was made superintendent of the Fayette plant and a year later was put in charge of the
land mine project at Court Street before being recalled to Fayette Street as superintendent. In
1952, he became assistant to the vice president for production and from there he went on to
manage the Household Tableware Division. In 1959, he joined the board and in 1961 was
elected a vice-president of the corporation.

SYRALITE
by SYRACUSE
U.S.A.

Syralite (1964–97)

For the second time since it had modified its standard flint china body with colors in the
late 1920s, the pottery in 1964 introduced a truly new clay body. It appeared first at the National
Restaurant Show in Chicago in May of that year. This body was *Syralite,* a high alumina body
that was whiter and thinner, and was claimed by the company to be more durable than any
other china then in production in the industry. *Syralite* was the product of years of research
begun by Edward Schramm. Known at first as X-15, it was completed in 1963 by Flemmon Hall
(who had come from Pass & Seymour to replace Schramm when he retired) and Ralph

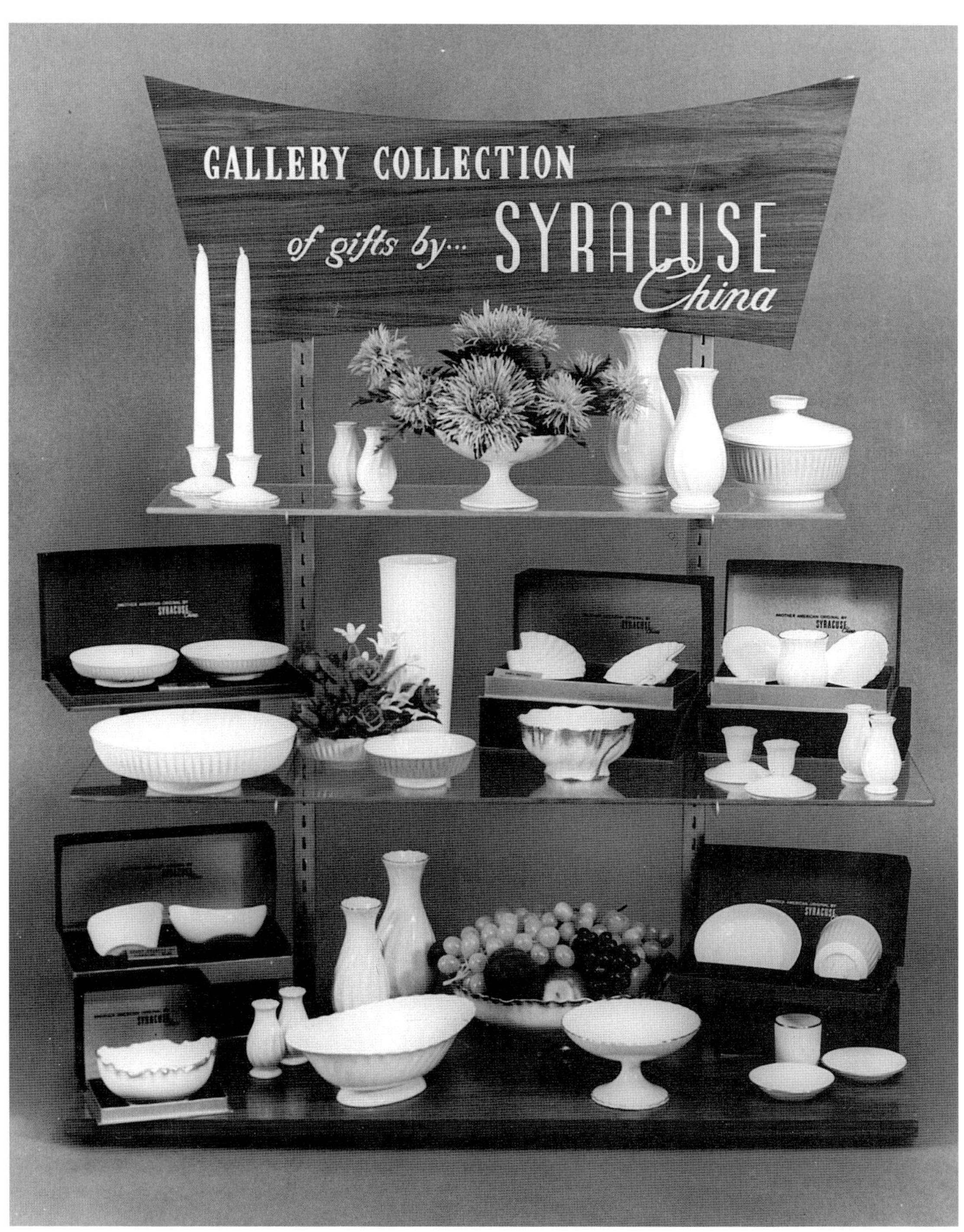

9.2. Convention display of the *Gallery Collection*, 1957.

Brigham. A stack of a dozen *Syralite* plates occupied half as much space as a stack of **Rolled Edge.** The body's strength made possible more elegantly refined shapes, but to show it off, the design department first created **American,** a straightforward shape with a vergeless, flared rim (see fig. A.8). Michael Szymanski created two new patterns, "Captain's Table" and "Versailles," to launch the shape. **American** was an instant success, and *Syralite* took its place as the company's standard clay body, only the second since the introduction of the pottery's first flint body in 1892. In 1966, the company offered **Tudor;** this coupe shape was a *Syralite* version of **Essex.**

In 1965, the company initiated a three-year, million-dollar expansion of the Court Street plant, increasing its capacity by one-third. State of the art manufacturing machinery arrived, most of it American-made, including a new biscuit tunnel kiln. Diagrams published in *China News* demonstrated how the plant had grown since its opening in 1922 (figs. 9.3 and 9.4). The expansion enabled the company to move many production operations from Fayette Street. The electronic data processing department moved in 1967, the color preparation lab in 1968. As the

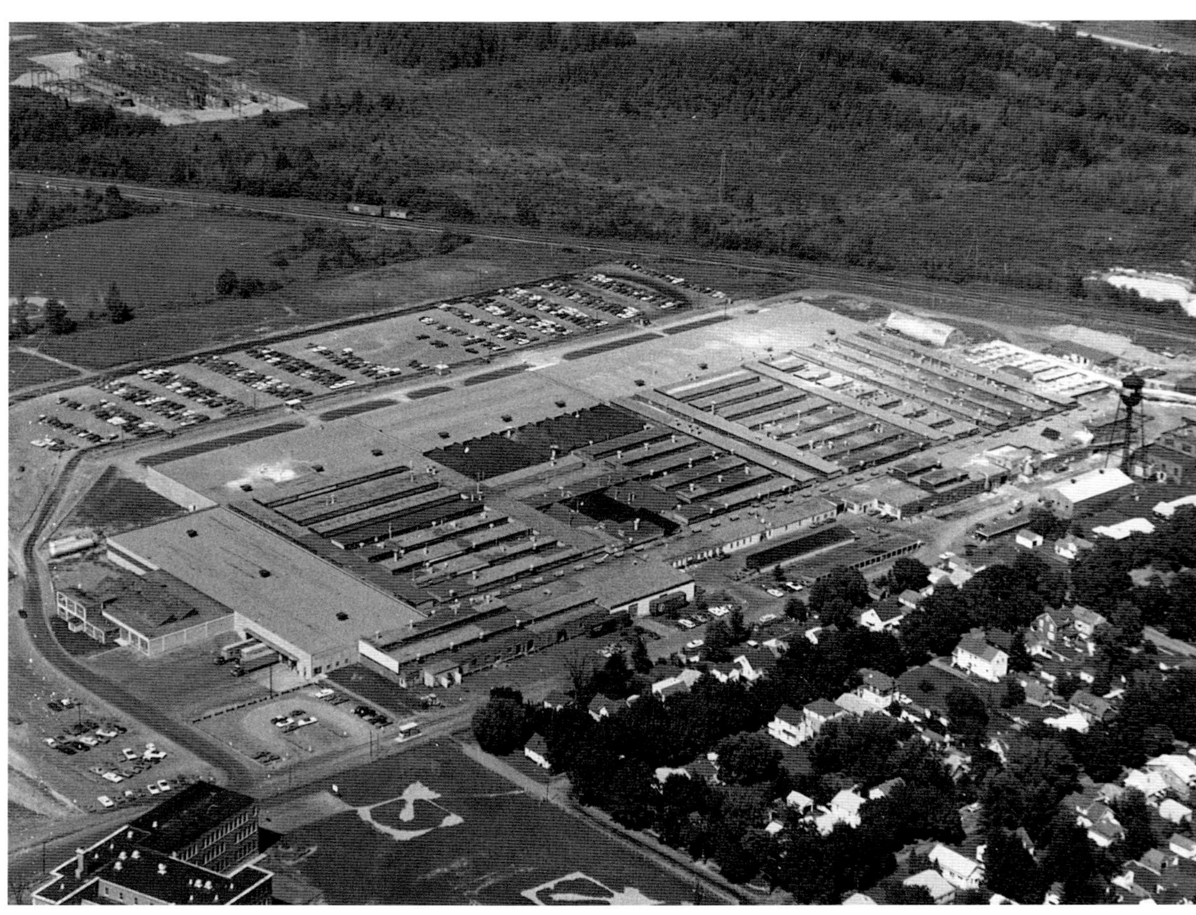

9.3. Aerial view of the Court Street plant, 1962.

Stages in the Development of Court Plant

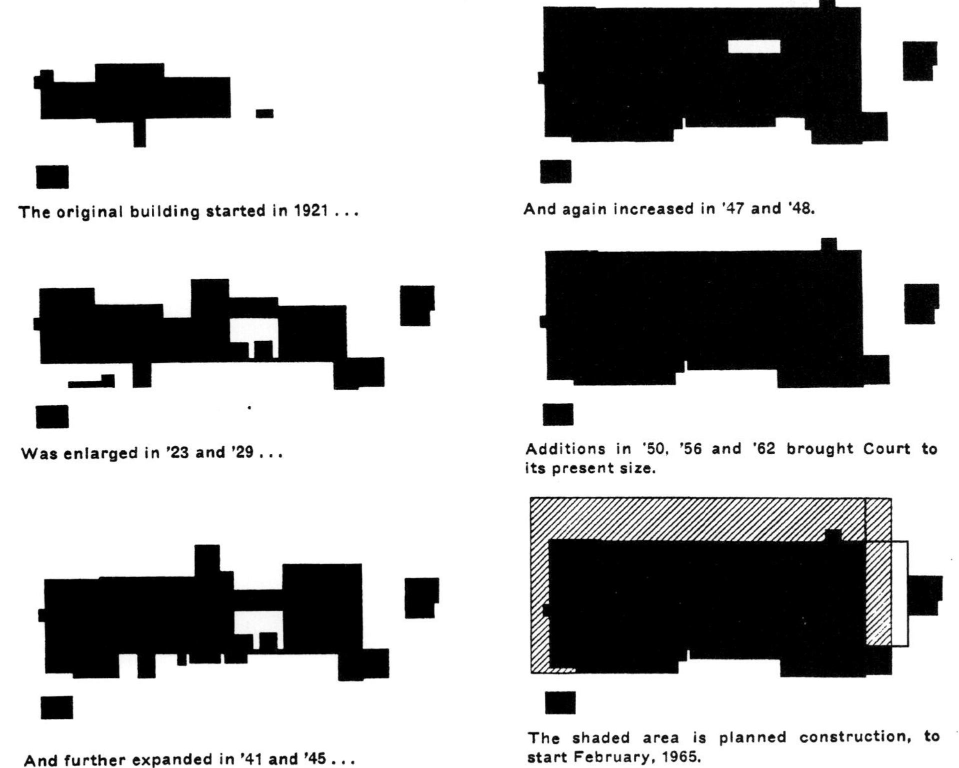

9.4. Stages of development of the Court Street plant.

factory that had begun to take shape on the Erie Canal in 1880 lost more and more of its vital functions to the new Court Street plant, doubt spread that the Fayette Street operation of Onondaga Pottery would long survive.

An important event concerning the identity of the company occurred in 1966. Over the years, the parallel use of the names Onondaga Pottery Company and Syracuse China had become increasingly confusing. The word "Onondaga" proved a stumbling block in pronunciation and spelling to many outside the local community. It signified little to anyone unfamiliar with its Native American origins. The company received letters addressed to Onondago Pott Col, Odagu Pottery, Onadagua Co., Onon. Potterie, etc. Correspondents increasingly addressed their letters simply to Syracuse China. Even the word "pottery" was confusing. By now the term was associated with flower pots and cemetery flower urns. In 1966, the company officially changed its name to the Syracuse China Corporation.

In June 1966, Syracuse China Corporation published the first issue of a new house organ titled *Chips. China News* became a quarterly concentrating on in-depth treatments of subjects of interest to the corporate family. *Chips* appeared in the interim months, reporting company and employee news and taking notice of the corporation's new products. In that year, an *Empire* dinnerware shape (fig. 9.5) was added to the already bulging household division offerings. Echoing nineteenth-century French Empire style, but modified to please twentieth century taste, **Empire** repeated the name of the corporation's early predecessor on Fayette Street as well as an earlier shape name, but the new shape had nothing in common with the old one. In style it stood in contrast to the **Carefree** line. **Empire** responded to the still-significant consumer demand for traditional formal dining service. It offered further proof that the predictions of the mid-1940s that the future lay in modern design had not taken into account the strength of tradition in dining in middle-class homes. The decorative patterns created for it by Syracuse China artists varied from elegance to stately simplicity.

Modern informality in dining was becoming a tradition in itself. This encouraged Syracuse China to keep its offerings fresh. In 1966, several new patterns were created for **Carefree X-L,** a *Syralite* addition to the **Carefree** dinnerware line. This streamlined shape used **American** flatware with slightly modified serving pieces and was guaranteed for three years against breakage.

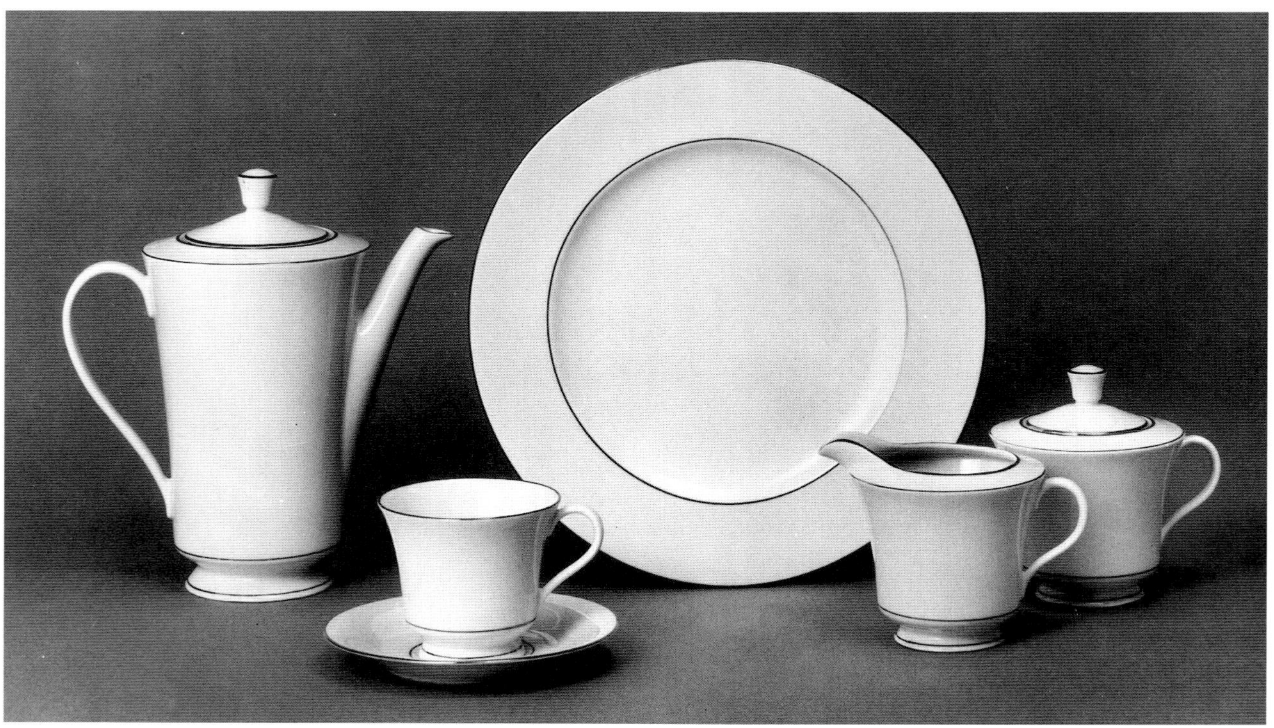

9.5. **Empire** shape with "Yorktown" pattern, 1966.

Two years later, the dinnerware division introduced another shape in the **Carefree** line, this one a distinct departure from past shapes. Promoted as a shape inspired by the "rhythms of the West Indies," **Calypso** (fig. 9.6) harkened back to the 1930s styles of Guy Cowan with its intaglio plate surfaces and moulded decorations. The rich, blended colors, applied by a new spraying technique called "color flooding," were meant to conjure up the tropical moods of its names: "Trinidad" (brown and blue), "Montego" (green and tan), "Aruba" (blue and gray), "Largo" (yellow), and "Surf White." This thicker china dinnerware, so seemingly casual in its styling, resembled nothing so much as earthenware, though it was in fact china. The firm's fine china still drew prestigious orders. In 1968, United Airlines ordered **Silhouette** dinnerware with a gold emblem and gold edge line for its executive Red Carpet 737 jet service. American Airlines used the **California** shape.

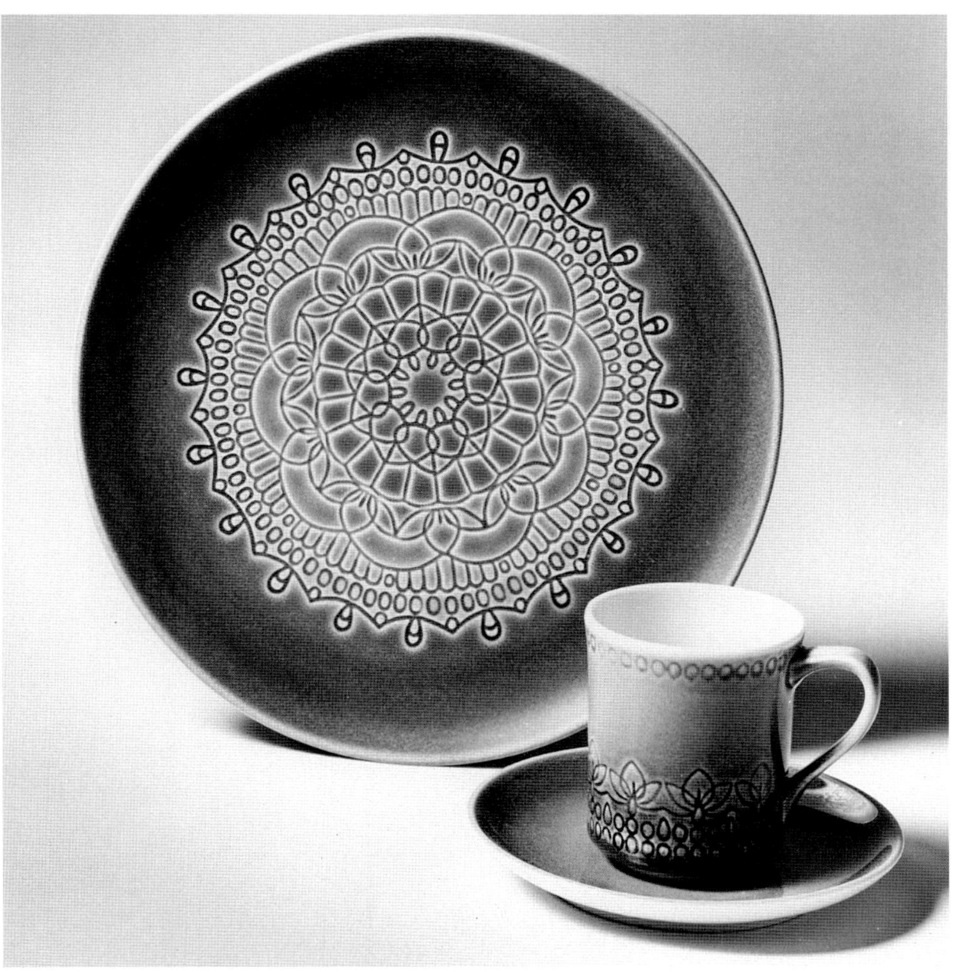

9.6. **Calypso** shape with "Montego" pattern, 1968.

In 1968, the company introduced a formal dinnerware shape designed by Richard Garvin: the **Wellington** shape (fig. 9.7). It was distinguished by an ivory color lighter than *Old Ivory*. A pedestal base supported its hollowware items. Three design consultants created patterns to embrace the late 1960s range of decorating trends: modernist, Spanish, or traditional.

As Syracuse China approached the one-hundredth anniversary of the founding of Onondaga Pottery, the household ware design coordinator, Andrew Petta, readied a prototype dinnerware shape to be called **Centennial**. It never went into production. In spite of heroic efforts to save the household dinnerware division, and despite the finest shapes, a vast selection of patterns, the most advanced marketing techniques, and outstanding customer service, the company could not in the end compete successfully with foreign imports. Since at least 1965, the household department had operated at a loss and was being carried by the still-flourishing commercial ware division. Part of the problem was that in recent years most of the firm's improvements in manufacturing technology had been invested in the Court Street plant, leaving at a disadvantage production operations in the increasingly antiquated facility that had grown on the site of Farrar's humble Rockingham pottery.

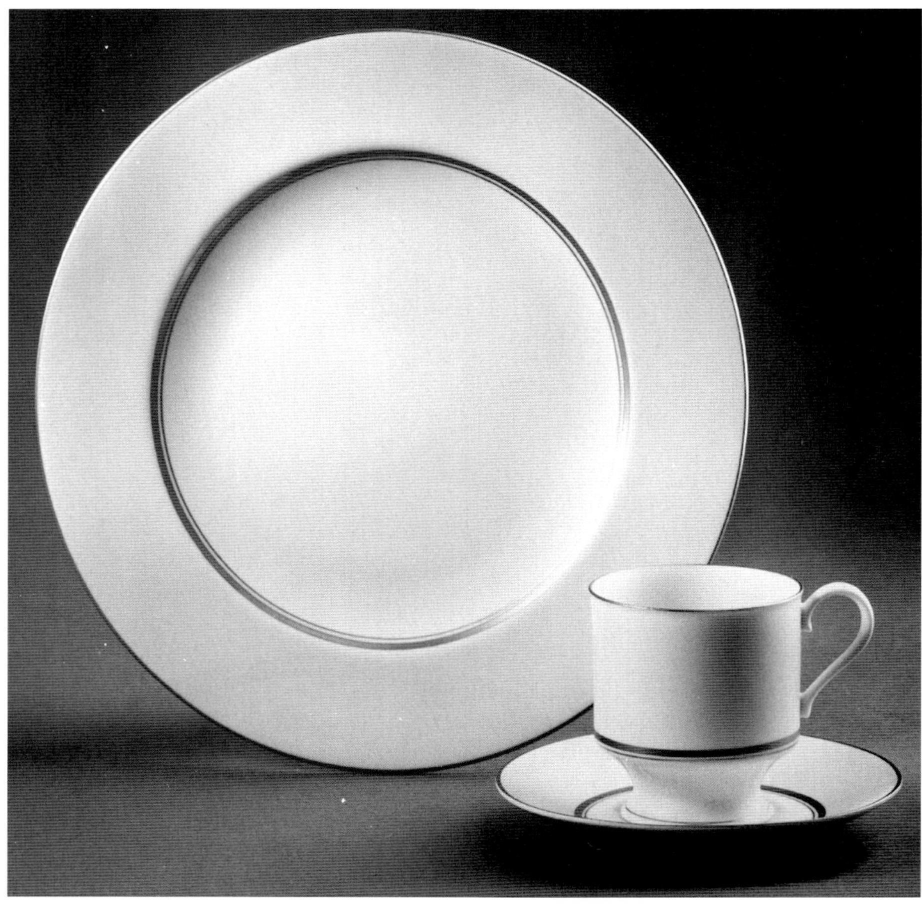

9.7 **Wellington** shape with "Kent" pattern, 1968.

Modernization was, as always, crucial to keeping up with competition. Advances in manufacturing technology now came not from the firm's own research and development efforts or from the American industry, but from Europe, where engineers had taken the lead in machine design. Nothing pointed up more clearly the extent of change that had occurred since the early 1950s than did the tour organized by *Ceramic Industry* to give American potteries a chance to see the latest developments in European manufacturing. Syracuse China's engineers had already taken this European tour on their own. They purchased a German Zeidler finishing machine for scalloped edge ware, a cup-making and handling machine, also German, and an English machine for turning footed cups. For the first time in the century, the American ceramic tableware industry looked back across the Atlantic for the most up-to-date technology.

The *News* reported that at his annual shop meeting in spring 1968, Salisbury divulged that some months earlier, Iroquois Industries of Buffalo had managed to buy 11 percent of the corporation's stock. At this point the Pass and Salisbury families held 59 percent. Iroquois Industries filed a lawsuit against Syracuse China Corporation charging illegal intervention with its tender offer. Iroquois and others saw that Syracuse China's still-sound financial condition meant that its assets could be used to borrow large sums for purposes other than tableware production. Syracuse China could serve as collateral for the purchase of other companies in other industries. Salisbury stated that he did not intend to allow the company to be swallowed up by any entity whose philosophies differed from those of Syracuse China's management, but it was unclear how he could prevent it. A year later, Salisbury reported that the federal court in Buffalo had rejected the Iroquois Industries suit. He also reported that orders for commercial ware had shown a gain of 10 percent. Orders for dinnerware were up also, especially for the **Calypso** and **Wellington** shapes, though not as high as he had hoped.

In the meantime, the company began to plan for its centennial celebration, to be held on 20 July 1971. A committee meeting on 15 July 1969 recommended that first priority be given the establishment of a museum at the Court Street plant to exhibit and interpret a historical collection of the firm's ware from its earliest days to the present. The ware had appeared "wherever food was served—from restaurants and private homes to the ships at sea and the planes that fly over our heads." The firm had actively collected ware for this purpose, believing that a factory museum would serve the rising interest in plant tours and, ultimately, sales. This never came to be, despite the building of new space for it.

On 11 June, 1970, Salisbury announced to the assembled employees of the Court and Fayette plants the company's decision to close the Fayette operation and to discontinue the production of fine china dinnerware. Production ended within seven weeks. Before the factory retail store closed to the public on 3 July, employees purchased dinnerware by the thousands of dozens in an unprecedented rush to own the product of which they were so proud. By mid-September they had purchased 7,334 dozens of ware. Salisbury explained what everyone already knew: that the dinnerware division had not been able to show a profit for some years and that the commercial ware division could no longer afford to carry the loss without seriously endangering its own future. He had hoped that all the highly skilled fine china makers could be phased into commercial ware production, but this did not happen. He developed a severance pay plan for some sixty skilled workers whom Court Street could not absorb. The Fayette Street factory complex was demolished in October 1971.

The corporation's centennial celebration in 1971 began one hundred days in advance of 20 July. Employees received an anniversary booklet and souvenir mugs. Over six hundred employees and guests dressed in period costumes recalling the past hundred years enlivened the annual company dinner dance on 17 April. The company decorated the Hotel Syracuse ballroom with historic photographs. A continuous slide presentation on large screens ran throughout the evening. On 14 and 15 May, the company invited the public to an open house at Court Street. Ten thousand guests arrived in a never-ending stream. On 15 July the biggest birthday party in town commenced when Salisbury cut the ceremonial first slice of cake before a gathering of invited guests and a large number of the Syracuse China "family" of employees and retirees. Seventy banquet tables lined the length of the main Court Street hallway; champagne punch flowed. In spite of the closing and razing of the Fayette plant, spirits were high.

In October 1971, *Chips* reported these festive centennial events without any hint that, unknown to nearly all in attendance, the celebrations also constituted a fond farewell to the old company as it had been known. *Chips* did not report that family ownership and management had come to a close and that a new Syracuse China Corporation had been formed as a public stock company in August, barely a month after the final centennial celebrations. On 30 September 1971, in one of the first leveraged buyouts in the nation, the new corporation had purchased the assets of the old one.

Ten

The New Company and New Challenges
1971–1997 (and Beyond)

> The sale will assure that Syracuse China will start its second century
> as a stronger organization and that continued growth will insure
> the security of our workers.
>
> —William Salisbury, 1971

FACED with the threat of an unfriendly takeover, William Salisbury had found an alternative that amounted to a friendly one. Robert J. Theis, a business executive based in New York City, had offered an attractive plan. His entrepreneurial credentials included extensive consumer product experience, first as general sales manager of Philco Corporation and then as president of GTE's Sylvania Home Products, where he had turned a losing business around in three years. Following that success, he had restructured ITT's worldwide consumer products business in two years, then had become group general manager of ITT's industrial products entity in the United States. A former colleague at ITT introduced him to Ed Gibbons, managing partner of Gibbons, Green and Rice, an investment banking firm in New York City. Theis and Gibbons began to look for a company in the United States in which Theis could become a major owner.

Over the next eighteen months they reviewed nearly twenty companies. They settled on Syracuse China, which had been brought to their attention by a Chicago consulting firm. Gibbons and Theis pulled together a group of investors, raised the necessary capital (7.7 million dollars), and in August 1971 formed a temporary public stock company in Syracuse, called Ryacuss, Inc. On September 30, the Syracuse China stockholders, by a narrow margin, voted their approval for the sale of the corporation to Ryacuss. The Syracuse China Corporation was reborn.[1] At a press conference, Salisbury commented, "The sale will assure that Syracuse China will start its second century as a stronger organization and that continued growth will insure the security of employment for our workers. Syracuse China will continue to be an integral part of the Syracuse community."

Theis became president. He brought with him two of his most trusted colleagues, the seasoned senior executives Charles B. Nairn, Jr., as senior vice president, and Martin B. Shellenberger as vice president of Advertising and Merchandising. Shellenberger took charge of new product planning and development. Stanley Campion, a long-time sales executive in the old company and right-hand man to Richard Pass until he retired, became vice-president of Marketing. William Salisbury became chairman of the board.[2] Two others from the old company remained at the executive level: Malcolm E. Kelley to oversee financial affairs as vice president and treasurer, and Richard W. Besse as vice president for manufacturing.

Syracuse China's corporate revitalization came from leadership that had little or no previous experience in the manufacture and marketing of ceramic tableware. Instead, Theis brought on board executives with a wealth of experience in contemporary corporate theory and practice, industrial design, human relations, industrial processes, and marketing. Still, the new company built on the best of the old.

Theis knew that to improve the company's profit margins, he needed to rebuild its approach to distribution, eliminate outdated items, and introduce new shapes and patterns. Foreign competition with low labor costs (70 to 80 percent lower) continued to reduce the nation's china manufacturing business. It became apparent in two months, however, that the new company could not grow without a major reduction in expenses. Since labor costs accounted for approximately 60 percent of the cost of china, the level of business Syracuse China was doing could not support its work force of thirteen hundred employees. Over one third of them had transferred from Fayette Street when that plant closed. Theis reduced this number at once. "Black Friday" (an event that was repeated more than once) was a sad day on which many old-timers were let go. The factory then became a union shop. *Syracuse China News* and many of the social and athletic activities the company had sponsored came to an end, but the endangered company survived and resumed its growth.

Theis had a genuine interest in making the firm succeed. His experience at ITT had been particularly instructive. For that firm he had led the search for, and acquisition of, more than a dozen manufacturers of industrial products. In the process, he had developed successful turnaround strategies, which he now implemented at Syracuse China. Central to his strategy was the concept of "no weak links."

Seeking a closer match between product and customer, Theis pruned the product line, eliminating 70 percent of the items the pottery had been making (as well the adobe body). As James Pass had given up earthenware production for china in the mid-1890s, Theis knew that the company now needed to limit its production to *foodservice dinnerware* of high quality and to fulfill customer orders with even greater speed than before. Foodservice was a new term. Since the turn of the century, the term *dinnerware* had applied within the company *only* to thin fine china made for the household division's sales to department and other retail specialty stores. The term *hotel ware* had applied *only* to the more durable chinaware made for the commercial division's sales to hotel and restaurant china distributors. Because most American potteries no longer made fine household china, and because markets were changing, the distinction in terms was no longer useful. By the 1970s, *foodservice dinnerware* described the durable hotel ware on which Syracuse China focused exclusively. Theis put aside all thought of returning to the manufacture of fine household china and issued a new company logo.

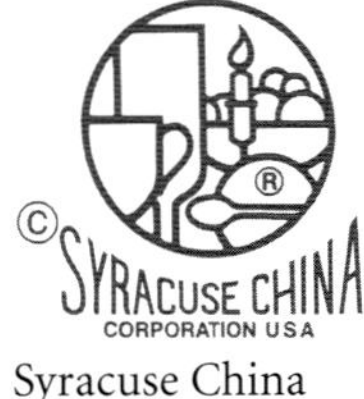

Syracuse China
Corporation USA
(1971–97)

For foodservice ware, upgraded and expanded computer operations tightened inventory control, production flow, and order tracking. By the 1980s, computers facilitated modern cost-accounting practices. The company was an early convert to bar-coded shipping containers. It made continuing capital investments to increase production and elevate quality; modernization of the pottery's manufacturing equipment became a critical requirement in the effort to reduce production costs and to decrease the advantage that low labor costs continued to give to foreign imports.

The corporation's engineering team traveled regularly to Europe to find the most advanced production equipment, always with the goal of testing the newest and adopting the best. The expansion program at Court Street initiated in the 1960s was accelerated and completed with the addition of seventy-three thousand square feet of space. This expansion included the building of a new kiln that fired ware on one continuously moving band instead of loaded in multi-layered cars, and it enlarged the glost production area while adding a new warehouse and a silo complex for advanced computer-controlled storing of bulk material and blending of raw materials. A leading force in the program to enhance manufacturing capability was Chester D. Amond, a manufacturing executive whom Theis had known at ITT. Amond assumed the post of vice president of Manufacturing when Richard Besse resigned late in 1972.

Determined to stay abreast of the fast-paced food-service industry's needs, and to keep competition off balance, the corporation moved ahead with a program of new product development. As a means of simplifying the complex challenge of selling chinaware to restaurants, the new company quickly reduced the traditional and bewildering number of "scale prices" for an item. Earlier, as many as two hundred different prices had been quoted for varying decorations for a single item; when the simplification process was completed, twenty scale prices covered the same range of decorations.

In 1973, it hired as director of design George Jensen, an internationally respected industrial designer who was a partner in the Latham Tyler Jensen firm of Chicago. That firm, and Jensen himself, had contributed prize-winning designs for some of the dinner services created a few years earlier by Rosenthal, the German producer of ceramic artware and fine porcelain dinnerware. For the first time in its history, a broadly experienced industrial designer who was also a gifted sculptor, rather than a practical potter or ceramic artist, took over the position as head designer, and this, too, reflected the spirit of the times.[3]

Under Shellenberger's direction, the design department became an integral part of the marketing operation. Jensen's first major project was the design of a non-ceramic product. He created **Country Ware,** a line of more than three hundred items of hand-finished, sand-cast aluminum alloy accessories (with only a few appropriate china coordinates) for dining and home decorative use (see fig. A.11). The company aimed **Country Ware** at both foodservice and retail markets. Jensen developed a new corporate logo type and new custom graphics and business document designs.

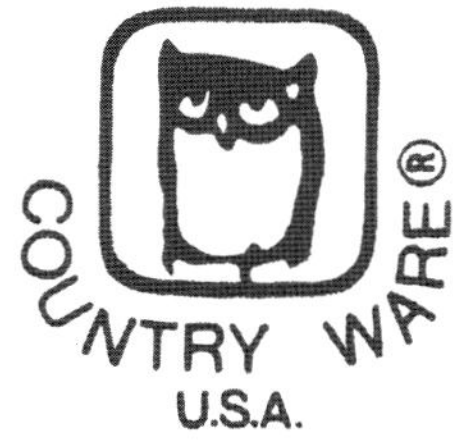

Country Ware (1973–78)

In the mid-1970s, a Jensen ceramic platter design, which the company trademarked as the "Great Plate," broke the production bottlenecks of conventional platters and had large-scale sales. Another Jensen success was the brown speckled and banded "Mesa Grande" decorative treatment on **Morwel.** This became one of the company's all-time best-selling and most copied patterns, perhaps only matched historically by the "Cardinal" lines and "Roxbury" print decorations.

In 1975, Jensen designed the ***Signet*** shape (see fig. A.10) in the premium *Syralite* body, with a distinctive carved intaglio well for upscale casual restaurants. (The pottery continued to make most of its ware in its standard flint china body, which, though modified slightly from time to time, was essentially unchanged since the 1890s.) Jensen also created a deliberately limited-item dinnerware service, the squared ***Olympus*** shape (see fig. A.11), for midscale casual restaurants. Jensen's last major shape program was the complete ***Gibraltar*** "Cook 'n' Serve" ovenware line (see fig. A.12), introduced shortly before his untimely death in 1978.

The result of a research program conducted to learn the nuances of the industry's usages and wants in commercial ovenware, the ***Gibraltar*** line capitalized on the appeal of healthful (baked, not fried) individual entrees served piping hot from the oven. Made to resist thermal shock, chipping, cracking, and abrasion, ***Gibraltar*** could go from freezer to oven to table. It was formed using a variation of the standard flint body developed for single-firing production. Designed to be the industry's most durable ovenware, ***Gibraltar*** was an immediate success. To help the company's distributor sales organizations to integrate their new ***Gibraltar*** ovenware line successfully with traditional Syracuse China dinnerware services, Shellenberger broadened the newly popular concept of "tabletop architecture" to highlight the firm's posture as a producer of both ovenware and foodservice dinnerware. New brochures illustrated visually appealing menu presentations incorporating fresh combinations of both kinds of ware.

Steve A. Unger, a gifted designer from Raymond Loewy and R. Latham & Associates of Chicago, succeeded Jensen. Several skillful designers and fine craftsmen-modelers worked on his team during his tenure, including Helen Zughaib, Jeanette Mattson, Joanne Capella, Lucie Wellner, Duane McCann, and Don Folley.

Under Shellenberger's direction, Syracuse China created a significant competitive edge in marketing by organizing its own proprietary national "conventions" for the top management of its distributors. These gatherings introduced new company products and marketing concepts to the distributors well in advance of the foodservice industry's national trade shows, at which the firm would be only one of many manufacturers. Syracuse China's meetings with its distributors included seminars to provide expertise of practical value in the merchandising of foodservice products. Exclusively for the company's distributors, these meetings reinforced a partnership relationship between Syracuse China and its chief customers in the trade. The meetings were held in such appealing venues as Las Vegas, Hawaii, Acapulco, and Mexico City.[4]

Shellenberger developed a number of creative merchandising programs during these years. One program, in 1976, offered a series of illustrated booklets on the theme, *The Perceived Value of Table Top Architecture.* The booklets included instructions for setting tables, designing menus, and arranging food on plates. For a single menu, they offered layouts on plates for "a standard," "a little more," "a little less," and "a dramatic" presentation. They ranged from bacon

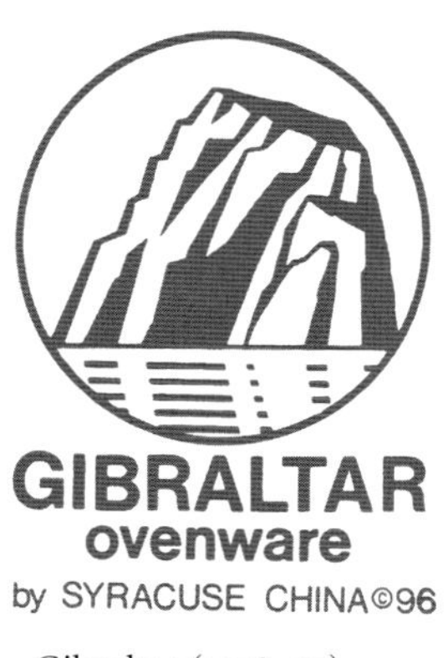

Gibraltar (1978–97)

and eggs to surf and turf, quiche, barbecued ribs, lasagna and garlic toast, a garnished sub, a club sandwich, and other common restaurant offerings. *Table Top Architecture* specified suitable selections of Syracuse China for each presentation.

The company commissioned Edith Gilbert, a leading author on household tabletop etiquette at the time, to create an illustrated fifty-four-page booklet, *Tabletop the Right Way*. This introduced the company's shapes by "getting down to basics" in the tabletop concept in foodservice. The contents page offered advice on "Using Flatware Creatively," "Waking-up Coffee Shops," "Tray Service & Cafeterias," "Beverage Service: Glassware, Bar Tips, a Word about Wine," "Napkin Mood: Folding Napkins for Fun and Profit," "Launch a Successful Lunch," "Merry Money-Making Merchandising [on good showmanship]," "Self Serve and Salad Bars," "Gourmet Service," "Hospital and Extended Care Facilities," "House Specials," "Dramatic Desserts," "Coordination," "17 Steps to Smooth Service," and "Metric Conversion Charts." Thin, elegant, translucent china played no part in this advice. Hotelware had been refined into highly durable yet relatively graceful commercial ware, fine enough for the new age of dining out that increasingly emphasized ethnic foods of various cultures, menu innovation, and fresh but casual tabletop architecture.

Early in 1973, Syracuse China bought the Will and Baumer candle manufacturing company of Syracuse. This gave Syracuse China combined sales exceeding twenty million dollars (its own approximately twelve million plus Will and Baumer's nine million). Will and Baumer turned out to be an unfortunate choice, however, because almost immediately after its acquisition paraffin costs went from five cents to twenty-one cents a pound owing to the Mideast oil embargo. At about the same time, Roman Catholic rituals changed, dramatically reducing the need for religious candles. The demand for candles for dining remained, but this was not enough to sustain profits.

In time, Theis appointed Chester Amond, his vice-president for china manufacturing, to head the candle company. William Fenn, a senior manufacturing process executive with Allied Chemical in nearby Solvay, assumed responsibility for manufacturing at Syracuse China. For more than twenty years he oversaw many manufacturing advances with the goal of steadily improving efficiency and quality. In the critical area of ware forming, the already begun process of roller-jiggering ware became the new standard procedure and eliminated the traditional blade-jiggering process. This procedure resulted in denser and more durable ware. Also, over a number of years, the claying-up support method of firing was successfully eliminated, greatly reducing airborne refractory dust in the pottery.

Throughout its decades of innovation as a pottery, Syracuse China had always been in the forefront of chinaware decorating technology. In the early 1970s it had stated with pride that it was armed with the greatest palette of underglaze colors and the greatest array of decorating techniques. Under Amond, and then under Fenn, it sought new solutions to the everchanging challenges of ceramic decoration. During the 1970s it abandoned a number of traditional decorating methods either as unneeded or prohibitively expensive, chief among them the varnish mounting of decals. Conversion to the newer uniflame type of decal permitted the sending of decorated ware directly to the glazing shop without a pause enroute in the hardening-on kiln to burn off the varnish. In spite of the uniflame decal's advantages, it was stiffer and not as easy to apply to many hollowware curves as were the old varnish-mount decals. In addition, the conversion process was time-consuming because almost every decal strike had to be recalibrated. Still, it succeeded, and was still in use in 1997.

Syracuse China discontinued traditional hand painting, which had also become prohibitively expensive. To satisfy customers who continued to want a hand-painted look, the design studio perfected decal techniques that effectively simulated a free, painterly look. Similarly, the *Syratone* method of creating solid-color decorative borders of finest quality had proved too labor intensive to sustain. For many years no effort was made to replicate the special effects made possible by this technique, though by the 1990s, decal mineral pigments were used to achieve similar dimensional effects. *Shadowtone* spray-decorating capability continued, but having fallen from fashion, there was little call for it.

Just as the company steadily improved its decal capability, so, too, did it persist in seeking better and more complex transfer printing capabilities. In the 1970s, the intricate and charming "Strawberry Hill" pattern was successfully applied with one-color transfer printing equipment, and it proved highly popular. By the mid-1980s, Fenn had broadened the firm's transfer printing capability to permit well-registered two-color transfer printing of the refined and delicate "Montlynn" pattern. By the mid-1990s, it was feasible to register four decorative colors in a transfer-printed pattern.

In the mid-1970s, Canadian Pacific, Ltd., which had begun to acquire companies in the United States, took an interest in Syracuse China. It knew about the company's financial successes from its annual reports and recognized its management strengths. Moreover, Canadian Pacific's, extensive hotel operations in Canada had been buying ware from Syracuse China's Vandesca subsidiary. At this time, Syracuse China was faced with the need to take on substantial additional debt to fund its modernization plans. The prospect of its sale to Canadian Pacific, Ltd. offered the benefits of having the financial support (in this and other needs) of Canadian Pacific, with its seventeen billion dollars of assets.

Canadian Pacific bought Syracuse China in 1978. Theis remained as chairman and president. A year later, Canadian Pacific asked him to assume greater responsibilities as president of its United States holdings. He agreed, with the proviso that his headquarters be located in downtown Syracuse. Theis remained as chairman of Syracuse China, though his major role was now as president and CEO of Canadian Pacific Enterprises (U.S.). Nairn retired; Malcolm Kelley became Theis's senior financial executive at Canadian Pacific' Syracuse headquarters. Theis ran Syracuse China from his downtown Syracuse office. When the corporation sold Will and Baumer late in 1978, Amond returned to Syracuse China as president. At about the same time, the firm sold its fast-growing **Country Ware** division to Wilton Armetale, a leading competitor, when Canadian Pacific decided not to undertake foundry operations in the United States.

In 1976, Charles Goodman had joined Syracuse China as vice-president of marketing and Campion had reverted to vice president of sales. Goodman had been president of Walker China of Ohio, a competitor in foodservice dinnerware. Goodman's experience in distribution had been a missing ingredient in Syracuse China's sales force. A dual-management system began. Amond as president was the "inside man," supervising manufacturing, finance, industrial relations, and general administraion; Goodman, as executive vice president, was the "outside man," with overall responsibility for sales, marketing, merchandising, and advertising. With the retirement in the early 1980s of Stanley Campion as vice president of sales, Dale Sutton assumed his role. In 1984, Theis retired from Syracuse China, remaining with Canadian Pacific Enterprises, the parent company. Amond became Syracuse China's chairman and CEO and Goodman took over as president. When Goodman resigned as president to pursue other business opportunities in 1987, Amond resumed day-to-day operating management duties as chairman and president.

One of Amond's goals in the 1980s was the acquisition of two other potteries, reflecting a trend in American industry to merge like businesses. In 1984, he negotiated the purchase of the Mayer China Company of Beaver Falls, Pennsylvania.[5] Syracuse China provided the company with technical assistance, made biscuit ware for it, and invested in its specialized decorating operation. Mayer at first prospered under the wing of its new owner. In 1988, Syracuse China bought another historically important company, Shenango Pottery of New Castle, Pennsylvania. This company proved to be a less successful investment than Mayer China had been, hampered as it was by inefficient and obsolete facilities.

Reinforcing the company's long-standing association with the Everson Museum of Art (formerly the Syracuse Museum of Fine Arts), Amond added another strong link between Syracuse China and the world of ceramic art. That association had begun in 1932, when Richard Pass and Cowan had helped launch the Syracuse National Ceramic exhibitions at the museum, to national acclaim. Three company presidents, Pass, Rhodes, and Amond had since served as museum trustees. The museum's collection of world ceramics grew steadily from the generosity of donors and Ceramic National purchase prizes (some funded by Syracuse China). In particular, the museum's collection of twentieth-century American ceramics, perhaps the largest in the world, had outgrown even its storage space.

While Amond served as the museum's president, he facilitated, in 1986, with generous support from Canadian Pacific, the renovation of a large space in I. M. Pei's handsome museum building into the Syracuse China Center for the Study of American Ceramics. With most of the ceramic collection at last accessible to museum visitors, the center attained an international prominence attested to in 1989 by the publication of the four-hundred page *American Ceramics: The Collection of Everson Museum of Art,* edited by the museum's curator of Ceramics, Barbara Perry. The collection included examples of **Imperial Geddo,** "Blue Plum," and other key shapes and patterns.

The 1980s was also a decade of new product activity. In 1984, Unger and his design team brought the broad-rimmed **Syrene** shape to market (see fig. A.13). The intaglio-carved **Syrene** took the concept of tabletop architecture to new heights. Reversing the concept of **Econorim,** it was a wide-rim, small-well shape conceived especially for nouvelle cuisine, suited to the vogue for dramatic presentations of small amounts of food. In 1988, *Encore* echoed the **Econo-rim** concept as a narrow-leafed variant of the *Syrene* shape.

In 1985 the design team developed the *A la Carte* program of universal serving items and accessories intended to complement all of the company's shapes (see fig. A.14). This line included such items as condiment holders, teapots, sauceboats, mugs, vases, after-dinner cups and saucers, sugars, creamers, and other shapes that could be decorated to coordinate with any style or pattern.

By then, the foodservice industry defined types of restaurants as midscale (in the mid-1980s, this meant eight to fifteen dollars per meal), upscale (over fifteen dollars), and low end or value priced (coffee shops, diners, and quick-service chains). In 1987, the company introduced the *Savoy* and *Turina* shapes to serve the growing needs of the midscale market. Both *Savoy* (fig. A.15) and *Turina* (fig. A.16) featured intaglio moulded decorations on medium-width flatrims. Most shapes offered a selection of as many as seven sizes of plates, great plates (but not traditional platters), cups and saucers (usually in two cup sizes or a mug), rim-deep soups and cereal bowls, fruit dishes, unhandled bouillon cups, and after 1991, large-diameter, rim-deep entree bowls and plates intended for popular pasta entrees and hearty ragouts, stews, and other lower-ingredient-cost provincial entrees. *Turina* proved to be one of the company's best-selling new shapes.

For its first fifteen years, with new general management, strong financial controls, effective human relations programs, modern manufacturing methods, and focused marketing practices in place, the new Syracuse China Corporation flourished. It experienced an average annual growth of 10 percent, sharing fully in the steady expansion of the food service industry in the 1970s. Indeed, it got good returns on its investments each year through 1987, even in an industry beginning to face market saturation. By the late 1980s, however, the corporation faced a growing threat from a new wave of offshore competition. Selling pressures rose significantly as Asian manufacturers began exporting higher quality foodservice dinnerware at selling prices that United States producers struggled to match, in spite of the latter's far greater sophistication in the technology of manufacturing.

Some American manufacturers, such as Buffalo China and Homer Laughlin, had already refocused production to meet the competition of imports at the low end of the market, the "coffee shop business." Syracuse continued to target the more rewarding middle to upper end. Distinctions in product quality narrowed, however, between the low end and the middle. Further, though Syracuse's "Mesa Grande" casual dinnerware pattern had captured a large part of the United States food-service market in the 1970s, by the 1980s, low-end manufacturers overseas had effectively copied it.

At the same time, at the very top of the dinnerware quality spectrum, leading and well-established European dinnerware producers found a small but receptive market in the United States in the 1980s for very elegant upscale ware. Such historically famous producers of fine porcelain and china as Villory and Boch in Luxembourg, and Royal Doulton and Wedgwood in England, now produced fine quality foodservice dinnerware, and this ware gained special distinction in an age fixated on prestigious brand names. Even America's own Lenox China, which had not earlier been a factor in foodservice dinnerware, moved into the competitive fray with a high-priced line of very white china imported as bisque ware from Japan.

In 1988, Syracuse China/Vandesca in Canada wanted a product to compete with Lenox's white-bodied china. Vandesca imported a similar white-bodied Japanese biscuit ware, which it decorated and glazed at the Canadian plant (except for some custom work decorated at the Mayer plant while it still operated, and then in Syracuse). The imported product was called *Royal Rideau* in Canada and *Luxor* in the United States, and served the upscale market. It was used for the **Parliament** and **Chateau** shapes and carried a *Royal Rideau* or a *Luxor © Mayer* backstamp, depending on where it was made. Problems developed however, because although the Japanese clay body was a superior product in most regards, the dimensions of the bisque ware proved inconsistent. In 1993, the ceramic engineers of the revived laboratory at Syracuse China, directed by Roger Markell, successfully developed its own "bright white" high alumina body with clay imported from England. The *Luxor* name was dropped in favor of *Royal Rideau* and this became the fourth china clay body type developed by the pottery in over one hundred years. When the Mayer Pottery closed, the manufacturing and decorating of **Parliament** and **Chateau** shifted to the Syracuse plant and occasioned the new backstamp, *Royal Rideau Fine China, Syracuse China Company*. The company had shown that American invention could still meet the best overseas competition with better quality.

During the 1990s the engineering team made other advances. Under the direction of ceramic engineer Joe Benoit, the company became the first American commercial pottery to form dinnerware successfully by pressure-casting in production quantities. It also developed a prototype kiln-pusher control system, making Syracuse China the first in the world to operate a fully computer-controlled tunnel kiln. Even the most advanced earlier kilns collected data, processed by computers, that required action by kiln technicians. The new computer-controlled system constantly collected information as the kiln fired and made needed adjustments automatically.

As the overall foodservice industry prospered in the mid-1980s, Syracuse China found itself challenged from above and below by more than fifty competing brands of dinnerware, nearly all of them imports. The number of domestic producers had been reduced from more than thirty in 1960 to fewer than ten in the late 1980s.

At the same time, new patterns of product distribution evolved to present a challenge of a different kind. Previously, two kinds of distributors had served the hotel and restaurant trade. One sold strictly foodstuffs, the other, "tabletop" supplies, including foodservice dinnerware and cooking equipment. Syracuse China had traditionally marketed through a limited network of the latter distributors. In the 1980s, consolidation among distributors led to the rise of giant regional and national food suppliers who also sold food-service equipment, including tabletop, and promoted "one-stop" shopping in the trade. To facilitate this consolidation, the inherently larger food distributors bought out less–well capitalized supply and equipment distributors.

The specialized knowledge of the product needed to sell Syracuse China as a quality product

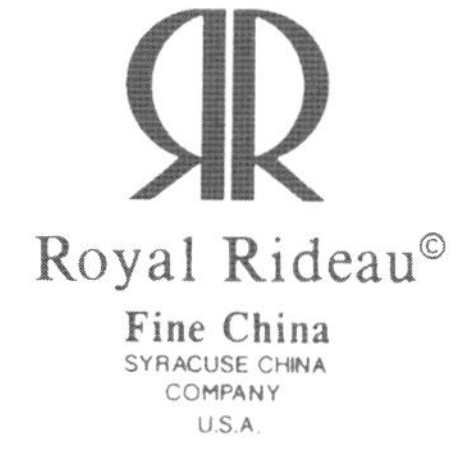

Royal Rideau©
Fine China
SYRACUSE CHINA
COMPANY
U.S.A.

Royal Rideau (1993–97)

largely disappeared as the army of professional tableware salespersons retired, lost their jobs, or were transformed into food salesmen. The industry went from several thousand expert tabletop salespersons to several hundred. As a result, at many distributorships the sale of ceramic dinnerware, which once involved a command of practical aesthetics, taste, technical knowledge, and an ability to discriminate among the fine points of competing products, was reduced to little more than elementary price-shopping.

In 1987 and 1988 Syracuse China experienced several other shocks. Shenango Pottery's difficulties began to drain corporate energy and resources. The 1987 stock market plunge and the ensuing recession meant that fewer people dined out and fewer hotels and restaurants were built. The restaurant industry, which had expanded rapidly in the previous few years, now complained that it had "too many seats and too few fannies to fill them." The growth of foodservice china sales halted. The industry could neither sustain capital investments for new equipment nor raise product prices. The inherent high quality of Syracuse China meant that it was a product that lasted and did not require frequent replacement, and although this was an advantage to the foodservice buyers who recognized it, it diminished production at the company.

The increasingly casual lifestyles of American consumers intensified the problem of decreasing demand. Less china was needed, rarely more than a five-piece place setting, and very few serving pieces. Demand was lessened also by the steady growth of a fast-food industry that created a boom for disposable products while it reduced the market for restaurant china by nearly 20 percent. More and more, Americans consumed meals that came wrapped in paper, poured into Styrofoam, and supported by pieces of cardboard.

By 1989, the foreign manufacturers of foodservice china for the the upscale American market were producing a durable, special-purpose new ware, usually decorated with elegant intaglio carving or tasteful, subtle, sophisticated decals. This ware closely resembled these makers' finest household china and porcelain, and performed as well as the less elegant, American-made, commercial dinnerware. Fortunately for Syracuse China and other American makers of upscale foodservice china, these imported, commercial-grade products from such companies as Japan's Noritake and Mikasa, Germany's Thomas, Schoenwald, and Rosenthal, England's Wedgwood and Royal Doulton, were expensive and affordable only to truly upscale restaurants. The sales of this ware were of relatively small volume.

To compete with the demand for high-end commercial dinnerware, Unger and his design staff created three new full-rim shape lines, **Belmont, Tremont,** and **Verona,** all in the whiter, thinner *Syralite* body. **Belmont** was a plain, unembellished shape (fig. A.17). **Tremont** (fig. A.18) and **Verona** (fig. A.19) were intaglio shapes, the former decorated with Art Deco motifs, the latter with a classical beaded rim. The sense of American affluence, which spawned a large number of casually elegant restaurants with celebrated chefs, seemed to reach a pinnacle in the late 1980s. In response, Shellenberger again expanded the tabletop architecture program, suggesting techniques to make more dramatic, formal presentations of gourmet entrees with the new *Syralite* shapes.

For the 1990s, Syracuse China planned to introduce new shapes at regular, measured intervals. This plan was interrupted by the pressures of absorbing the Mayer and Shenango lines. Unger and his team had designed the **Castleton** line to initiate this series in 1990. Aimed at the upscale restaurant market, this elegant shape needed to wait until 1992 to go into production.[6] In 1993, **Justine,** an old fashioned shape in *Syralite* (with a Mayer backstamp) responded to a renewed interest in the more elaborate Victorian shapes that European competitors had always made.

In 1989 Canadian Pacific, feeling the same pressures of economic recession, started to redirect its assets into its defined core businesses, and put the Syracuse China Company on the market. In 1989 the family-owned Susquehanna-Pfaltzgraff Company of York, Pennsylvania, outbid more than twenty investors for the highly regarded pottery along with its Mayer and Shenango properties. Pfaltzgraff Pottery was the oldest continuously operated, family-owned pottery in America. It had become a major producer of household casual stoneware, and in 1988 had become the nation's sole producer of fine bone china. With the acquisition of Syracuse China, the company entered the middle-to-upper end of the commercial foodservice market.

Amond continued as president and CEO. Jere Riggs, a Pfaltzgraff financial executive, was brought in as vice president/treasurer to provide liaison with the new parent company. All other management remained in place. The company closed Mayer in 1989 and moved its operations to the Shenango Pottery in nearby New Castle, Pennsylvania, but because of union hiring practices, it was unable to transfer Mayer's experienced decorators, some of the best in the country, at the pay level they required. This was the case even though Shenango had earlier lost its own fine decorating staff. The decision was reached to close the New Castle operation in 1990–91 and move to Syracuse the brand names, decals, moulds, and paperwork needed to make the Mayer/Shenango lines at the Court Street facility. This consolidation presented its own problems. The Mayer and Shenango lines required the diversion of significant resources to service the large clientele of these two companies. At the same time, new environmental concerns about the use of leaded glazes in ceramic manufacturing, required a thoroughgoing review of past and present practices for all brands.[7]

The company experienced some of its most difficult times during this period. Amond retired in late 1991. William Simpson, president of Pfaltzgraff, served as interim chief executive until April 1992, when the company invited Charles Goodman to return as president. Goodman quickly refocused sharply on what the company had always excelled at, high quality ware and excellent service. He set out to further improve the ware, as well as customer relations, and to develop a better-field service corps. He upgraded an old Syracuse China practice into the innovative concept of "Hospitality Express," a customer-delivery service that guaranteed thirty-six hour delivery for seventeen dinnerware patterns. This remarkable level of service was an instant success. Goodman also saw that the company's best chance to gain a competitive edge was to anticipate market shifts and quickly generate original shapes and patterns to meet them.

Sutton retired in 1991, Shellenberger in 1992, and Unger in 1994. After Unger's retirement, for the first time in the company's history a team of women headed the design department. Joanne Capella and Lucie Wellner continued to design custom patterns and crests, and also created displays for national hotel and restaurant shows. In 1995, Wellner designed the **Barista** shape, a series of various-sized stacking coffee cups aimed at the growing and trendy expresso and cappuccino market.

In 1995, Libbey, Inc. of Toledo, Ohio, purchased the company from Pfaltzgraff. Libbey, the largest commercial glassware manufacturer in the world, was now able to strengthen its position as a supplier of tabletop glassware by adding quality foodservice china to its product array. Its chairman and CEO, John Meier, combined the Syracuse China management team with his own. At Syracuse China Company, A Unit of Libbey Inc., Goodman remained as president and Fenn as vice president of Manufacturing. The company immediately issued a *Hospitality Express* catalogue to its food service distributors offering thirty-seven patterns available for delivery in thirty-six hours, many coordinated with Libbey glass products. The catalogue's color photographs showed elegant presentations of food on Syracuse China with Libbey glassware at hand, all in designer settings. Among the shapes was Cowan's **Econo-Rim,** which had been introduced more than sixty years earlier and had been in constant production ever since, but was now pared down to seventeen items. The catalogue also offered **Morwel, Winthrop, Essex, American, Savoy, Tremont, Verona, Turina, Justine, Castleton,** and **Syrene** shapes. **Gibraltar** ovenware, specialty entree bowls, and oval platters, were carried over. The catalogue illustrated several successful Mayer and Shenango shapes, including **Stylus, Staffordshire, Carlton, Parliament, Fanfare,** and **Cord-edge.**

Once Libbey assumed ownership, it continued and accelerated the ambitious capital investment program for advanced manufacturing capability begun under Pfaltzgraff. Libbey/Syracuse China made its new presence known in the trade by bringing its combined ceramic and glass products together at the 1996 National Restaurant Association show in Chicago. For this show, Wellner designed a new shape, **Cantina,** decorated with incised "sgraffito" lines and solid colors, inspired by "the painted deserts of the American southwest." As had now been the case for some years with new shapes, **Cantina** (plate 30) consisted of a carefully limited number of items, essentially platters and plates of several sizes, variations of two mugs, and three bowls (including one for salsa and one for chili). These items were all that were required for the popular and casual dining style of Mexican and other theme restaurants.

When Syracuse China reached its 125th birthday on 20 July 1996, there were no public celebrations, no dances, banquets, or parades. The Court Street plant's halls were quiet, void of the normal daily roar and clatter of the making of thousands of dozens of pieces of china. Except for a skeleton office staff, most of the work force of 350 was on its summer vacation. The vision, practices, and policies of the "new company" had proved to be valid, durable, and, for the most part, still operable.

The number of Syracuse China employees in 1996 was less than a third of what it had been a quarter of a century earlier (fig. 10.1). This was so not only because of greater efficiency

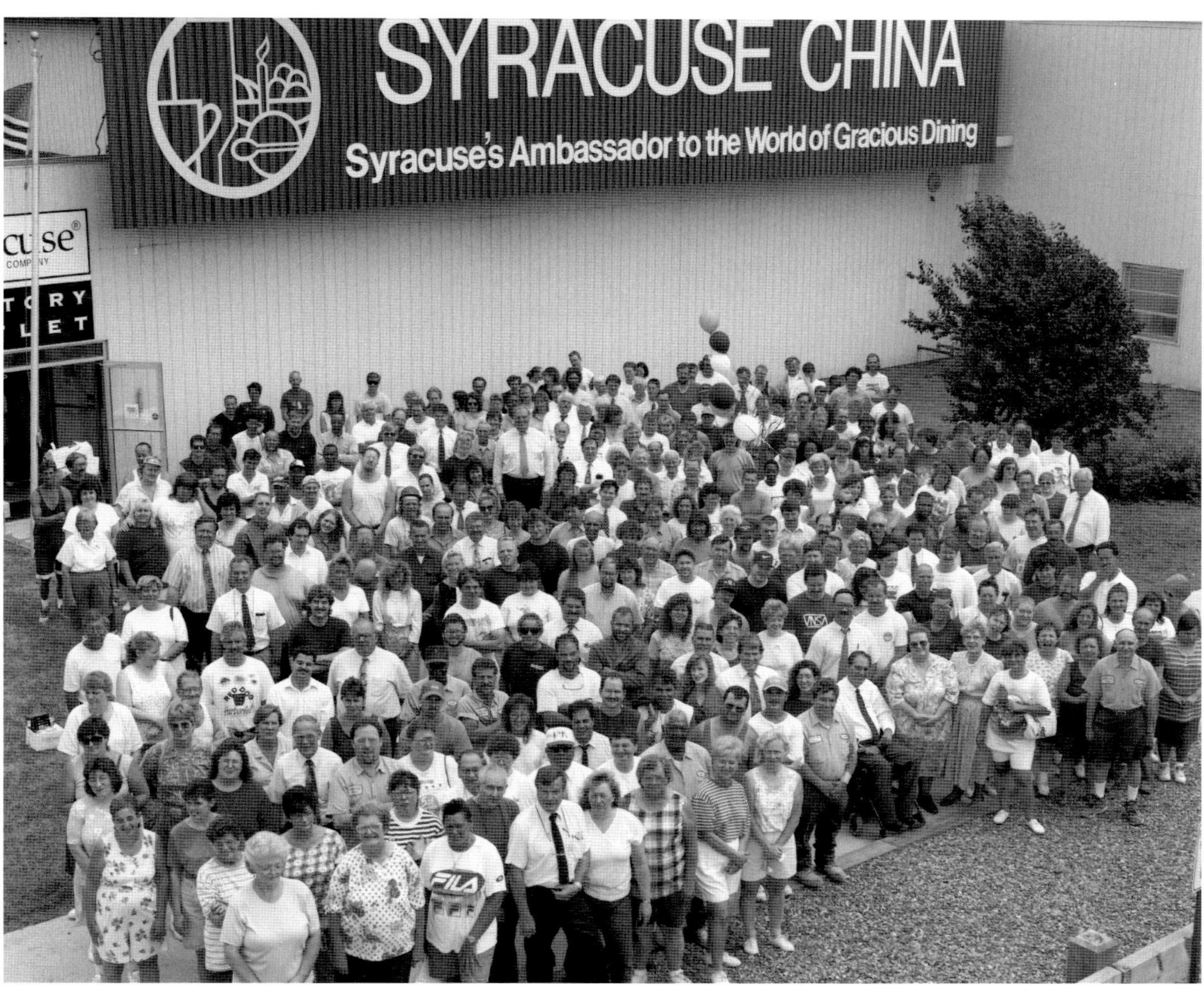

10.1. Employees gather to celebrate the 125th anniversary of Syracuse China, 1997.
Photo by J. William Courtney, Klineberg, Inc., Syracuse, N.Y.

throughout the company but also because a once labor-intensive industry now used advanced equipment and computers to accomplish many tasks. William Fenn, who in the 1980s and 1990s had put modern manufacturing equipment and processes in place, retired in 1997. He was the last of his executive generation to retire, except for President Goodman. And so, as the employees gathered in October with the corporations' management to celebrate the 125th anniversary of the founding of Onondaga Pottery, there was cause for new, if guarded, optimism, a feeling that the company that had survived 125 years of challenges was still a leader and again on the rise.

Eleven

The Story of a Piece of China

Tʜᴇ success of Syracuse China as a product always depended on expert execution and the vigilant control of every step in the china making process. Even before 1900, each step occurred in a highly specialized department. The number of departments grew or diminished as the china-making process changed over the century following James Pass's introduction of American china to the world of ceramics. The account of china production that follows describes the practices in place from the mid-1920s to the mid-1940s.

Shape Design

Before a piece of china was made, a small staff of designers, modelers, and pattern artists had created models and studies for the shape of which it was a part and its applied decorations. From initial design to readiness for manufacture, a new shape sometimes required as much as a year of preparation. After the *head designer* developed a concept for the shape and made preliminary drawings for each item, approval was needed from those who would make it and those who would sell it. The head designer turned over to the *modeler* the job of creating a full-scale model for each item. Sometimes the head designer modeled the shape himself; less often, the modeler created an item for a shape. Once approved, the models went to the mould shop where the necessary moulds were made.

Mould Shop

Clay shop potters used moulds[1] to produce ware of uniform dimensions. These were: the *block mould,* the *case mould,* and *working moulds* (fig. 11.1). They made moulds from plaster of paris because it absorbed moisture from clay, diminishing its plasticity and enabling the un-

Adapted and with the addition of introductory and concluding paragraphs from the series "The Story of a Piece of China" which first appeared in serial form in the *Syracuse China News* in 1925 and 1926 and republished in revised form in 1946 and 1947. Most of this "story" takes place at the Fayette Street plant. At Court Street, assembly lines made possible a linear and much more efficient production flow (see fig. 11.28).

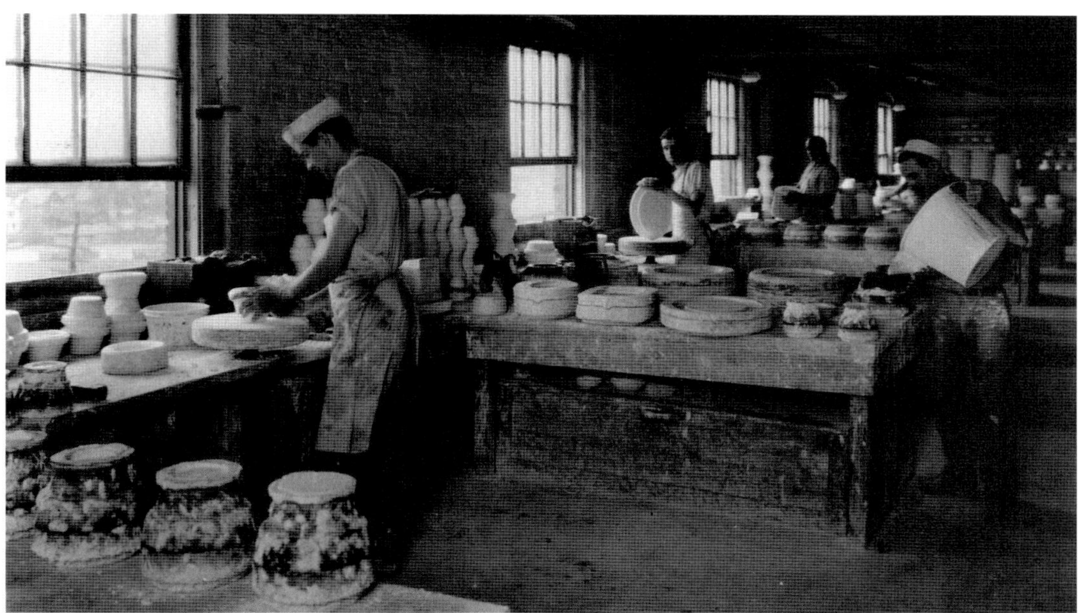

11.1. Mould shop, Fayette Street plant, 1925.

fired *green ware* to keep its shape. Each item in a shape required a series of moulds. The block mould, or master mould, was a reverse impression of the model. The case mould was made from the block mould. From the case mould came the many working moulds from which ware was made. Through the wear and tear of constant use, working moulds deteriorated, making it necessary to replace them from the case mould. When, in time, a case mould showed signs of ware, a new one was made from the block mould. Moulds were made larger than the finished piece of china to allow for the 10 to 15 percent shrinkage that occurred during drying and firing of ware.

Clay Preparation

The production of a piece of china began in the *slip cellar* with a formula for the clay body. Three general types of materials constituted the flint body used to make Syracuse China. The essential component was a high-quality *clay,* such as English china clay or another kaolin, some from America. Clay's main ingredient, hydrous silicate of alumina, becomes plastic when wet, making it possible to form the ware. The clay body's two other components were *inerts,* such as flint, to hold the formed body together and give it strength, and *fluxes,* such as feldspar or whiting, that melt in firing to fuse the clay and inerts together. Because the properties of each of these three ingredients varied according to the site where they originated, and even within the same deposit, a close monitoring and cleaning of ingredients was required to ensure consis-

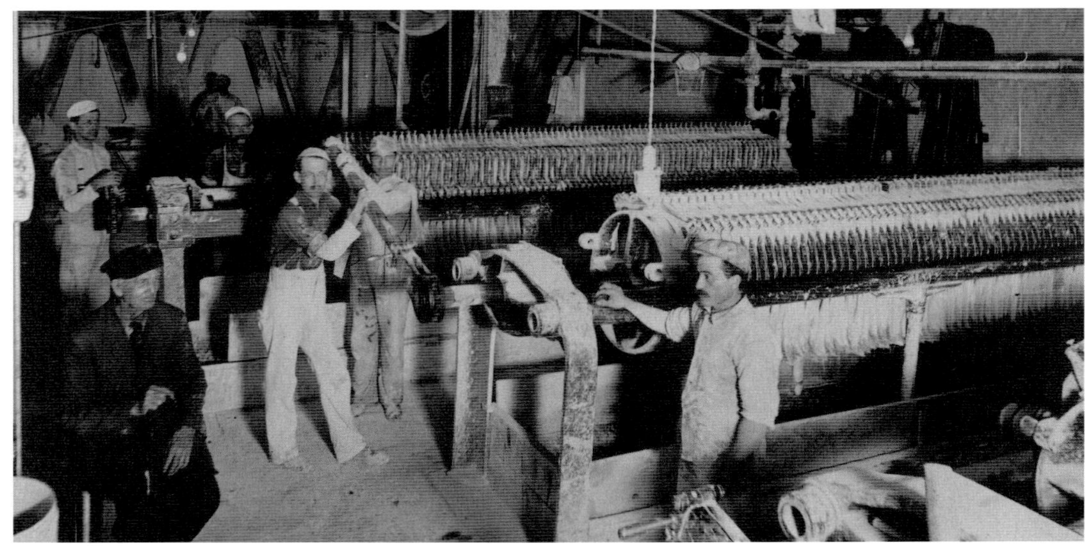

11.2. Filter presses in the clay prep shop, Fayette Street plant, 1915.

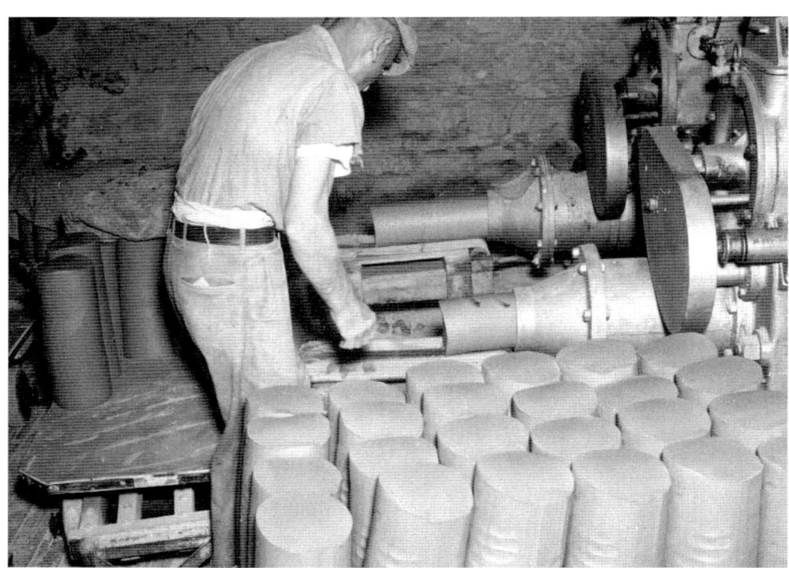

11.3. Pugging clay in the clay prep shop, Fayette Street plant, 1946.

tency. In the early years of the pottery, raw materials first came to the factory via the canal systems of New York State, and then by rail. After the mid–twentieth century, they also arrived by truck.

Because clay formulas were highly guarded trade secrets, the dry materials were weighed on "blind" scales and then deposited in a tank mixer, or *blunger,* with several tons' capacity of dry materials. Water was added to the formulated dry clay body and mixed until it reached a thick, milk-like or *slip* condition. Although the dry materials had been processed to remove impurities, the slip was further purified. First it was run through fine metal screens, called *lawning,* to filter out small pieces of rock and other foreign materials. Then, the slip passed over magnetic separators to remove iron particles that would discolor the clay body when fired.

After these purifications, the slip was continuously mixed in an underground tank called an *agitator* until it was pumped into a *filter press* (fig. 11.2). The filter press, an accordion-like series of iron disks and canvas filters, used hydraulic pressure to extract excess water from the slip, leaving it with the proper moisture content and consistency. When the clay was removed from the filter press, it was stored in zinc and slate vaults that preserved its condition and kept it contaminent-free until it was ready to be put in the *pugmill* (fig. 11.3). The pugmill kneaded the clay through a series of augers (similar to a meat grinder) while vacuum pumps removed excess air. Then the pugmill produced long, cylindrical pieces of clay called *puggings* that now were ready for distribution to the clay shop.

Forming the Ware

From the mould shop, the working moulds were sent to the *clay shop* for use in making ware (fig. 11.4). In its early years, Onondaga Pottery produced its ware in a time-honored way by hand-pressing the plastic clay body into the moulds, either by hand or by hand-operated tools. Later, the factory maintained steam, and then electric-powered, machinery for use with the moulds.

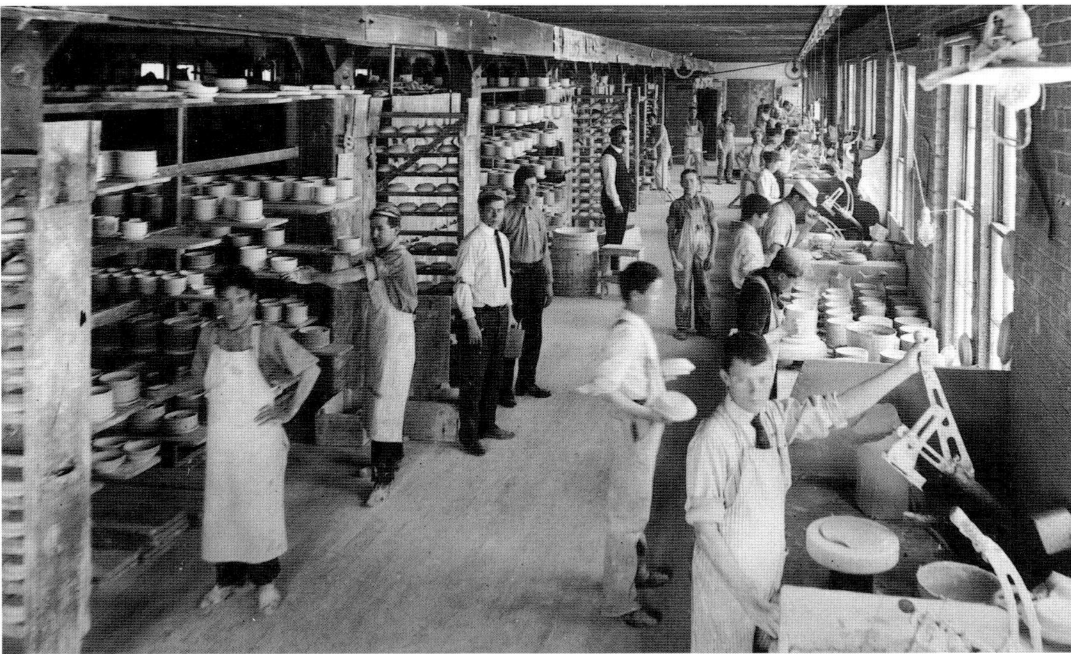

11.4. Flatware forming (jiggering) in the clay shop at Fayette Street plant, 1910.

The working moulds for flatware were disc-shaped and used on a machine called a *jigger*. This was a rotating vertical spindle with a ring or cup head at the upper end into which the mould fit. The face of the mould expressed the face or upper surface of the piece. The *jiggerman* created the back of the piece by pressing a precision-cut cast iron tool into the clay on the mould while the jiggerhead revolved.

The jiggerman, a potter who possessed a high order of skill typically learned through years of apprentice and journeyman experience, was assisted by two other workers: the *batter-out* and the *finisher*. The batter-out cut a slice of clay from the puggings, worked it into a ball, and then placed it on the *spreader* to form a clay disc, or *bat*. The batter-out then placed the bat on the mould for the jiggerman. The jiggerman carefully set the mould on the jiggerhead and, with the tool, formed the back of the piece (fig. 11.5). After removing any scrap clay from the mould, the jiggerman removed the mould holding the finished piece from the jiggerhead. The batter-out collected these moulds on a *slide* (a shelf that slid into the dryer) and moved them to the *stove room* to dry.

After the ware had been in the stove room long enough to dry, the *stripper* separated the pieces from the moulds and piled, or *bunged*, them six to twelve high (depending on the item) on a *setter*. In the green state, ware was too fragile to handle or support itself. A setter was a heavy dish made of refractory clay, a form of purchased kiln furniture, that supported the bung of ware in this fragile green stage until it came from the kilns and could be handled with little danger of its breaking. The *bungs* were set on a *whirler*, a turning circular surface, where the finisher removed the scrap line and smoothed the edges with an edge sponge. The finisher also smoothed the face of each piece with a very fine sponge and cleared bits of scrap clay from the back. All flatware pieces were dusted with a very fine calcined clay, which prevented them from sticking together when bunged and fired. The large pieces of flatware also received a *cob*, a carefully measured pile of dry calcined clay placed in their centers to support the bunged pieces and prevent the centers from dropping during firing. Finally the bunged ware went to *clay-up*, where *calcined*

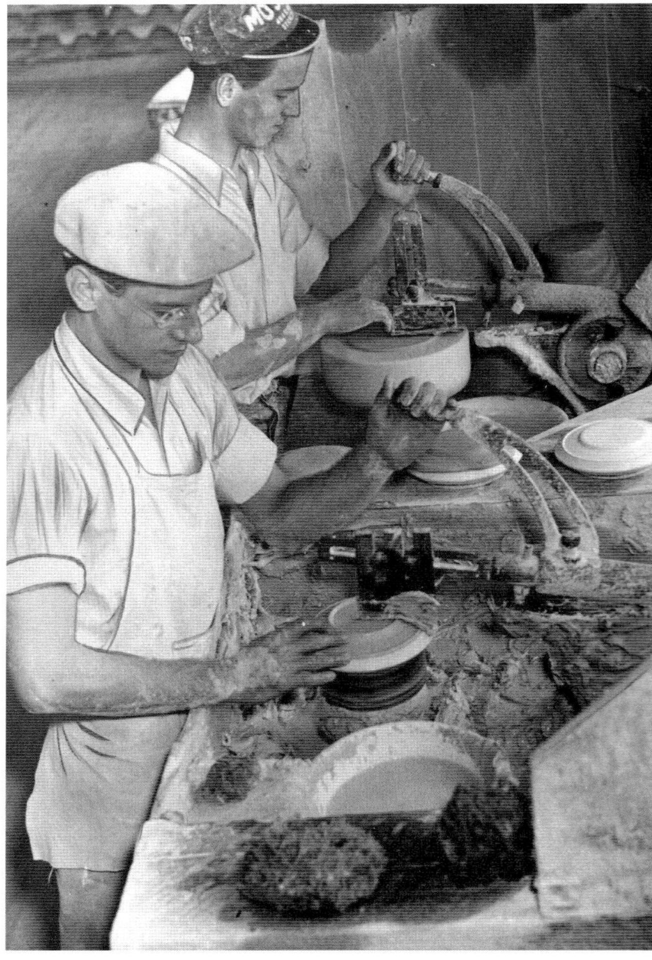

11.5. Jiggermen forming flatware in the clay shop, Fayette Street plant, 1938.

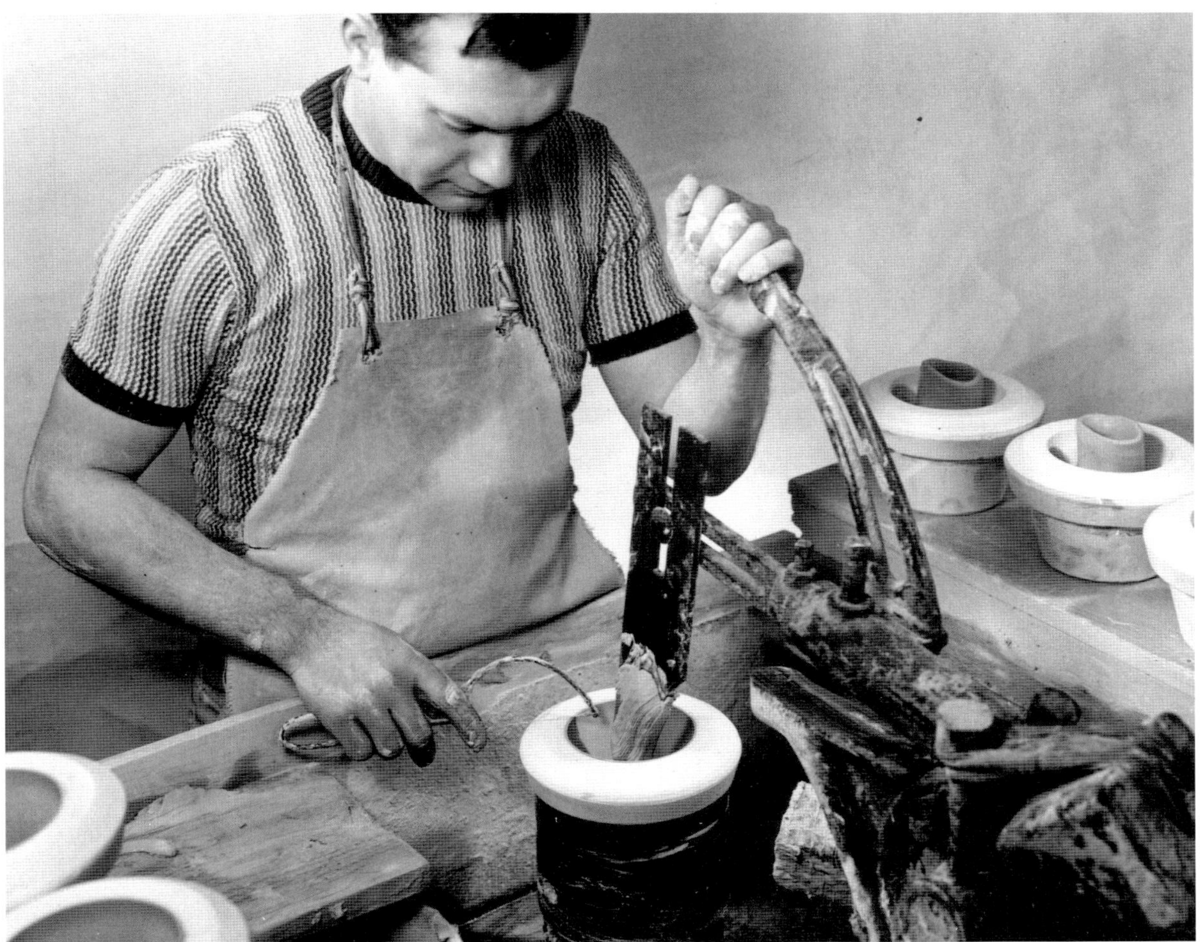

11.6. Ted Lukowski jiggering cups, Fayette Street plant, 1947. Linings are in place on the right.

clay was blown into the space between pieces to ready the bungs for firing. Calcining is the process of prefiring clay.

Beginning late in the nineteenth century, the making of ware went from a strictly hand oper-ation to a mass-production assembly line process that increasingly depended on mechaniza-tion. Nonetheless, the process always required certain hand steps. For example, to make cups, a team of workers performed the several tasks in three stages. First a two-man team consisting of a *cup maker* and his assistant, called a *runner,* worked together to form the cups. The cup maker threw cup *linings* (rough cylindrical cup forms) on a potter's wheel until he had a good supply, then moved to his jigger to shape the linings into cups. (Unlike flatware pieces, which were formed on the top of a mould, a cup was formed on the inside of its mould.) He positioned the cup mould in the jiggerhead, pressed a lining into the mould and pulled the jigger tool down to form its inside (fig. 11.6). His runner (who earned his name by being constantly on the move)

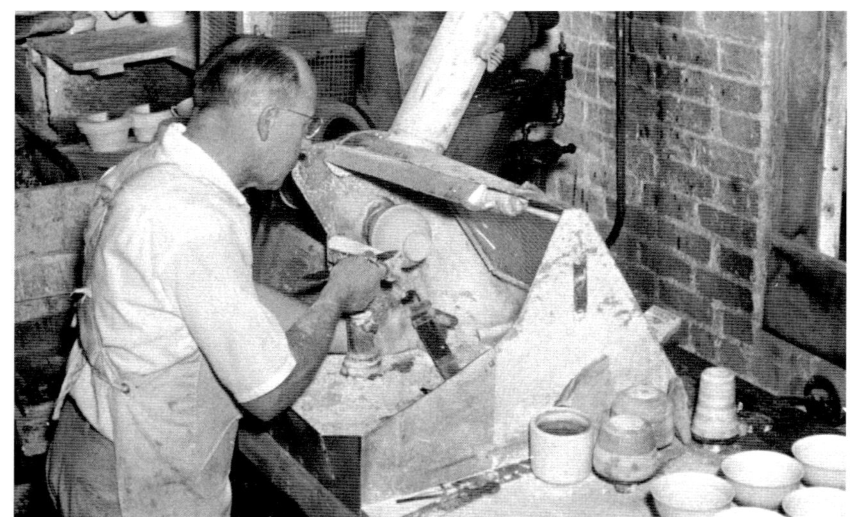

11.7. Turning the feet of cups, Fayette Street plant, 1947.

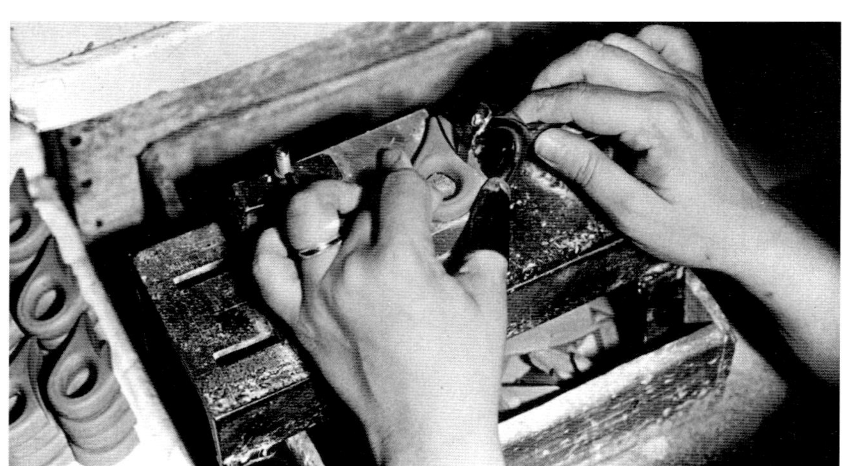

11.8. Cutting cup handles, Court Street plant, 1947.

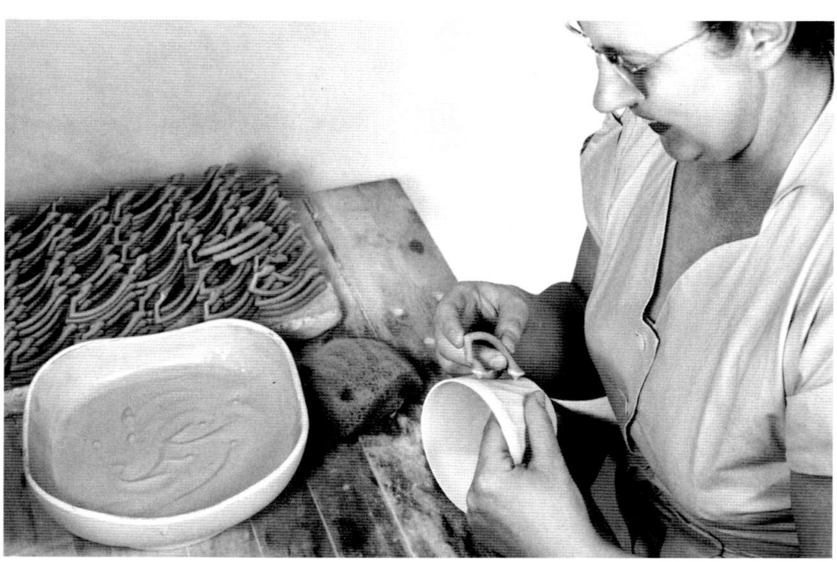

11.9. Mary Kubrak putting handles on cups, Fayette Street plant, 1947.

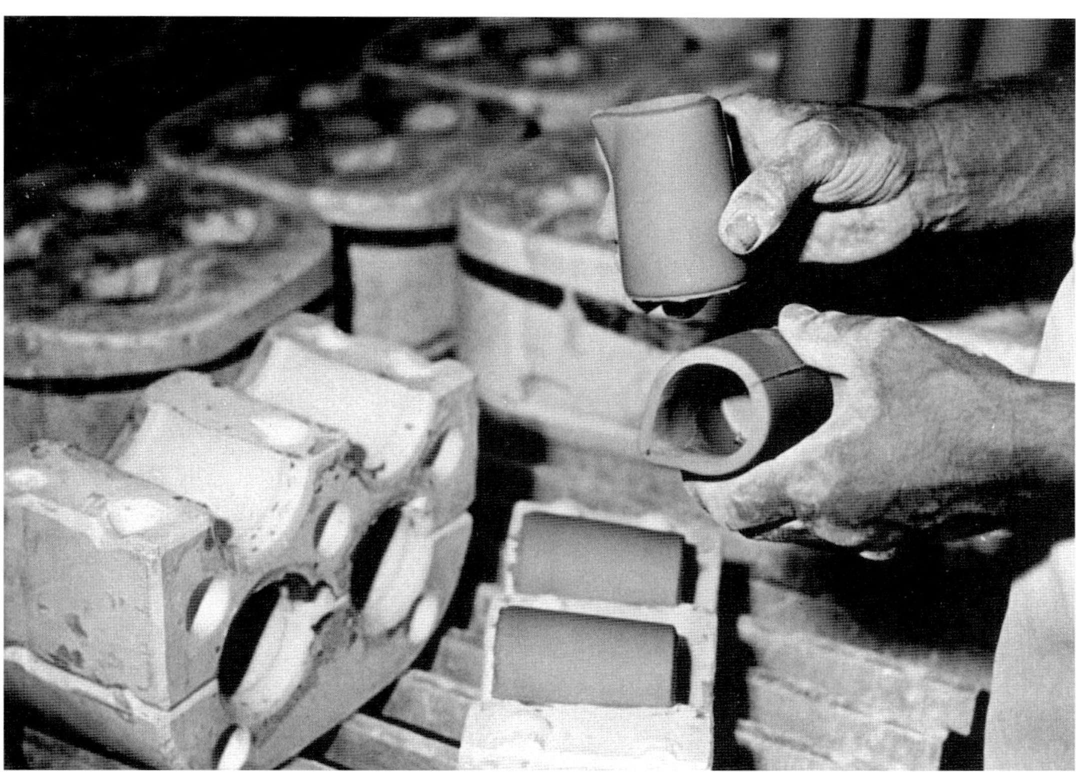

11.10. Casting **Boston** creamers, Fayette Street plant, 1947.

brought clay and moulds to the cupmaker, *snatching* first the linings and then the moulds away when the cups were complete, placing them on a *slide* or board and into the stove room to begin drying. The runner removed them from moulds when they were hard enough to release easily and placed them on boards, and removed them to *coolers* to keep them damp. A *trucker* then took cooled cups to a third worker, the *cup turner*. Before turning, cups were rough at the top and solid at the bottom. On a *cup lathe* the cup turner shaped the *foot* on the cup (fig. 11.7), trimming the excess clay left from jiggering, and rounded its rim (not all cups were turned). The turned ware was returned to the cooler to await *handling*.

In a separate operation, handles were pressed or cast in moulds, turned out, trimmed, and put in cool storage. A *handle cutter* carefully cut the handle to fit the cup (fig. 11.8). The *handler* then dipped the handle in clay slip (which acted as a glue) and attached it to the cup (fig. 11.9). A finisher sponged the cup smooth, removing any excess clay or marks left by the process before placing it on a rack to dry.

Some types of ware, such as jugs, sugars, and teapots, could not be formed on a revolving jigger, but instead were made by the casting process (fig. 11.10). A jug was made in a three-piece mould, consisting of two sides and a bottom. These parts were set together, a strap put around to hold them in place, and the slip poured in from the top. The slip adhered to the inside of the mould and gradually solidified against the side of the mould as the plaster absorbed moisture from the slip. When the required thickness was obtained, the surplus slip was poured out and the cast piece of ware remained in the mould. After the item was removed from the mould, it dried to a "rubber hard" stage on a board before final finishing and sponging smoothed out the seams on the top, bottom, and back.

Kilns

Once a piece of greenware had dried sufficiently, it was ready for its first firing in the kiln. Originally, the pottery fired its ware in brick *bottle ovens* or *beehive kilns* of English design (so-named for their shape). They rose high above the roof of the factory, creating a distinctive skyline. They were *periodic kilns* because they fired in periodic cycles of loading, firing, cooling, and unloading. Beginning in the 1930s, the pottery began to install *continuous* or *tunnel kilns* in which cars loaded with ware moved slowly on tracks through the constantly firing kiln. Tunnel kilns fired in zones, warming up gradually, reaching peak temperature, and cooling through firing cycles that imitated those in a periodic kiln. The last bottle kiln was fired in the late 1940s but was taken down a few years later. Maintenance of the kilns, placing of the ware within their ovens or on cars, firing, and unloading were each separate and distinct skills. Like the skills of the clay room, where potters formed the ware, the skills of the kiln room reflected specialized knowledge gained through long experience.

The first firing at 2250 degrees Fahrenheit, called *biscuit* firing, hardened the ware. After the biscuit firing and cooling, the ware had to run through a series of handling procedures. It might be processed immediately or put into storage to await further handling. When coated with liquid glaze it was returned to another, *glost* (glaze) kiln. This second firing at a lower temperature of 2050 degrees Fahrenheit gave the ware its protective surface of glass-like hardness. Sometimes the ware received a third firing at 1450 degrees Fahrenheit to fuse certain decorations. Any of the pottery's kilns could in principle accommodate all three types of firing: biscuit, glost, and decorating, but each kiln had distinctive properties, and each type of firing required different firing cycles.

Before the ware was fired in the kiln, it was placed in box-like fired ceramic containers known as *saggers,* which protected it from the direct action of the fire. At first, ware going through tunnel kilns was placed in saggers; later it was set open on refractory shelf structures designed for tunnel kilns. Saggers were made at the factory of clays, flint, and *grog,* previously fired saggers that had been ground up and reconstituted (fig. 11.11). They came in a variety of sizes to accommodate the diversity of the items of ware. Placing the ware in saggers or on shelves required great skill, for the greenware broke easily. Ware bunged up (piled) on *setters* was *clayed up* (packed firmly with dry clay) to further help maintain its shape and prevent crooking and sagging (fig. 11.12). Cup rims were dipped in oil and refractory clay and then *boxed* two together, rim to rim, to keep them round and to prevent them from sticking together during firing.

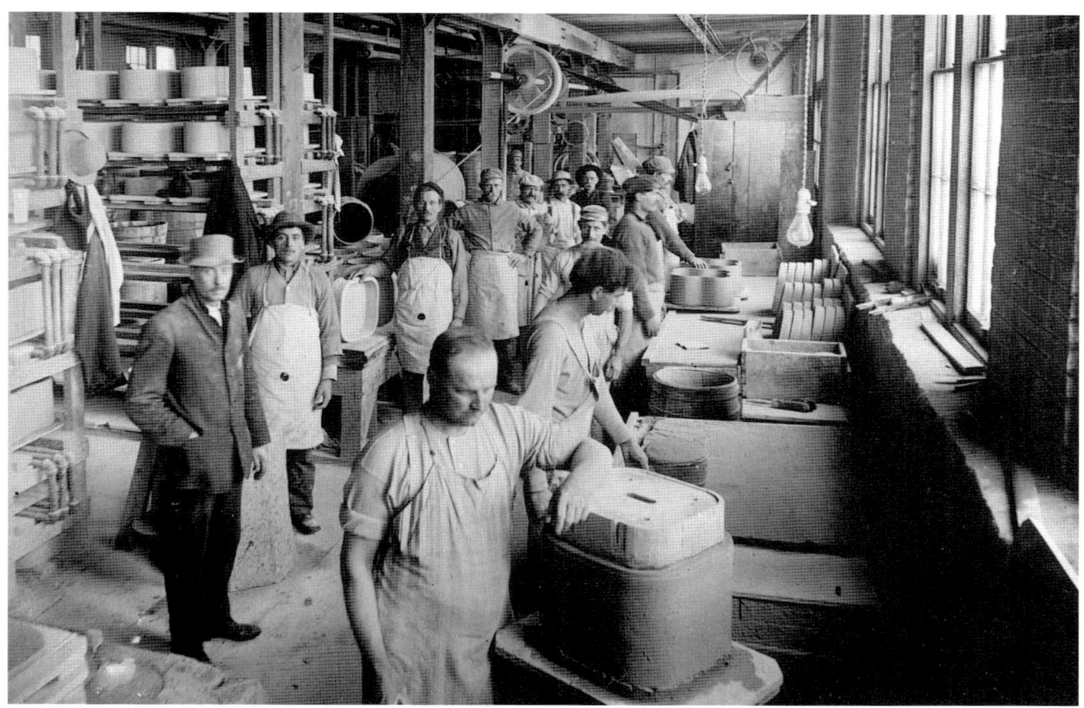

11.11. Sagger-making shop, Fayette Street plant, 1925.

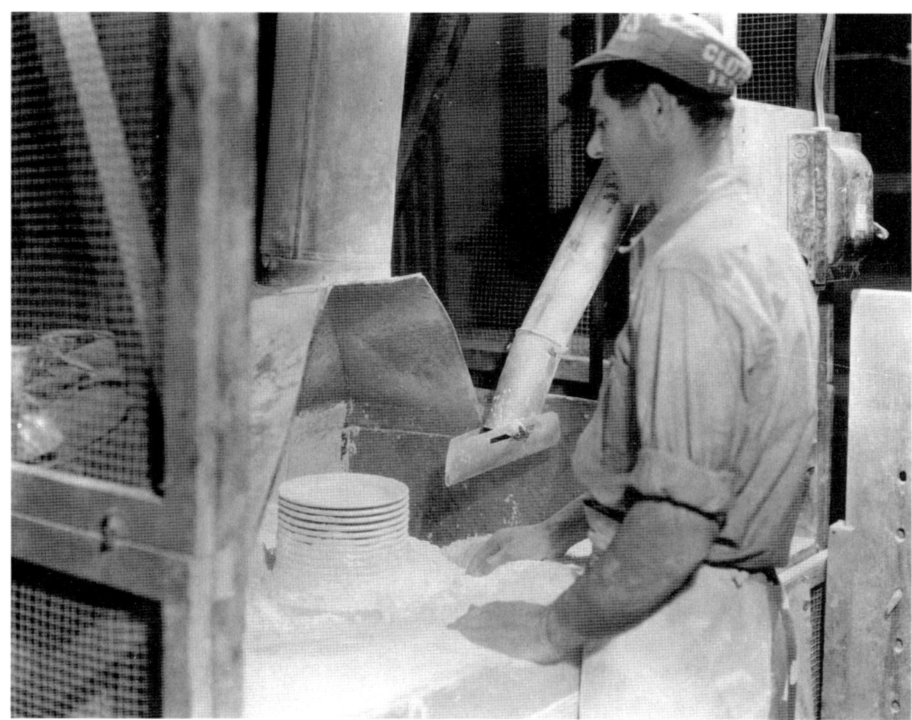

11.12. Claying up a bung
of plates for biscuit firing,
Fayette Street plant, 1947.

In periodic kilns, the *placers* carried the loaded saggers, which often weighed as much as seventy or eighty pounds, on their heads, cushioned by a roll made of ladies' stockings positioned under their hats (fig. 11.13). They stacked the saggers in *bungs* arranged in *rings* within the kiln according to the type of the ware and the "behavior" of the heat at various locations within the oven (fig 11.14). Different types of ware required different intensities of heat.

Bottle kilns were filled to capacity for greatest economy. One often held as many as 3,600 dozen pieces for biscuit firing and 3,000 dozen for glost firing. When the oven was full, the door opening was bricked-in and sealed with clay and sand. The large *firing chamber* (*oven*) had fireboxes around its perimeter that the *kilnmen* stoked night and day with coal. (Tunnel kilns were fired with gas.) The kiln temperature was raised over time through a carefully monitored succession of stages. Once the final stage was reached, the kiln was allowed to cool slowly. In tunnel kilns, the *kiln cars* loaded with ware moved through the same succession of monitered stages of heating and cooling (fig. 11.15). A typical cycle of setting a periodic kiln, firing it, cooling it, and unloading it, took about a week.

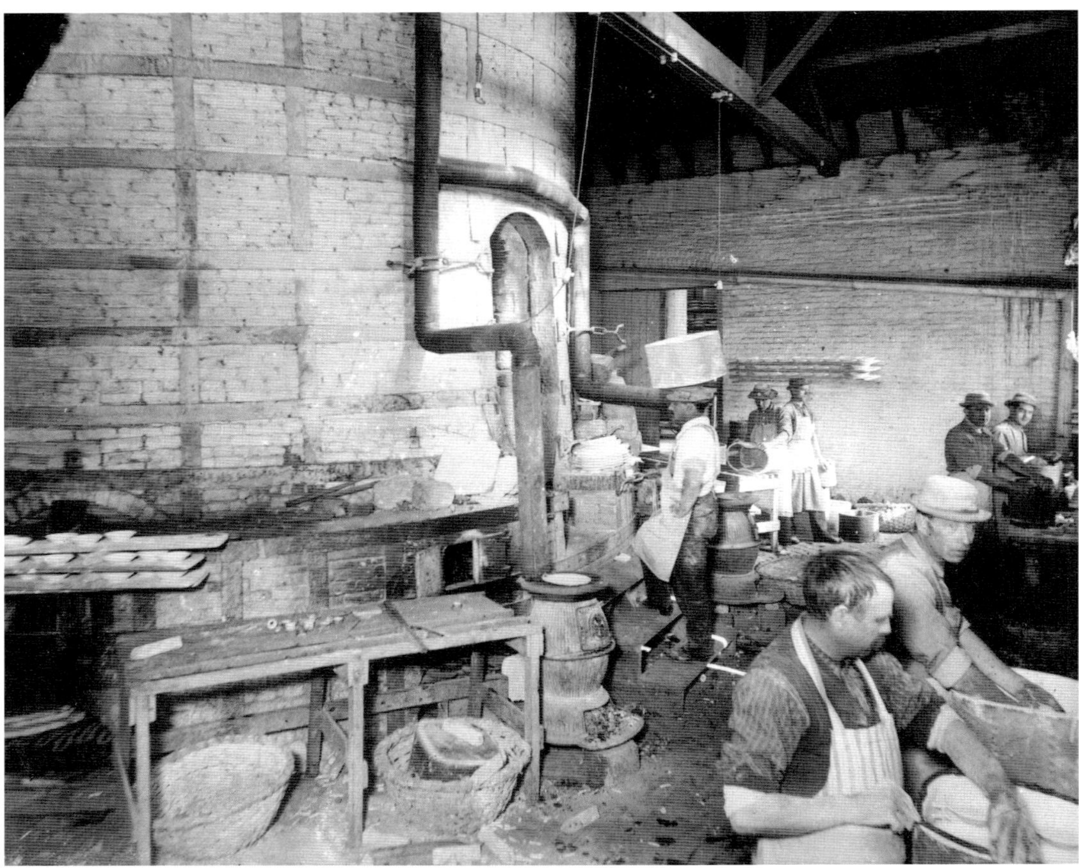

11.13. Loading a periodic glost kiln at the Fayette Street plant, 1910. Workers in the foreground are loading saggers. A placer carries a sagger on his head into the kiln.

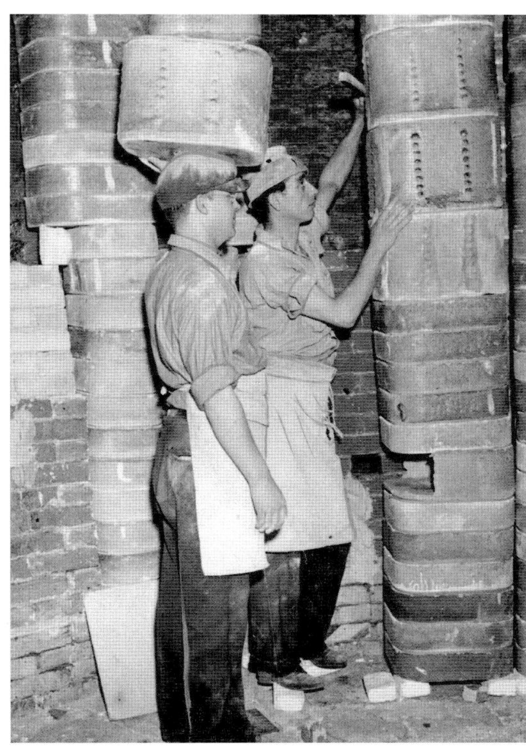

11.14. Placing saggers inside a periodic glost kiln, Fayette Street plant, 1938. Large saggers with pin marks are for flatware; smaller saggers are for cups.

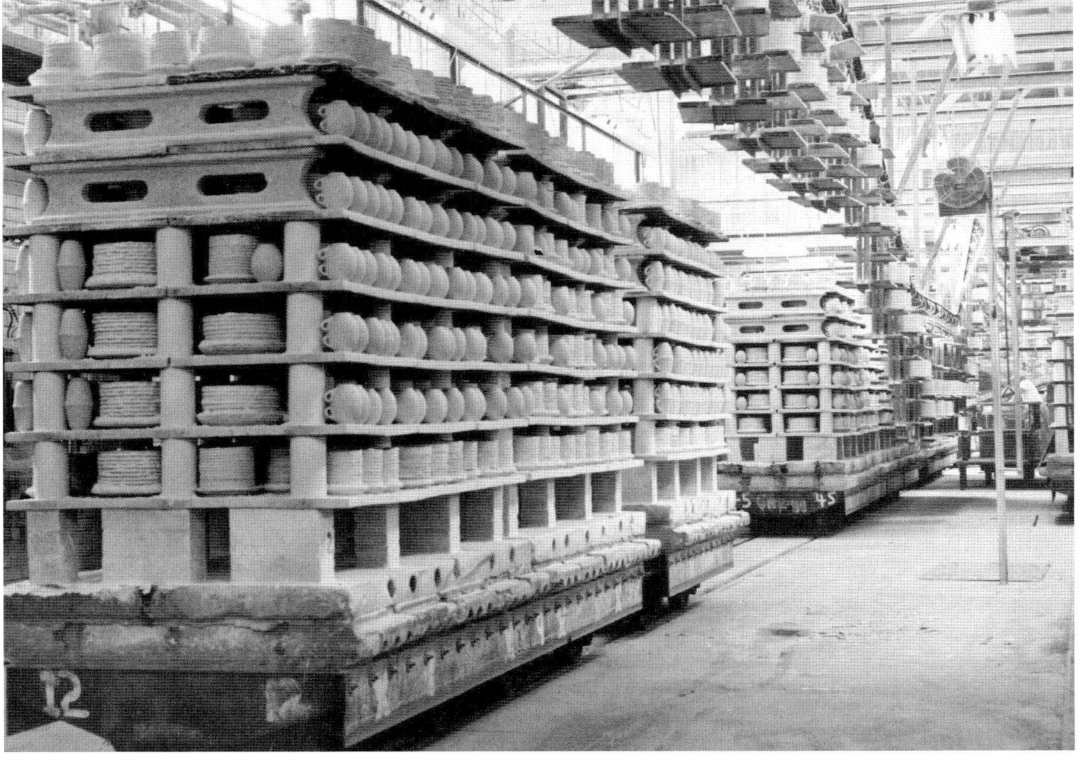

11.15. Tunnel cars loaded with biscuit ware, Court Street plant, 1963. When first introduced, tunnel kiln cars were loaded with saggers. The cars move on tracks through the kiln.

Biscuit Firing

Biscuit firing transformed the greenware into stone-like hardness. The process consisted of the *smoking-out* or *dehydration* stage, which removed any excess water that remained in the clay. The *oxidation* stage slowly burned impurities out of the clay, and was followed by the *heating-up* and *shrinkage* stages. The final stage of firing for china was the *vitrification* period, during which the components fused together to produce a ware of great compressive strength, denseness, and impermeability. Timing and temperature control were critical during all stages of firing to prevent spoiling the ware, for ill-timed or poorly controlled firing could cause blistering and cracking of the ware, black spots, gassing, crooking, or discoloration. Rapid cooling caused cracking.

Biscuit Ware

Once cooled, the biscuit ware was removed from saggers. Flatware went onto skids, hollowware into baskets, and then was moved to the *biscuit cleaning room* for cleaning and *selection*. Here the ware was sorted and faulty pieces removed. The flatware had to be scraped of the layer of excessive clay surrounding it. The ware was then carefully placed in *tumblers* so that it would not be nicked or cracked while very small pebbles or bits of broken biscuit ware were tumbled gently against the ware to act as an abrasive and give it the smooth, satiny surface necessary for the application of underglaze decorations (fig. 11.16). When this work was completed, the ware was inspected for flaws, such as crookedness, cracks, nicks, or impurities such as discoloration from iron spots. The usable ware was backstamped with the trademark and sent to the biscuit stock room (warehouse) to await decoration orders.

Before decorating, all ware was washed in hot water to remove dust particles. This was accomplished by hand until the 1940s, when washing machines came into use. Ware that was to be used for plain white orders or for overglaze decorations was taken directly to the *dipping rooms*.

The Color Lab

Technicians in the *color lab* prepared colors to decorate china. The color was made from three ingredients: *oxides*, *fluxes*, and *pacifiers*. The oxides determined what color would be produced. For instance, chromium oxide produced greens; iron oxide mixed with chromium, brown; cobalt, blue. Fluxes caused the oxides to combine and pacifiers cut the intensity of color, giving it shades. The ingredients were weighed, screened, and spread in saggers. They were fired in the biscuit kiln and emerged as hard cakes. These cakes were crushed into granular form by a jaw crusher and mixed with water. They were then put into porcelain-lined revolving *ball mills* and ground by flint pebbles for a week (fig. 11.17) into a milky *slip*. The slip was poured through a magnet into a huge kettle where the color batch was washed of all soluble material. The color residue settled to the bottom of the kettle and was dried into a solid cake, then screened into a

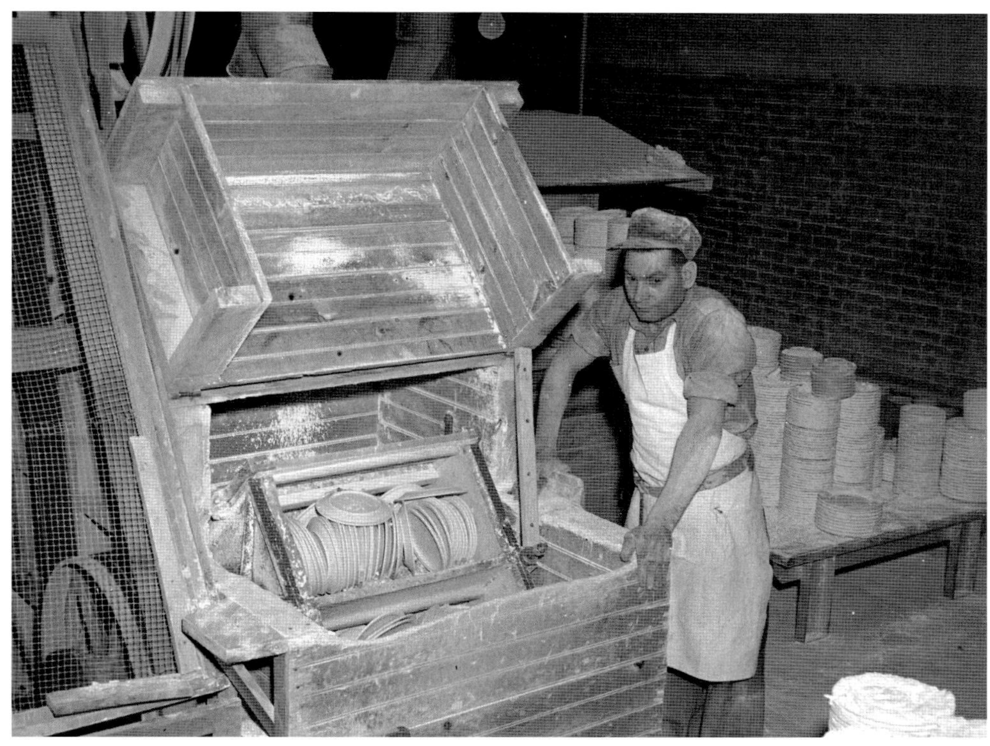

11.16. John Kutzera work-
ing at a ware-tumbling
machine, Court Street
plant, 1946.

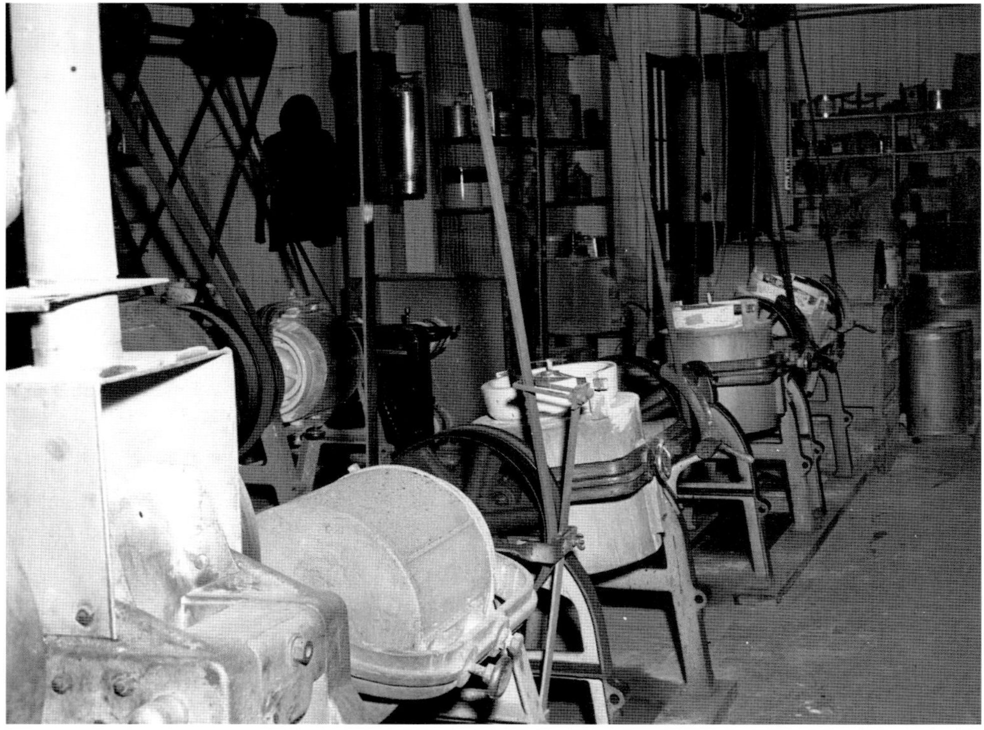

11.17. Colors are finely
ground for a week in
porcelain-lined, revolv-
ing ball mills, Fayette
Street plant, 1946.

fine powder. From this point on, the color was prepared for specific decorating jobs: it was mixed with turpentine into a thick paste to use for lining and hand-fill, ground with linseed oil into a heavy paste for printers, or ground into small mills to make suitable slips for spraying.

Decoration

There were two kinds of decorations, *formed* and *applied.* Formed decoration in relief or intaglio carved in the mould became an intregal part of the piece during biscuit firing. Applied decoration went on the ware's surface after biscuit or glost firing. Decorations of prints, decalcomania, lines, or other treatments applied to biscuit ware before it was glazed was known as *underglaze;* decoration applied to glost ware after it had been glazed and fired was *overglaze.*

From the earliest days of the pottery, decorators had many methods of applying decoration to ware. At first, all decorations were overglaze because the mineral-based colors then used were not formulated to mix with the glaze composition, nor could they withstand the high temperatures of the glost firing. The disadvantage of overglaze decoration was that, lacking the protective coating of glaze it was prone to wear away after repeated use, especially under the pressure of utensils. A major innovation after the turn of the twentieth century allowed many types of decoration to be applied underglaze. Research into mineral-based ceramic colors ultimately produced a palette compatible with the glaze composition that allowed many colors to withstand the intense heat of the glost firing.

Lining and *hand-fill* were two of the oldest types of applied decoration. Originally overglaze, they were later used either over or under the glaze depending on the material used. (Gold, plat-

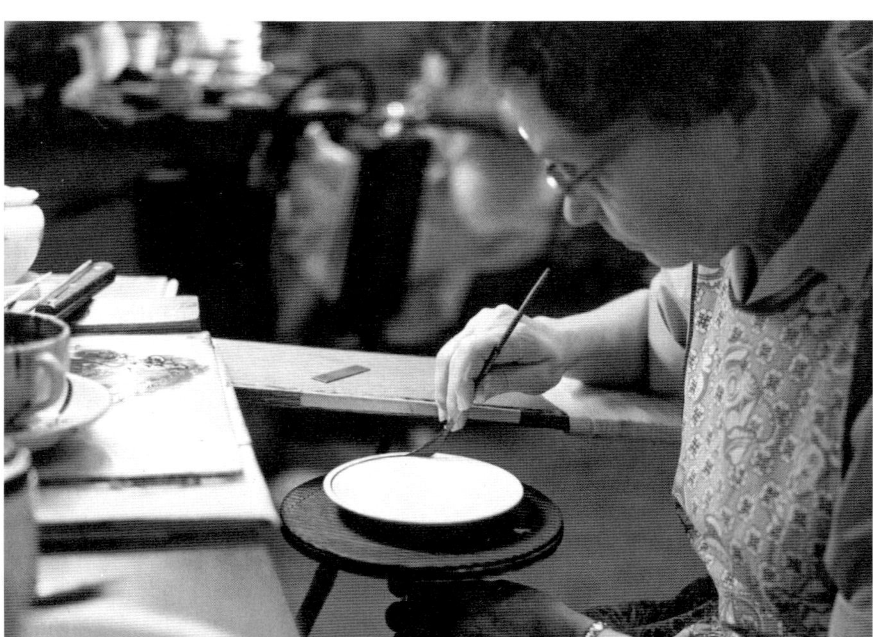

11.18. Mary Sears lining a plate, Fayette Street plant, 1946.

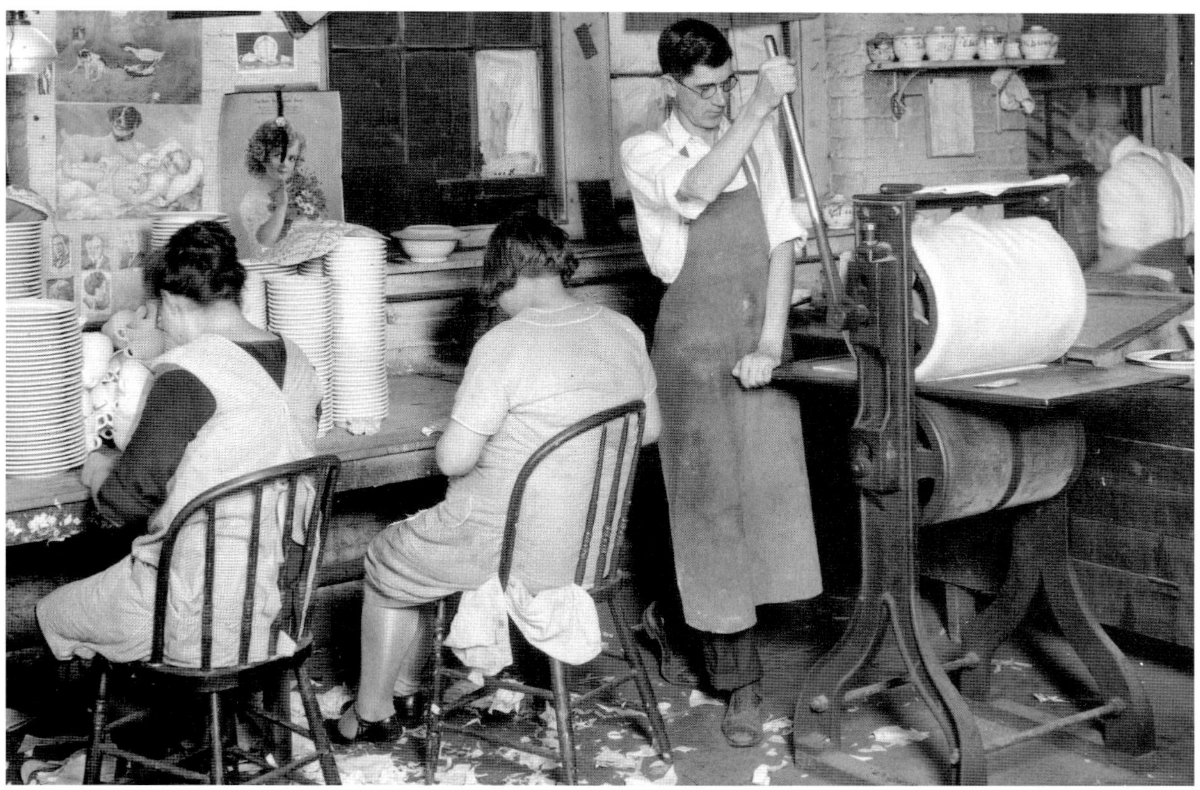

11.19. Two women decorators apply transfer prints while pressman prints on a
hand-operated flat-bed press, Fayette Street plant, 1938.

inum, and certain delicate colors have always required overglaze applications.) Lining was done
on a *liner's wheel* mounted on a bearing; with one hand the *liner* turned the wheel, while holding the brush against the piece of ware with the other hand (fig. 11.18). The liner had to center
the ware perfectly on the wheel for flawless application and needed keen hand-eye coordination. Liners also did freehand work on handles or irregularly shaped ware. *Hand-fill*, or *hand-
painted* decoration, was freehand work, sometimes guided by printed outlines (or by charcoal
dusting of stenciled outlines) that burned off during firing. Many liners and hand-fillers were
women.

The English *transfer printing* process was an important early type of applied overglaze decoration suitable for large quantities of ware. In this process, a prepared design was drawn and
engraved on a flat copper plate, usually about twenty inches square. The plate typically contained several versions of the design, sized to fit different items. The printer first inked the plate
with color (ceramic pigment mixed with oil), wiped the surface clean, leaving the color only in
the engraved lines, and printed it on a hand press (fig. 11.19) on dampened tissue paper (rice
paper imported from England).[2]

After printing, and before the color dried, a team of three women transferred the design to the ware. The first woman cut the printed pattern for a specific item from the tissue; the second placed the print color-side down on a piece of ware and rubbed it gently but firmly with a short bristle brush to ensure that the color adhered to the ware (fig. 11.20); the third placed the ware in water, floating the tissue off but leaving the oil-based printed pigment on the ware as an exact reproduction of the copper engraving. Decorators then often hand-filled the printed outlines with ceramic pigments. Finally, the ware was fired in a decorating, or *hardening-on,* kiln at 1450 degrees Fahrenheit.

The limitation of the transfer printing process was that it allowed the printed application of only one color (except for a short time in the early twentieth-century when a two-color transfer print called *duo-tone* was used); other colors had to be applied by hand. The arrival of the French *decalcomania* process in the 1890s enabled decorators to apply a multicolored overglaze *decal* produced lithographically.

Decalcomania required a large commitment of factory space for presses, storage (of limestone printing blocks, colors, and papers), and the large sinks needed to "grain" the stones (to prepare their surfaces with fine sand in water), as well as printing rooms free of dust and extremes of temperature and humidity.

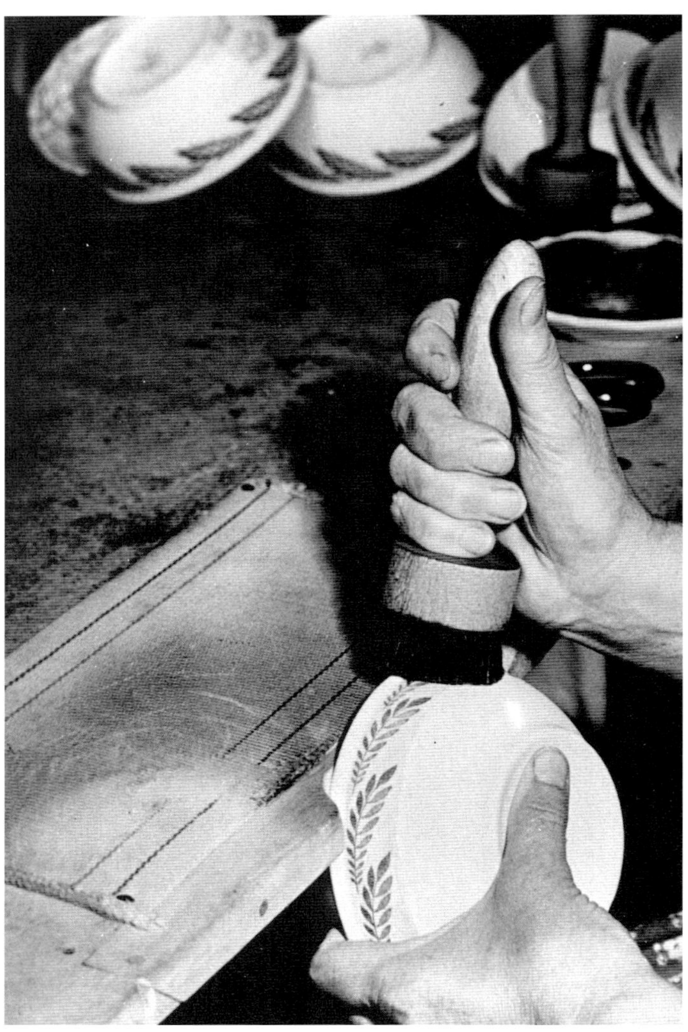

11.20. Rubbing "Willow," a transfer print decoration, with a stiff brush, Fayette Street plant, 1938.

Each color in a decal required a drawing with an oil-based ink or crayon on a separate stone. Each stone was inked with a single oil-based ceramic color which adhered only to the drawing. The pressman printed the stones in a lithography press in succession on a piece of transparent film bonded to a specially prepared paper with a water soluble glue.[3] Each stone needed to be *registered* (printed in exact alignment) to the other colors. Each color needed to dry before the next was added. The result was a multiple-color image of high quality. A single printed sheet of decals held patterns for several items. In applying them, a decorator cut the appropriate design from the sheet, wet it, and applied it to a piece of ware that had been prepared with a size to receive it. She rubbed it onto the ware to ensure the proper transfer of the decal ware (fig. 11.21) and then floated off the backing paper. This

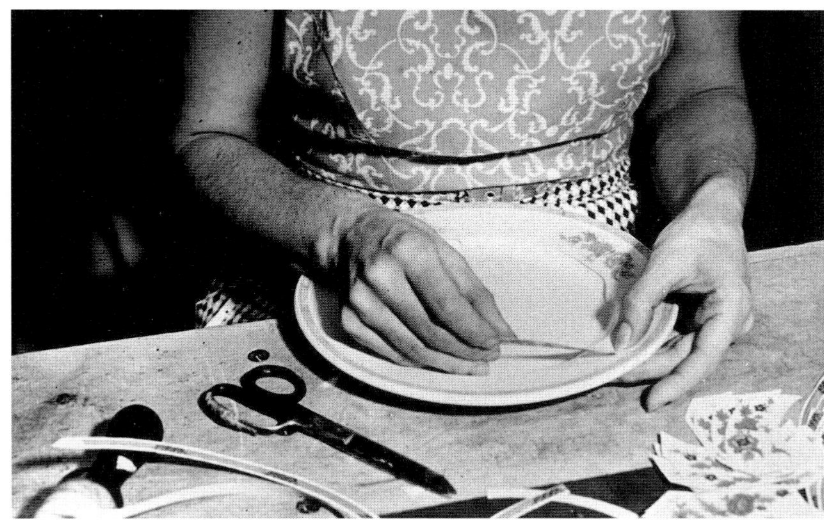

11.21. Applying decals to a plate, Fayette Street plant, 1938.

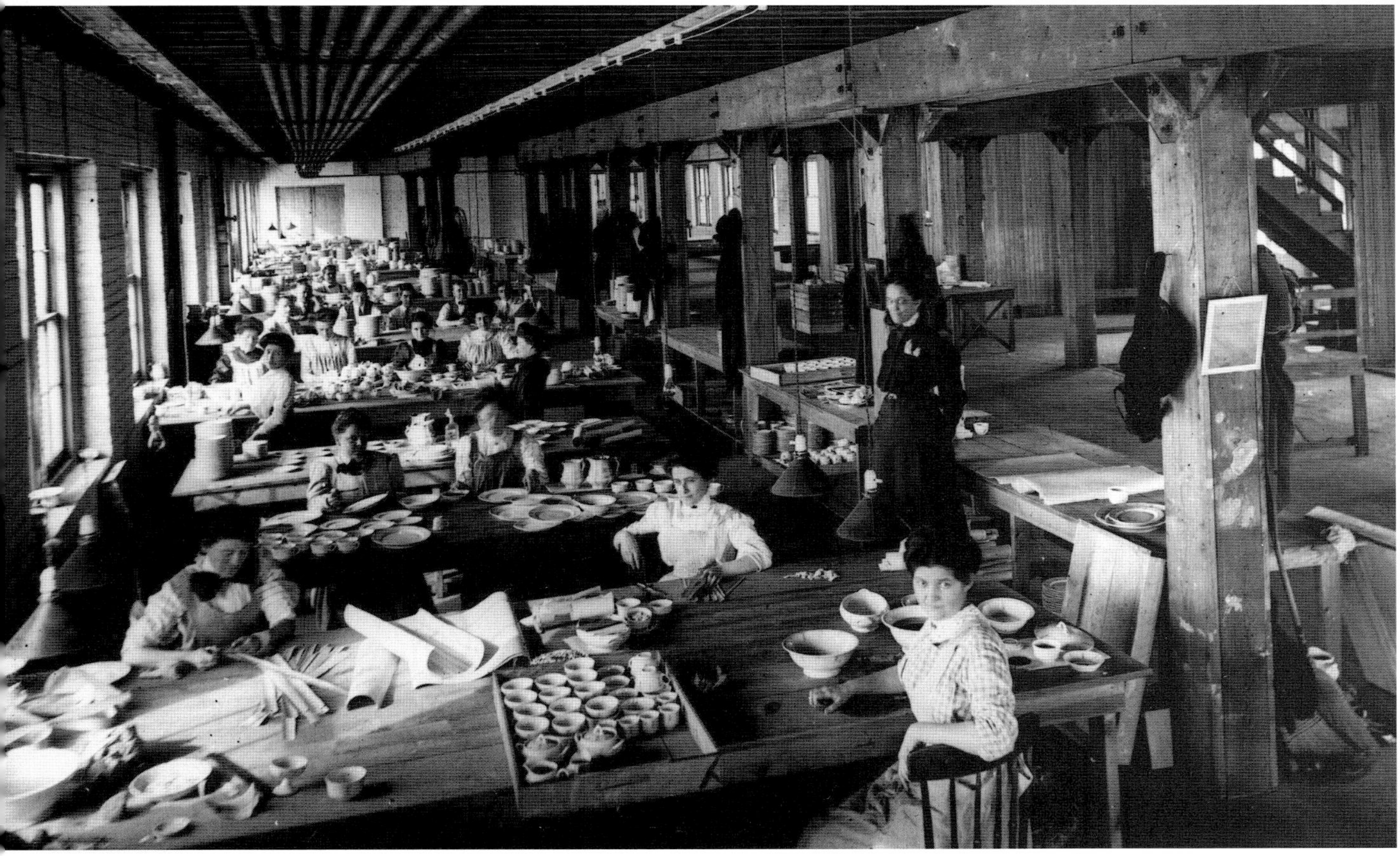

11.22. The decalcomania department at the Fayette Street plant, 1910.

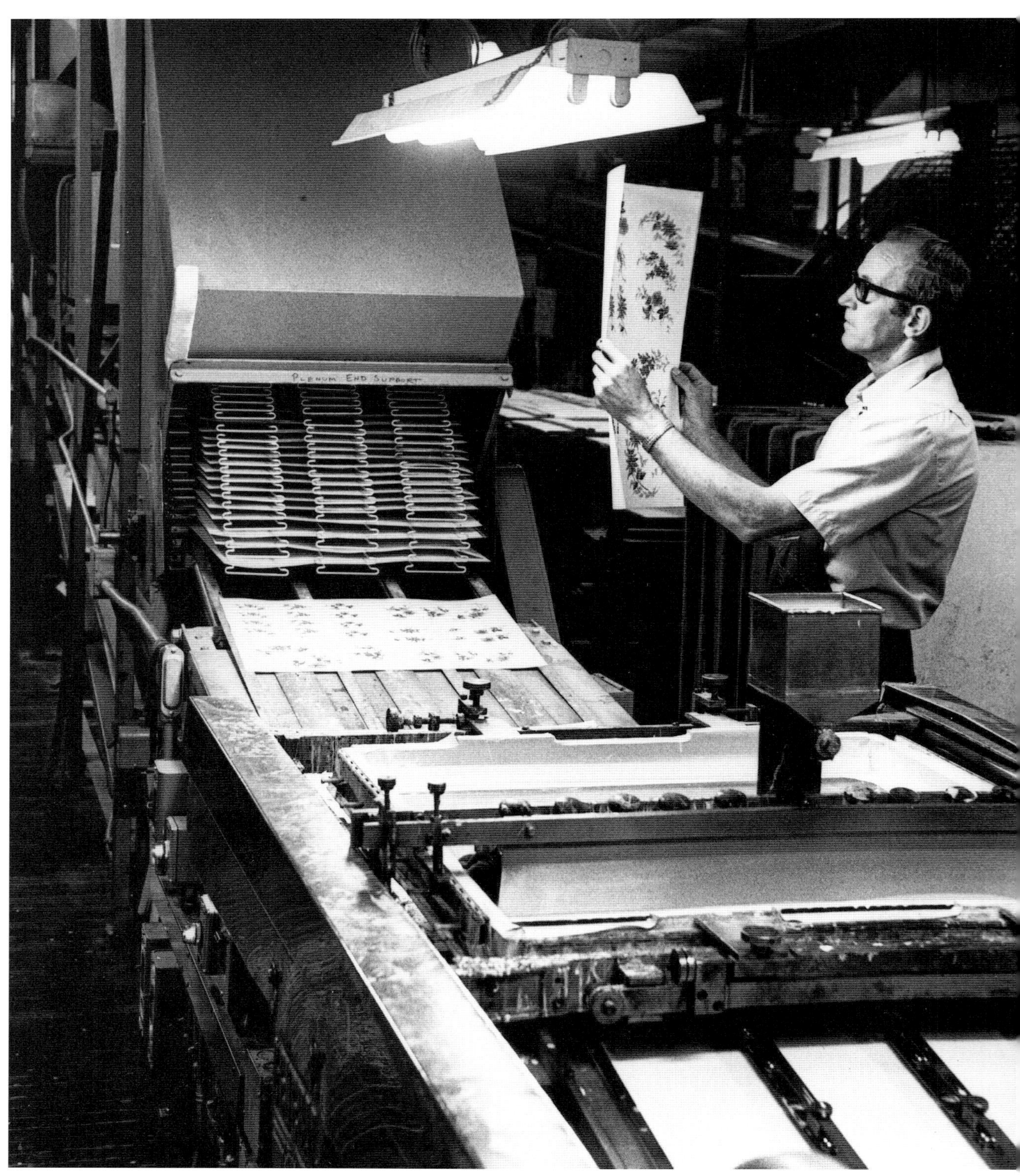

11.23. Jim Warren inspects a sheet of decalcomania printed on a modern press. Court Street litho plant, 1960s.

left the film and printed design in place and ready for firing in a hardening-on kiln. It took many decorators to transfer print ware and they were usually paid on a piece-work basis (fig. 11.22). After World War II silk screen printing replaced stone lithography (fig. 11.23).

Underglaze decalcomania made decorations practically indestructible. Other developments in underglaze decoration included the sprayed *Vitritone, Syratone, Artint,* and *Shadowtone* processes. Vitritone involved an application of color to greenware before the biscuit firing. Syratone, a process developed at Onondaga Pottery, permitted a uniform application of sprayed color in bands on the rim or as all-over color on biscuit ware. Artint was a sprayed band of color on the intaglio pattern of **Econo-rim** ware. Shadowtone was color sprayed through and around stencils, plus delicate hand-shading. Once underglaze decorations had been applied, those decorations that required hardening-on went to the hardening-on kiln, then all ware was sent to the dipping rooms for glazing.

Coin gold decorations were always overglaze decorations. They included lining and hand-fill, stippling and raised paste, as well as *encrusted gold* decorations, and were sometimes used with some type of crest design, mostly for household ware, special services, or commemoratives.

Encrusted gold decorations required an exacting and highly skilled application. First, an acid-resisting compound (a type of wax used in place of printing color) was spread on an etched steel plate, from which the pattern was printed in a press on prepared paper. The pattern was transferred to the ware from the paper, carefully rubbed down, and the paper floated off. In the next step, a wide band of *asphalt lining* (or red wax) was applied to the ware on all areas of the ware (except on the transferred design) until the entire piece was covered. The ware with its protective coating of wax was submerged in acid, which *etched,* or ate away, the unprotected pattern. Once this was completed the ware was washed in cold water to remove the acid and then in hot water to remove the wax.

Now the ware was ready to receive the first of two applications of gold (to achieve the proper thickness), each followed by a firing in the hardening-on kiln. After the second firing, the gold appeared brown and lusterless. To restore the luster, the gold surfaces were *burnished,* or polished, by machine or by hand. In either case great care had to be taken to prevent the scratching of the surface of the piece. Finally, the burnished ware was wrapped in color-coded paper and sent to the stockroom.

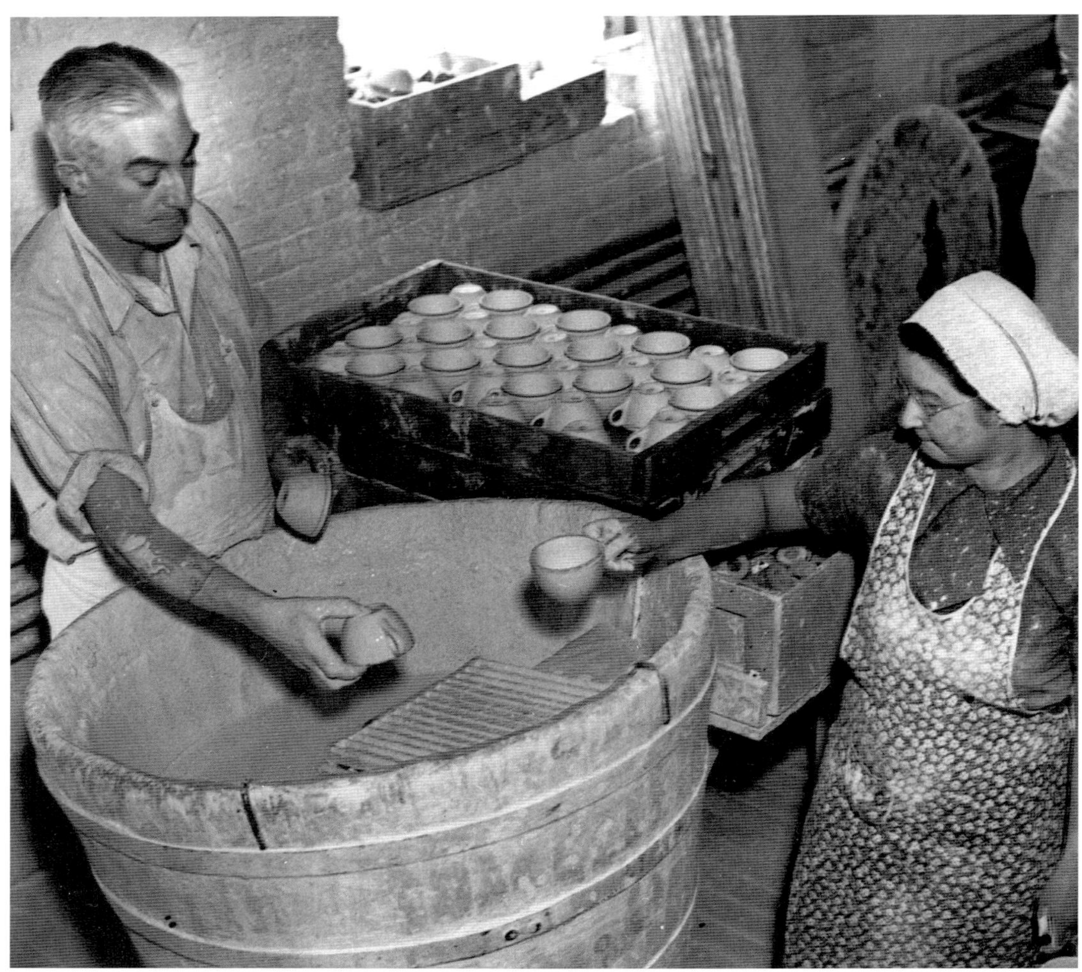

11.24. Dipping cups in liquid glaze, Fayette Street plant, 1938.

Glaze

When fired onto ware, glaze gave a protective surface of glass-like hardness. Glaze formulas contained silica (glass-forming), flux (melting), and inerts (strengthening). Glazes were formulated and compounded in the *color lab*. The wet glaze mix was placed in a closed jar filled with smooth pebbles, called a *ball mill*, and milled by turning for hours until it became smooth and creamy. The liquid glaze was poured into large wooden vats into which the ware was immersed by *dipping* (fig. 11.24). Successful dipping required experienced *dippers* who knew the properties of each glaze and how each type of ware and underglaze decoration needed to be treated. For pieces larger than a hand could hold, the dipper used a special hook attached to his thumb to extend his reach. After the ware was dipped, it was ready for the next step: placing in the glost kiln.

Glost Firing

Like biscuit placing, glost placing in a periodic kiln had two stages: arranging the ware in saggers and placing the saggers in the kiln. Glost saggers were coated with a glaze wash on the inside to prevent the absorption of glaze from the ware. A layer of *bitstone,* finely ground calcium flint chips, was placed inside the saggers. This allowed cups and hollowware to be placed on their feet without fusing to the sagger during firing.

The next step in glost placing was *pinning up* the saggers with three-sided pins made of fired clay (purchased refractory *kiln furniture*). These were placed in triangular holes that had been punched in the inside walls of the saggers when they were made. Each plate rested on three of these pins. Smaller flatware rested on two pins supported by the sides of the sagger and a third pin held by the *star* (fig. 11.25), a cylinder in the center of the sagger used to support pins (fig. 11.26). Great skill was required to place as many pieces as possible without allowing any of them to touch others or the sides of the sagger. In later years, in tunnel kilns, glazed ware was placed on *stilts* (kiln furniture) and fired on *racks* of refractory shelves.

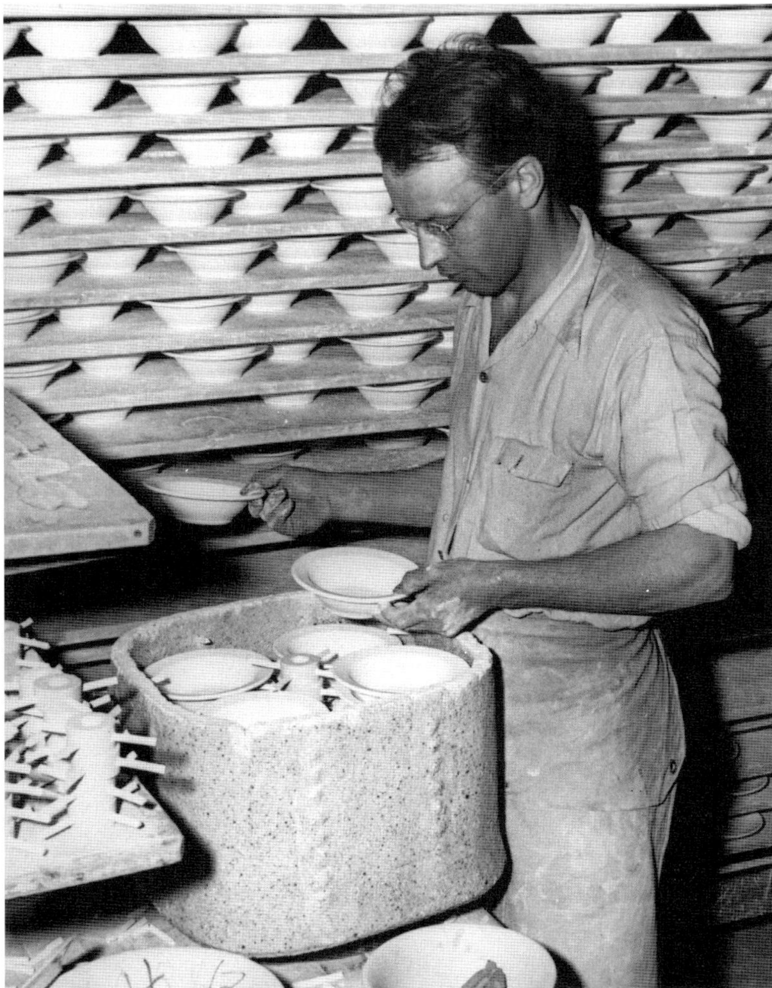

11.25. Arnie Crook places ware in saggers using "star" plate supports for glost firing, Fayette Street plant, 1946.

Once the saggers filled with glost were placed in the kiln and sealed with clay wadding, the kiln was ready for firing. The kiln atmosphere had to be kept free from dust, for it could cause blemishes on the glazed surfaces. Undesirable gases in the atmosphere could lead to dull, scummy glazes, changes in colors, or even the disappearance of the glaze. In extreme cases, the glaze could blacken or bubble. After the glost kiln had been fired and cooled, the glazed ware was carefully removed from the saggers and transported from the kiln room to the glost wareroom.

Glost Ware

In the glost wareroom, the *kiln sorter* separated the ware into types and distributed it to the grinders. First the *back grinders* used grinding wheels to remove the marks left on the backs of flatware by pins that had supported it in the saggers. A *back polisher* then gave the area of the

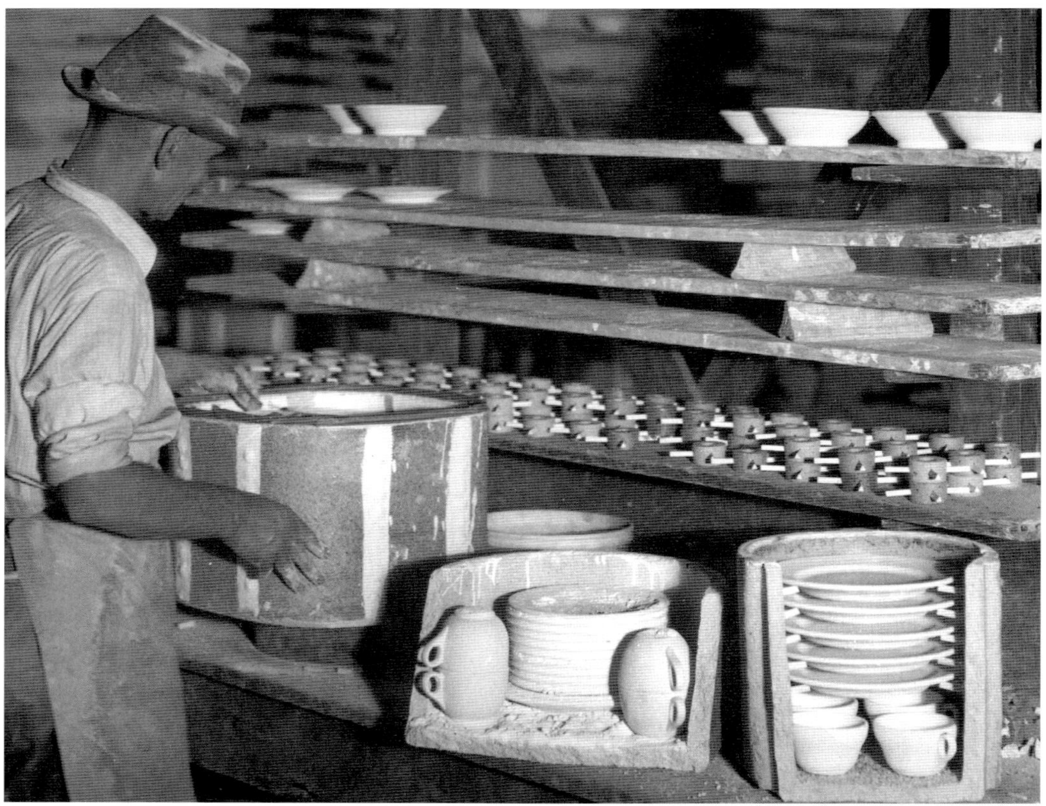

11.26. A worker demonstrates different approaches to placing ware in saggers for biscuit and glost firings. Biscuit-ware cups stacked rim to rim and clayed-up plates share the same sagger for economic use of space. In the glost sagger (*right*), cups are separated from one another, and plates are pinned so they will not touch.

pin mark on finer ware a lustrous polish. Next the ware was selected according to three classifi-cations. *Firsts* went directly to the stockroom. *Grinders* were sent to *face grinders* who removed surface defects with a grinding wheel and a buffing wheel. *Seconds* and *thirds* went to the re-claim room where they were sorted further to distinguish ware that could be reprocessed or sold as thirds; unsalvageable ware was destroyed.

Packing and Shipping

Once production was complete, all ware was assembled in stock bins. *Packers* were responsi-ble for ensuring the safety of the ware in transit to the customer. From the stock bins, the ware was counted out piece by piece to fit the specifications of orders. After each item was counted out and transported to the packing room, packers used a content sheet to determine the size and number of barrels or (later) cartons needed. In barrel packing, all shapes and sizes of china were fit into a barrel container with oat straw used as filler and padding (fig. 11.27). The ware was first *strawed-up,* stacked and nested between layers of straw, and then it was placed in

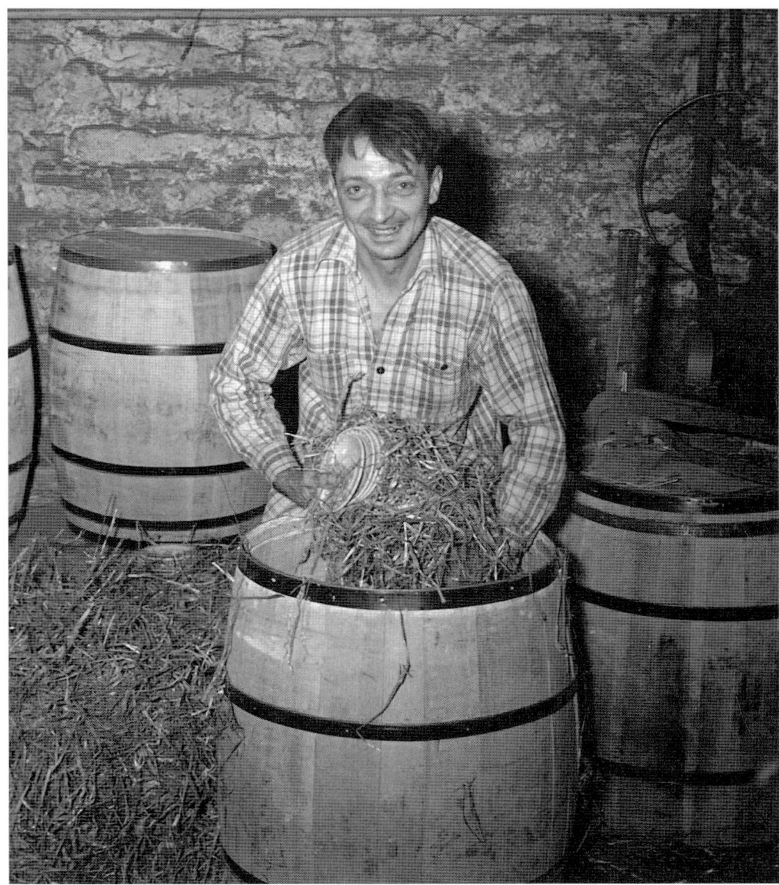

11.27. Vic Gilette packing ware into a barrel, Fayette Street plant, 1950s.

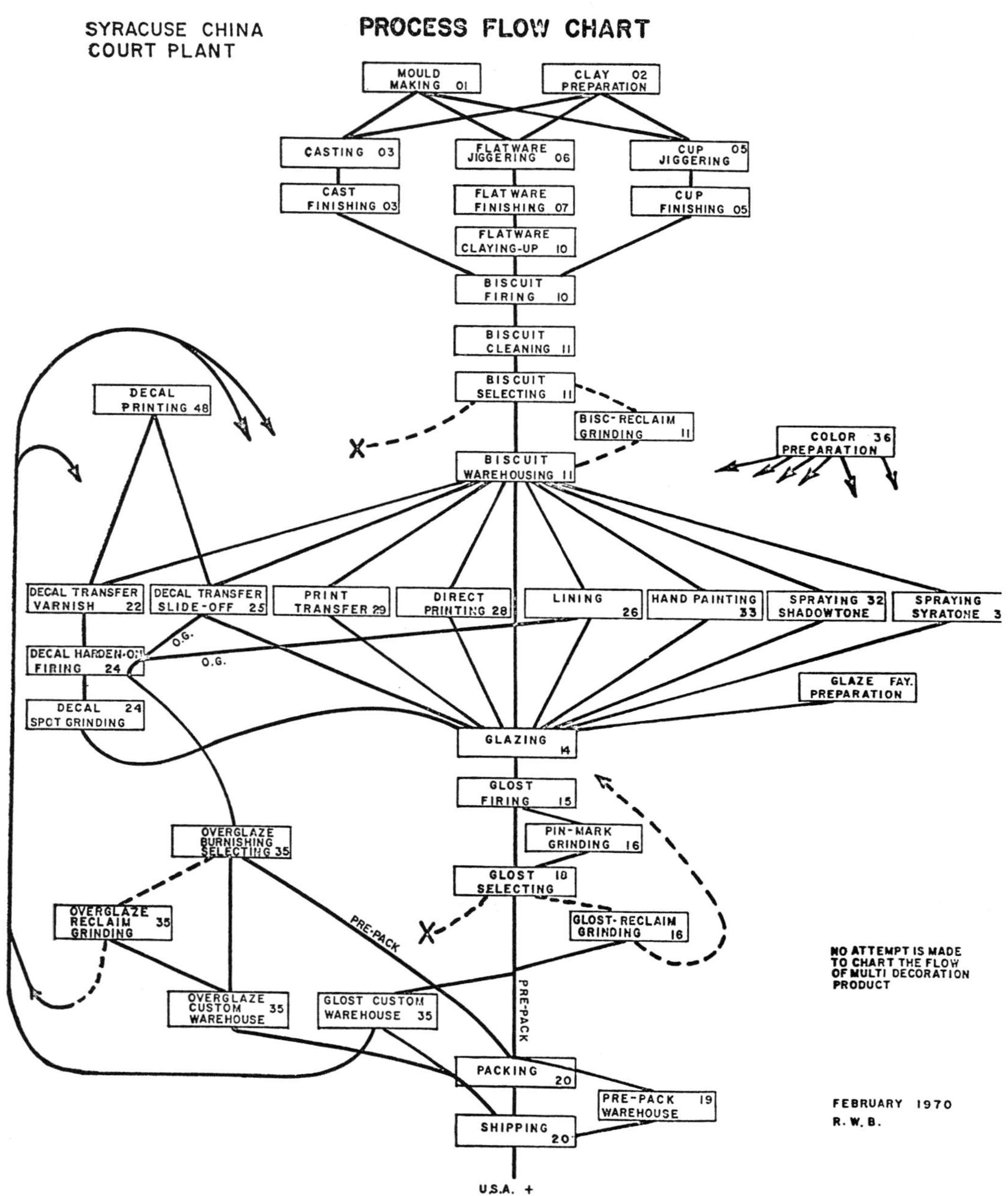

11.28. This chart shows production flow at the Court Street plant in 1970.

the barrel in ringed layers separated by straw. Cartons, used universally after the late 1950s, contained corregated cardboard liners and separators to cushion the ware. As soon as ware was packed and weighed, the order went to the shipping clerk, who determined the best method of shipment to the customer.

Engineering, Power, and Maintenance

The engineering department worked behind the scenes through all the stages, providing power in the boilerhouse for heat and electric generators, building and maintaining machinery, making capital improvements, and protecting the plant from fire. The maintenance departments serviced all parts of the factory as electricians, carpenters, yard crews, maintenance, and watchmen. All this was required to produce the handsome American china that became so prominent a part of the material culture of American dining.

Research and Development

From the days of James Pass, the pottery was established as a leader in the progressive development of the ceramic industry. After 1913, university-trained *ceramic engineers* carried on his work. The company took great pride in its completely equipped *laboratory,* built in 1923 across from the Fayette Street facility. In this building, they conducted experiments and tested raw materials. They developed new clay bodies and glazes and monitored the color lab. Gradually, an engineering heirarchy developed, headed by the manager of ceramic engineering, a specialized field that combines mechanical and chemical engineering as applied to ceramic manufacture. Most of the ceramic engineers who worked at Syracuse China were educated at Alfred University. They constantly tested the ware to maintain and improve its quality.

My Pottery Heritage

Ruth Pass Hancock

Generations of pottery families molded the history of the Syracuse China Company. This is the story of one of them, the Pass family, my family.

My great-grandfather, Richard Pass, was a Staffordshire potter. He was born in October 1824 in Burslem, "mother" town of the six adjacent Staffordshire towns long known as "The Potteries." Blessed with the Midlands' ample supplies of coal and clay, potbanks had popped up along every hedgerow; bottle kilns burned in every backyard. Generations of potters had made their livelihoods forming and firing clay. A true cottage industry, pottery-making and its related specialities consumed the community. With the success of Josiah Wedgwood's Queensware in 1765, England's pottery-making was elevated to a fine art and the industry grew. I don't know how many potbanks existed in Burslem in 1824, but I do know that there were thirty-eight pubs and ninety beer houses.

Pottery families intermarried and passed on their skills, their knowledge, and their lore. Family tradition holds that Richard's future wife, Ann Greatbach, born in 1821, was carried in a box by her mother to the potbank each day where, close by the work bench, the baby slept lulled by her mother's voice. This cradle song has been passed down through the generations:

> Hi chee-up for Burslem!
> Hi for Hanley Green!
> Two cups a penny,
> The like was never seen.
> Chee-up my little boxer.
> Chee-up my little boxer.

Ann came from a family of engravers in the pottery trade. Richard and Ann were married in 1848 in the Methodist Church. Their daughter, Sarah Ann, was born in 1850 and their son, James, in 1856.

Among these early potters developed a rebellious spirit of free-thinking independence, and a degree of humanism that turned them away from the established Church of England and drew them to the teachings of John Wesley. The Burslem Methodist Church was one of Wesley's first missions. He converted many potters.

In 1844, when Richard was twenty years old, he too was converted and joined the Methodists. His obituary in 1880 stated, "soon after his conversion he was moved by the Holy Ghost to preach the gospel . . . and [was] appointed to regular work on the Burslem circuit for sixteen years performing his ministerial work with great acceptability and usefulness and at the same time supporting himself and his family by his labor as a potter."

In the early 1860s, something prompted Richard to leave his native land for America. Was it a calling to spread the mission of Methodism or, perhaps, changing conditions in England's pottery industry? No one knows. But Richard could see opportunities abroad. He sold his interest in a Burslem pottery, packed up his family, and booked passage to New York on the steamship *Great Eastern.*[1] Years later, a worn piece of paper with a formula for burnished gold written in Richard's hand was found secreted in a pocket of his son James's diary. Perhaps this was to be Richard Pass's "open sesame" to successful pottery-making in the New World.

The *Great Eastern* was the talk of its time, an iron behemoth, 693 feet long, 120 feet wide, with 58-foot side paddles, five stacks, and five masts carrying 6500 yards of sail. She gained fame laying the Atlantic Cable in 1866 before meeting an ignominious end in 1889, ravaged by souvenir hunters and scrapdealers. When the Pass family crossed the Atlantic in 1863, the finely appointed *Great Eastern* was far from being a typical immigrant ship.

The family went directly to Cincinnati, then the fifth largest city in the United States, where, five days later, they found themselves on the front line of the "rebellion." Still a subject of Queen Victoria, Richard was trapped in a war that did not concern him and which went against his Methodist principles.

It is generally believed that in Cincinnati Richard was associated with the Tempest, Brockmann Pottery, newly reorganized in 1862. Although the city's directories in 1859 listed twelve potteries, by 1863 the number had shrunk to seven, and by 1864 to five, of which Tempest, Brockmann was the most significant. It was not a prosperous or promising time for Cincinnati potters. Nonmilitary shipping had been suspended on the Ohio River. I do not know how long the Passes remained in Ohio, but by 26 October 1865 Richard was in Trenton, New Jersey. On that date he renounced his allegiance to Queen Victoria in the Inferior Court of Common Pleas, Mercer County, New Jersey, and stated his intention to become an American. Three years later he gained citizenship in the United States.

At the close of the Civil War, Trenton was the industrial pottery center of the nation, though most of its operations were small by modern standards. Over the next ten years, Richard worked as foreman or superintendent at several of the potteries as they changed hands, changed names, merged, and regrouped. He may have worked at Cook Pottery, at Greenwood, and at Ott & Brewer. A Brewer Pottery also existed in Cincinnati; Richard may have taken advantage of this connection. Wherever he worked, he established a reputation as a capable potter who was also an able superintendent. This reputation reached Syracuse.

The four-year-old Onondaga Pottery Company was still struggling when, on 5 June 1875, its board, "on motion of Mr. Van Dusen . . . [resolved] that the president is hereby authorized to engage the services of Richard Pass or some other suitable person as superintendent at such salary and in such conditions as he may deem best to the interests of the company." Richard accepted the challenge, moved to the village of Geddes, and started work at a salary of $25 a week commencing in June 1875. The company had one large kiln and a workforce of forty-five persons.

By 1877 the company was showing a profit. Three years later, in June 1880, Richard Pass died. He was fifty-six years old. His obituary indicates that "for a few years his feeble constitution would not allow him to do much preaching, though he continued to serve as a steward, class leader, and Sunday School teacher." His energies had been directed towards assuring the future of the Onondaga Pottery Company.

There are no family stories about my great-grandfather. I look at his photograph, at the thin, worn, forbidding face and see more the upright bearing of a Methodist preacher than the sensitivity of a man who created things of clay with his hands (fig. 12.1). His wife's face also reveals the ravages of a hard life.

When Richard Pass died, he left with his son, James, his vast lore of pottery-making and his secret formulas, his devotion to the Methodist Church, and the care of his mother. James had been born in Burslem, in the heart of England's pottery district, on a bright summer day, 1 June 1856. I learned the measure of the man and the events of his life around our dinner table. My father found in every story of his father a lesson for me and my four younger sisters, and inspired in us the same love and respect he had for his father.

As the story goes, at seven James crossed the ocean to America with his family. When he was only thirteen his parents yielded to his pleas to allow him to leave school in Trenton and follow his father into the pottery. Shy and sensitive, he was slow in his studies and not considered a promising student by his teachers. He was teased by his schoolmates, who made fun of his foreign manners, clothes, and speech. Our father held this up to us as lesson number one: "This early trial is significant, for it may explain largely his championship of the 'underdog': his sympathetic understanding, and his fine sense of justice, which were pronounced traits of his character."

James had no further formal schooling. "The little limey" was apprenticed to a potter who was asked to make of him the best clay worker in Trenton. As was the custom with all beginners, he was put to work wedging clay, a heavy task. In one way or another, clay remained his work for forty-four years.

He learned every aspect of potting. When his father was called to Onondaga Pottery to be its superintendent in 1875, nineteen year old "Young Jimmie," as he was affectionately called around the pottery, was hired too. Pay ledgers in the Syracuse China archives show James Pass's signature in receipt of his weekly wages in 1876. At first $1.84 a day, a few months later he earned $2.00. One old timer, Will Shearer, recalled that "Mr. Pass used to be around the plant in knee

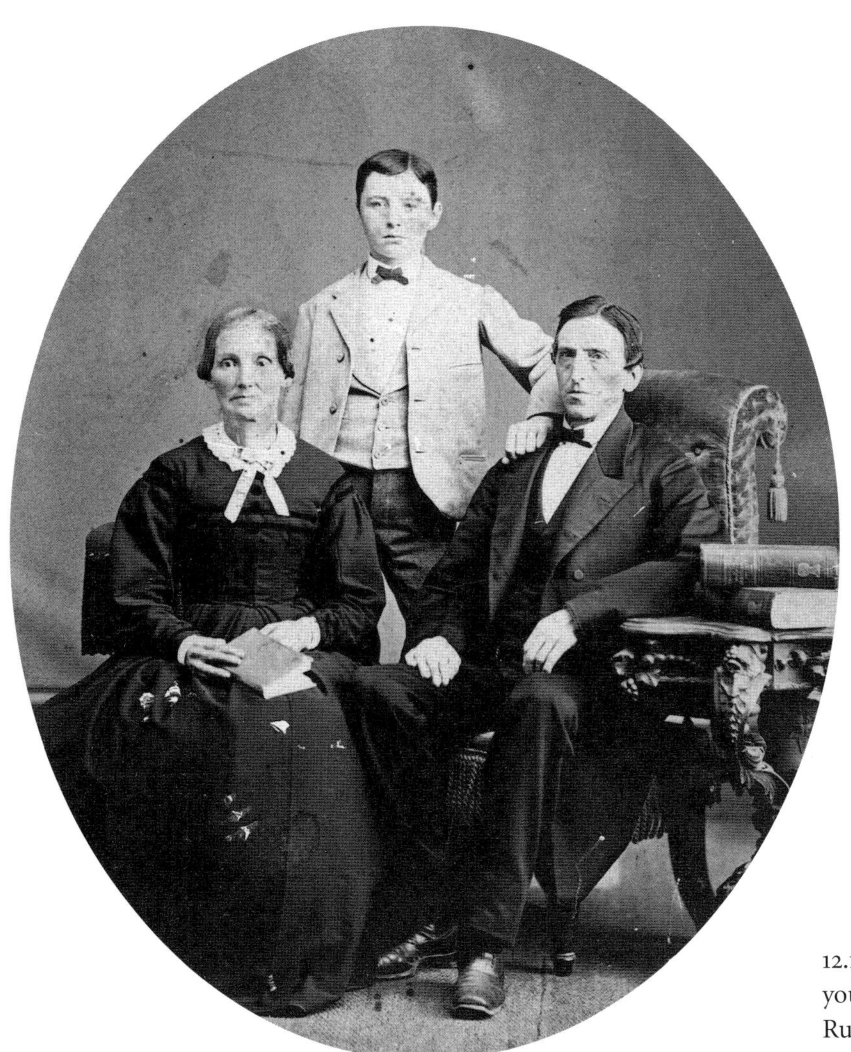

12.1 Richard and Anne Pass with young James, c.1870. Courtesy of Ruth Pass Hancock.

pants helping his father take the time of the men."[2] James soon rose to foreman and supervisor.

Although the study of science and fractions had been a struggle in his early school days, James found that he had a talent for chemistry. He was certain that it held the key to many problems in ceramic manufacturing. In 1876–77, he enrolled in an evening course in analytical chemistry, taught by Professor John Jackson Brown at Syracuse University. He earned his tuition fee by instructing students in the elementary chemistry course. He set up a small laboratory in the cellar of his home. After a day's work, he experimented late into the night with his chemistry lessons. The family still has the mortar and pestle shown in a photograph of him taken at that time (fig. 12.2).

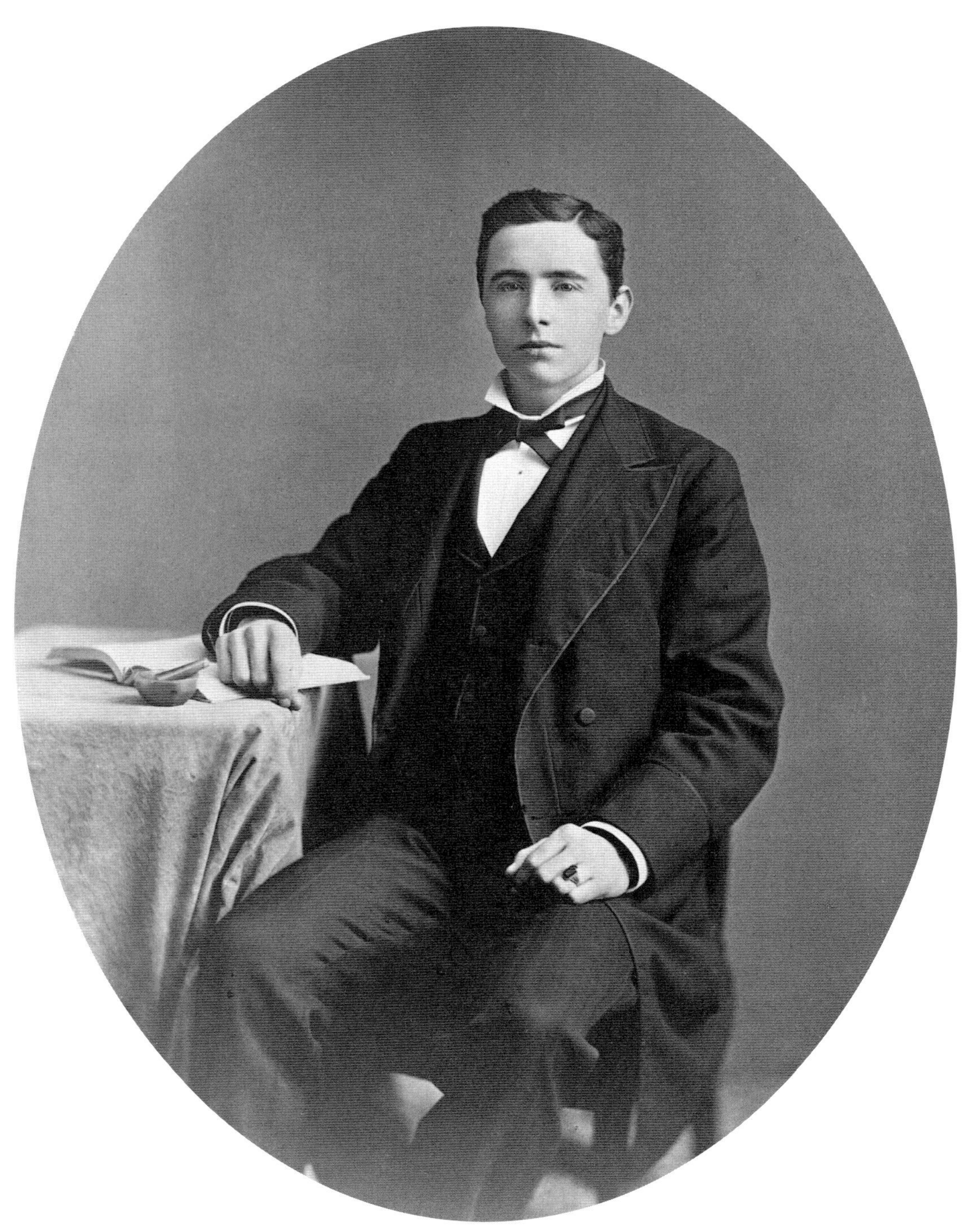

12.2. James Pass, at age twenty-three in 1879. His mortar is still in the family's possession. Courtesy of Ruth Pass Hancock.

With the confidence gained from his potting experience, and fortified with his newly acquired academic credentials in chemistry, James left Syracuse in 1879 for Beaver Falls, Pennsylvania, to manage a pottery begun there in 1878 by the Economite Society, also known as the Harmony Society, an intensely religious sect from Germany that had sought asylum in America from persecution and oppression. Sadie McGowan worked under James that year, the first of her seventy-three year career in the pottery industry. She wrote of those days: "Under the supervision of the Economites, a large bell tolled when it was time to report for work. Inside, two old upright kilns, one a bisque and the other a glost, were fired by coal. Eight girls labored over the two kilns for ten hours a day."[3]

With the death of James's father, Richard, in 1880, responsibility for his family fell on his shoulders, along with the mantle of the pottery tradition. In 1881, the brothers Joseph and Ernest Mayer, English potters, acquired the Economite Pottery. Whether they offered James an opportunity to stay on is not known.[4]

In 1881 he settled his mother in Trenton near his sister, Sarah Pass Mellor. James then took the inheritance he had received from his father and made his way to Cape Girardeau, Missouri, one hundred and twenty miles south of St. Louis. Louis Houck, prime mover in the early development of that area, wrote in his memoirs, published in 1969, forty-four years after his death, that he recalled

a young man by the name of Pass . . . drifted into Cape Girardeau from somewhere and began to talk and discuss the possibility of establishing in Cape Girardeau a pottery, urging that the large beds of kaolin found near [here] . . . would greatly help the business . . . He talked so plausibly and seemed to have such good reason for urging the establishment of a pottery that I became interested in the subject with several others . . . We organized the Cape Pottery Company and acquired the property where at one time a small distillery to make whiskey had been operated . . . in what was then, and is still, called "The Happy Hollow." We put up the money and built a kiln and Mr. Pass made the molds and after spending several thousand dollars finally burned a kiln of stone china ware. . . . We sold what we had made of course at a loss, but that was explained by the fact that we were just beginning the business and did not burn enough to make the work profitable. So we built another kiln and to make our ware more valuable, decorated some of it. Since our ware had a faint ivory color, we adopted an elephant as our trademark and stamped this elephant on the bottom so as to make sure of our future rights. . . . After working one or two years, and spending a comparatively large sum of money, Mr. Pass resigned, no doubt discouraged, and we got someone else to run the pottery for a while; but finally the enterprise collapsed. . . . [We] lost four or five thousand dollars in trying to establish a pottery here; but we had the idle pottery with the dead kilns standing in the "Happy Hollow" as an object lesson for a long time.[5]

James was twenty-seven when he left Cape Girardeau in 1883. He traveled to Trenton, where he took a job setting out periodic kilns, working ten hours a day, six days a week, to support himself and his mother.

Undeterred by his failure, he attempted to interest Trenton potters in pooling their resources to develop a cooperative laboratory for the study of problems common to them all. This was too radical a proposal for the "old conservative rule-of-thumb potters who turned a deaf ear on his arguments."[6] Eventually his idea came into being when in 1898 he helped found the American Ceramic Society, dedicated to this very purpose. He told his son, "I experienced many failures and from each learned a lesson which proved of immeasurable help to my later success."

James was gaining a reputation for himself. When George Oliver, manager of the Onondaga Pottery Company, went to Trenton looking for a replacement for his superintendent, John Brewer of Ott and Brewer Pottery reportedly recommended James to Oliver, who, knowing him, must have been amused by Brewer's recommendation: "Well, we got a fellow down below working for us," he told Oliver. "He's a young fellow, but he's a good deal of a dreamer. He has this idea that a pottery should be based on chemistry. He urges me to get a chemist in here to examine all materials before we use them, but I think he's a nut. His name is James Pass. You can take him and if you can make anything out of him you're welcome to him."

In July 1884, this "dreamer" assumed the job of superintendent that his father had filled less than ten years before. He had returned to the pottery where he would stay for his remaining twenty-nine years, devoting his life to its success. Starting with his first day at work, he made daily entries in diaries, recording experimental formulas for new clay mixes, carefully disguised in a complicated system of "hieroglyphics." He was soon able to make use of the formulas he had developed at Cape Girardeau.

James and his mother set up housekeeping in Geddes, close by the pottery. In the back of the diaries, he kept a daily record of his personal expenses: the rent to Mr. J. W. Hooper ($12.50 a month), insurance on the furniture, a weekly allowance to his mother ($1), a suit of clothes ($40), buttons ($2), collars and cuffs (50 cents), a necktie, a dress for his mother ($28.48). The entries tell a good deal about life in those days: kindling, coal, potatoes, butter, soap, a poker, an ice cream freezer, smoked glasses, a subscription to the *Daily Journal,* a looking glass, a mustache cup, hair oil, socks and braces, artics and rubbers, carfare, *The Judicial County Convention,* a lawn mower, repair of the carpet sweeper, and papering the hall. There was money and time for some pleasures as well: a day at the fair, a ticket for the theater, an excursion to Niagara, a case of Bass ale, fishing tackle, oysters for six. There were day trips to Montezuma, South Bay (Oneida Lake), to Jack's Reef, and longer trips to Chautauqua. In 1888 he bought a flannel shirt, a toilet case, a straw hat, and a satchel, and took a berth on a boat for a trip through the Thousand Islands on the St. Lawrence River. At Christmas time there were gifts for his cousins Annie and Mabel in Trenton, a hat for the minister, a shawl for his mother, and "presents for the tree." James took good care of his mother for the rest of her life.

His systematic effort to improve his mind began with the purchase of the *Encyclopaedia Britannica* on the installment plan. He bought *Plutarch's Lives, The Religion of Philosophy* by Raymond S. Perrin, added Ruskin and Trollope to his home library, and purchased a globe, a micrometer ($6), a microscope ($1), a tuning fork, a drawing book, boxing gloves, and a fly rod ($10). He took music lessons and sang in the Choral Society. Although in later years his son Richard was to describe his father as not very gregarious, James paid dues to the YMCA, the

Century Club, the Elks, the Odd Fellows, and the Republican Club. A Century Club history describes him as "intellectually the most pleasant of companions."

He rejoined the church of his youth, the West Genesee Methodist Episcopal Church. He made weekly donations to the church of one dollar, one-twelfth of his salary, and supported many of its activities as well, including someone to pump the organ for Addie Mae (Adelaide) Salisbury, who accompanied the choir he directed.

James was a handsome young man with the high cheekbones and high coloring of an Englishman, my father used to say. His eyes were blue and his hair dark. Tall and erect, he was well-dressed and well-mannered. In 1890 he married Addie Mae, a schoolteacher whose father had been superintendent of the salt works just down the hill from the pottery behind Star Park. Although Addie volunteered to postpone the purchase of a new house so that the money could be invested in the newly founded Pass & Seymour Company, she got her house in 1892, when James had risen to general manager at the pottery. They built the house on the hilltop property that her father, Oscar Salisbury, had bought in 1872, where he tended his bees. After the children, Eleanor, and Richard Henry, and James Salisbury, were born, the Passes enlarged and embellished the house. In its final form, West Rising was a handsome English Tudor house with formal rooms on the first floor, five bedrooms on the second, maids' rooms and a billiard room on the third. My father grew up there and at the family summer home, Swiftwater Point, on the St. Lawrence River near Thousand Islands Park. At times we lived at West Rising. The house and its big barn were enchanting places. Later we lived in Orchard House, the smaller English Tudor residence built on the property by my grandmother for my father when he married. In 1941, we sold the hilltop house and moved across town to be closer to the Court Street plant. West Rising survived until 1948; Orchard House is (in 1997) a home for the disabled. When my grandmother, Adelaide Pass, had leisure and financial security, she became actively involved in many of the social programs in her community. She was a founding member of the Huntington Foundation, which was dedicated to making life better for young factory women. She possessed a fine mind and unusual organizational and leadership abilities. When James Pass died, his wife took his place on the company boards. My father often reminded his five daughters that there was nothing a man could do that a woman could not do as well. He had seen his mother's example.

The diaries do not lend much insight into the Passes' life together. Their important function was to secret the clay body formulas he was developing. James was threatened more than once with the "theft" of his formulas. In 1888 his boss (Oliver), feigning difficulty mixing one of James's clay bodies, tricked him into divulging more details of the composition than he intended. "Was there ever such a fool?" James wrote of himself in his diary that day. Another incident nearly destroyed his association with the pottery. When in the early 1890s one of the pottery's officers died, Pass was sent to the bank to collect the contents of the deceased's safety deposit box. In it he found a contract between the officer, claiming to represent the company, and a worker in the shop. The contract specified that the worker was to ferret out Pass's formulas and bring them to the officer. Directors of the company disclaimed any knowledge of the contract.

By the time Bert Salisbury was president, clay formulas were company property, but still carefully guarded secrets. I remember having dinner with my family one night when my father was called to the phone. Uncle Bert had suffered a stroke. My father's first thought was that he must get to his bedside in time to learn where the formulas were kept. In 1941, news of Pearl Harbor reached an inn in Northhampton, Massachusetts, where I was having lunch with my father, then president of the company. He left the table abruptly and headed for Syracuse, concerned for the security of the formulas.

Along with his inventive genius, James Pass had a fine artistic sense. He created several patterns in 1903, ranging in style from a delicate gold filigree to a bold art nouveau motif. In the pocket of his 1891 diary he tucked a finely detailed pencil drawing of a rampant dragon labeled "Imperial Geddo." He felt it was important to bring together a group charged with the responsibility for developing the very best designs and decorations for his product.

His work was demanding and exhausting. Self-discipline saw him through. Even in his later years, when plagued with chronic asthma and occasional pneumonia, nothing kept him from his appointed rounds of his two companies, visiting the potting shop, and personally performing most of the laboratory work. Business increased. He had proved that the American potter could make true china, in quantity and at a profit. Under his leadership, Onondaga Pottery became the largest producer of vitrified hotel china in the country. It was debt free; and it provided employment for over seven hundred people.

My grandfather earned the respect and trust of a workforce whose welfare was of utmost importance to him. He had sometimes resorted to going to their homes on a Monday morning to rout them out of bed after a long weekend in a bar. The Mutual Benefit Society, begun in 1888 for the potters' welfare, and the Potters' Club, established to provide social activities and educational opportunities, were only examples of his care and concern. The company furnished the quarters for the club, and the members paid dues to run its activities. The company managers were active members. His son wrote after his death,

> His most important service to his fellow man was the influence of his sympathetic, generous, helpful nature. Nothing in life gave him so much happiness as the love of those who worked for him; yet the happiness which sprung from their affection was not the motive which prompted his kindness. The relationship which existed between [Father] and his workmen is evidenced by the fact that when his men had learned to know him well they never felt the necessity of securing the assistance of outside agencies in the adjustment of differences between employer and employees.[7]

If James Pass never faltered in his devotion to his workers, neither did he finish experimenting with clay. Many times I have heard that in the pan of his balance scale at the time of his death in 1913 was the formula for a new clay body, made once again entirely of American materials. We grandchildren never tired of the stories about the life of this consummate potter.

Richard Henry Pass, James's eldest son and my father, was born in 1895. Eighteen years old and only halfway through Harvard University, he was ill-prepared for the loss of the man he considered his best friend and champion. Finding little comfort and no acceptable explanation in the Methodist Church of his upbringing, he commenced his own life long quest for the meaning of life.

The mantle of his pottery forebears fell heavily on Richard's shoulders. He had spent long hours with his father at the pottery and had traveled with him to Europe to visit and observe European potteries at work. Two weeks after he graduated from Harvard, he went to work at the pottery's laboratory. His first real project took him to North Carolina to look again into the possible use of the "uniker" kaolin that his father had used in making ***Imperial Geddo.*** He developed a process to refine this clay before he headed south. He boarded at Mrs. Harris's Jarrett House. She was the wife of C. J. Harris, who had refined the clay for James Pass in 1888. My father often reminisced about those days in that wild mountain country of moonshiners and revenuers where he lived closely in the out-of-doors with people so different from any he had known before. Sometimes he would travel all night on horseback over the mountains to Asheville. It must have been especially disappointing not to have been able to improve the clay's yield profitably. It would take nearly thirty years until the best of that clay, refined by his method, was used again in one of the pottery's body formulas.

With the onset of World War I, my father was caught up in the thrill of the new engines on land and in the air, especially flying machines. He enlisted in the United States Naval Reserve Flying Corps. After ground school at the Massachusetts Institute of Technology and flight training at Pensacola and Key West, Florida, he received his wings and commission and was assigned to an HS-1, the first flying boat put in service.[8] My father took to flying naturally; he loved it.

Always the inventor, he started tinkering with aeronautical improvements. He developed a bomb-sight for moving targets and a new method of aerial navigation by dead reckoning.[9] Although the naval air fleet did not see wartime action overseas, it learned a great deal about flight. As the war ended, my father was in line for a position as navigator on the first navy transatlantic flight. Now he needed to make the most important decision of his life—to follow the future of aviation or to return to the responsibilites of the family companies.[10] Discharged as a Lieutenant, Junior Grade, in May 1919, and bouyed by the new-found confidence his Navy experiences had brought, my father returned to the pottery.

He was immediately made assistant superintendent, and in 1922, superintendent. My father supervised the design and construction of the new laboratory building on School Street across from the plant. The jobber's notebook indicates, "Mr. Pass came by, adding 2 percent lamp black to the floor and requesting brick work changes as follows 'red headers set in Flemish bond above soldier course.'" This gem of a building still stands today, the only surviving presence of Syracuse China on the west side of Syracuse.

At the same time, there was tremendous excitement across town over the construction of the new Court Street plant, a model one-story linear building. On June 7, 1922, my mother and father, newly wed, led the torchlight parade of potters around the outside perimeter of the new facility and then a Grand March down the long interior hall of the "Great Plant," decorated with flags and bunting for the occasion.[11]

My parents' marriage was a happy union. They were both West Siders. Ruth Pennock was the daughter of the Harvard-trained chief chemist of the Solvay Process Company, John D. Pennock. Her older brother Stan had roomed with my father during their first year at college. Mother was beautiful, bright, and very caring. After graduating from Radcliffe College in 1918, with the country at war, my mother enrolled in an intensive summer nurses' training course at Vassar College. She shared my father's interest in the people at the pottery and was always at his side for the company parties, picnics, and dances. She loved it all. At the same time she was a devoted mother to her five daughters. I was the first, followed by the twins Adelaide Salisbury and Eleanor, then the twins Ann Greatbach and Eunice Amelia.

In many respects my father was frustrated by the limitations on what he was able to accomplish and by what he considered his lack of ability to put his ideas across. In the late 1920s, he proposed the construction of a biscuit tunnel kiln he had planned with Philip Dressler of the Dressler Corporation, but his presentation to the board of directors was rejected.[12] (Eventually Dressler would install many kilns in both plants.) He found himself eager to innovate but faced leadership much his senior that moved ever more cautiously.

Enlarging the lives of others would be the essence of his leadership. In a 1928 essay, he described his mission:

> If my task is to lead others, I found that I could not do so without becoming responsive to their thoughts. This involved getting outside of myself and into their lives, learning to know them. As I did so, my interest in others increased surprisingly. I found in their characters noble qualities which, before, I had ignored. I learned that they had much to teach me. A real sympathy with and fondness for my fellow workers began to develop. At last I was becoming objective, beginning to find myself through . . . other lives. I cannot overemphasize the importance of this change from living in one's self to living in others, for I am convinced that herein lies the basic truth on which all ethical leadership must be founded.

My father assumed the presidency of Pass & Seymour in 1929, while continuing to serve as vice president of Onondaga Pottery in charge of research and production.[13] The next months brought the financial disaster of the Great Depression. Although Pass & Seymour had adequate reserves, it was losing money. He was advised, "Close the business. Take the cash. In fairness to the stockholders this is the only thing you can do." But like his father, who in 1890 had taken over the faltering pottery, asking for two years to turn it around,[14] my father met the challenge of preserving the jobs and livelihoods of the people at Pass & Seymour. He built back the business on the strength of the new line of Despard switches. During his presidency, which lasted until 1955, sales increased 500 percent.

Through these years, the pottery kept people working at both plants. Although at times the company went to four-day weeks, it never shut down and never let anyone go. As the Premier Egg Cup had saved the company during the crash of 1893, **Econo-Rim** came to the rescue forty years later.

As the economy picked up, the pottery sustained its reputation for being a good place to work. Ray Ross, who with forty-two years service was still (in 1997) working at Court Street along with his wife, Pat, recalled that "They always treated you good. The pay was good and there was no union. If you needed something they would help you. And every holiday Mr. Pass would come around and ask how you're doing, how're your kids, and shake your hand. It didn't matter if your hand was covered with clay." The door to my father's office was always open. Often, late into the night, he would walk through the plant, sitting a while to share thoughts and experiences with a worker at his lonely job on the night shift. In those days management *boasted* of the number of people employed. Another old timer remembered, "You looked forward to going to work each morning. It was like a country club. You always had time to chat with your friends in the halls." There were many layers of management. If it wasn't an efficient way to do business, it was an enlightened way. And the company operated at a profit.

My father had just been made president of Onondaga Pottery in 1941 (while continuing as president of Pass & Seymour) when World War II broke out. A way had to be found not only to keep production going at the plants but also to win the war. He teamed the companies to produce a nondetectable ceramic anti-tank mine with a chemical fuse, and shipped the first consignment of the weapon in November 1942. My father wrote to me, "Have to keep going like a buzzsaw these days. There just aren't enough hours in the days and days in the weeks."

The mines, made with extreme tolerances of temperature and pressure, were thoroughly tested at a proving ground in Highland Forest, south of Syracuse. The tests were awesome to witness. One cold fall afternoon I watched an exploding mine hurl a huge steel plate into the sky and heard my father's urgent yell, "Run north everyone, run north!" Everyone scattered every which way. Which way was north? Luckily we all managed to stay clear of the errant missile.[15] "If there are chances to be taken," my father said to Walter Haswell, an engineer about his age, "you and I will take them.[16] The rest of the group is younger and they are the future of the company." Young Bill Salisbury (another third generation at work), who helped steer the mine project, told this story:

> At one time in the proving, the explosive mechanism failed and it was necessary for someone to approach the testing site and repair the damage or correct the situation so the explosion could occur. Richard refused to let anyone else come within a hundred yards of the land mine. He himself crawled to the spot, corrected the mechanical difficulty, retreated, and we were able to detonate the land mine. This is an example of the courage he had. Richard took those chances.

There is a tale that one of these mines shared a berth with my father in a Pullman car on the overnight train enroute to an appointment at the Pentagon.

In January 1944 my father wrote to me, "Recent weeks have been rather hectic as usual, ordnance and more ordnance, Army, Navy, china, District 50, NLRB, taxes, finances, wiring devices, regulations and more regulations, and always weak bretheren to be coddled and nursed and carried along." With the close of the war, the company and my father received citations for outstanding service to the country.

With a workforce and production space triple the size of his father's day, and with a barrage of government paperwork and regulations his father could not have dreamed of, my father carried a heavy load. Some twenty-seven hundred people worked in the three plants. "Adding to

this number those who are dependent on them for a living makes quite a group whose livelihood comes from these factories," my father wrote. It was becoming more difficult to maintain
close relationships with the people. "If anybody thinks he wants to try to run a business in these
days, he is welcome to it." He was beginning to wear out. Like his father, he was a perfectionist
and he found it difficult to delegate responsibility.

To help lighten his load, in 1942 he called upon his brother, James Salisbury Pass, for help.
Uncle Jim, eleven years younger than my father, had graduated from Harvard University with a
degree in geological engineering. He had been pursuing a career in the nickel mines of Canada
with great success. With characteristic family selflessness Uncle Jim came to Pass & Seymour
and served eventually as president and chairman of the board until 1976.

When at last he became president of Onondaga Pottery, my father had the opportunity to
initiate changes and institute some of his own ideas. Over the entrance to the Court Street
building he raised a motto cribbed from a doorway of his school days: "Enter to Be and Find
a Friend." Now buried under layers of "new company" signage, the words are preserved on a

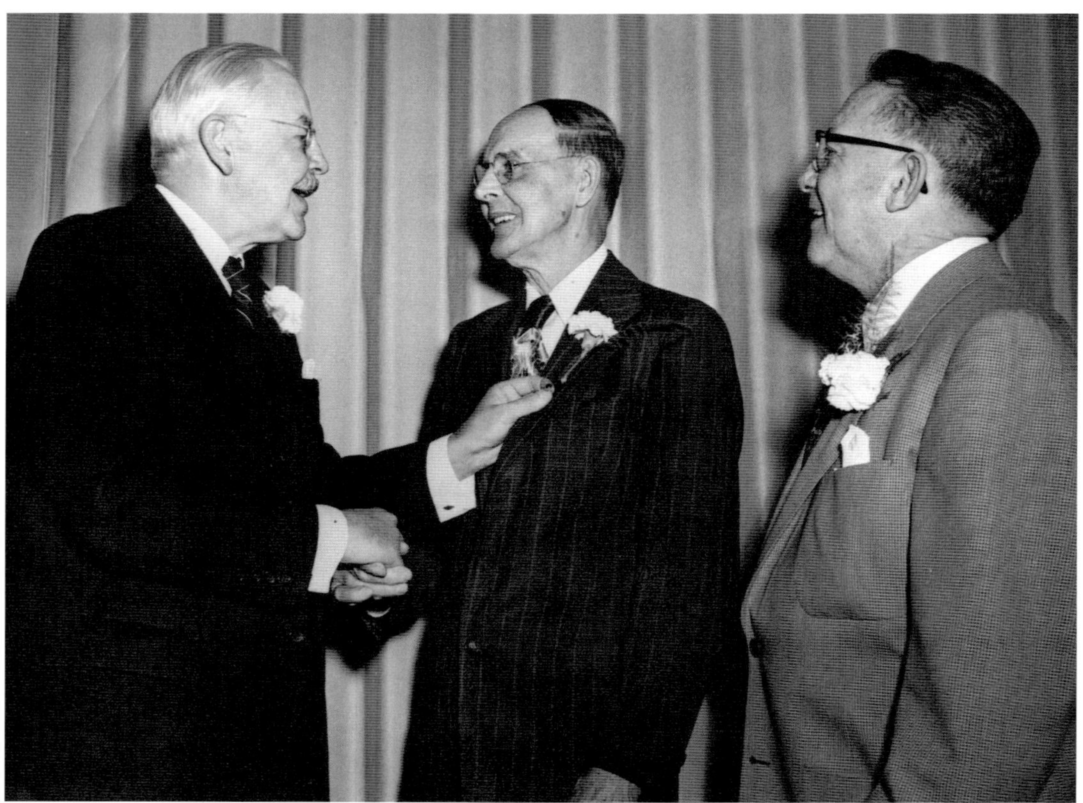

12.3. Richard Pass giving service awards to Freeman Craver and
William Cummings, 27 October 1947.

sticker in the glass window of the office of the litho department. The motto summed up suc-
cinctly the first two paragraphs of the company's statement of policy:

> The first principle of management of the Onondaga Pottery is that this business exists to be of
> service—to the public—to the people working in this organization—to the stockholders; that
> the interest of all three groups are in the end the same; that the interests can best be served well
> only as we serve each other well.
>
> The policy and the practice of the management is to maintain a friendly, cooperative rela-
> tionship with each other and every member of this organization on the basis of mutual respect,
> confidence and good will and the mutual service inherent in our work here together.

Foster Rhodes, in his remarks at my father's funeral, talked of discussions long into the night
between himself and my father regarding individuals who needed help because of unexpected
troubles that had come to them or their families. They had coined a word for this approach to
corporate management: "Industrial Humanics." It paid off with a close-knit and intensely loyal
work force.

My sisters and I heard about the people of Onondaga Pottery and Pass & Seymour around
the dinner table. They were our greater family, as they were my father's. We appreciated that
their concerns came first. Their names became so familiar that to this day I recognize the name
of a pottery or Pass & Seymour person in the obituary page. We were taught to cherish and care
for the Syracuse China we ate from as the handiwork of these skilled craftspersons. To break a
piece of china was tantamount to breaking trust with its maker. At home, we revered the pieces
of **Imperial Geddo** in our china cabinet. When my mother died my sisters and I divided up the
five piece tête-à-tête set. I have a cup and saucer. When "out," we patronized only those restaurants and hotels that carried Syracuse China. Sometimes after Sunday dinner we would go with
my father to call on a retired, "service reward" (pensioned) employee.

I remember going to see Mr. Cowan in the little fieldstone studio on Woodchuck Hill Road;
we were always paraded to the Ceramic National Exhibitions to see his work. And I remember
the awe surrounding a visit with my father to pay respects to an ailing Adelaide Robineau before her death in 1929. We have no Robineau ceramics, but I have a tiny, pink, heart-shaped ashtray (made at his Ohio pottery) that Mr. Cowan gave to my mother. She always kept it on her
bureau.

Along the way my father found time to guide our education, aimed at preparing us to lead
"useful lives in service to others" according to his own philosophy, proven at the pottery. He
helped pick our schools and courses and methods of study. He wrote letters of encouragement,
chiding us ever to do our best, to never give up, to stand squarely on our own feet, beholden to
no one, to be proud of our own self reliance and economic independence. Each of us had summer jobs in local factories. It is probably not surprising that my college thesis was "Industrial
Democracy." Although I worked one summer as an overglaze liner at Court Street and my
sister Eunice joined the marketing team for several years after college, none of us assumed our

father's mantle. Our cousin, Richard Besse, son of James Pass's daughter, Eleanor, did follow the family pottery tradition in the management of the company.

From early days sailing his boat on the St. Lawrence River, or wandering the North Carolina mountains, or flying through the clouds of Pensacola, my father had sought solace from the pressures of the work-a-day world in the restorative powers of the wonders of nature—the color of the fall leaves in Pumpkin Hollow and the spring blossoms in Cedarvale—the majesty of mountain peaks and sunsets. He knew the birds' calls and the names of wildflowers. If he could snatch an afternoon casting for brook trout or raising a pheasant he was sublimely happy. He wrote to me, "We had a wonderful time in the woods and swamps, as near Heaven as I ever expect to be." Wanting to share these experiences, he made it possible for the pottery to acquire a large secluded tract of land in the Adirondack mountains with streams and a lake. Named Pine Lake Preserve, it was furnished with boats and cabins and a caretaker and was available winter and summer for family vacations by all the people of the pottery, free of charge. My father retained Star Sapphire, a smaller lake up on the mountain, for his own use. (Once he was caught in an early winter snowstorm while camping in a little pup tent alone with my mother.) He made a gift of Star Sapphire to the pottery people on the Christmas of his retirement.

My father had completed his mission. "Industrial humanics" was a success. The pottery was providing jobs for more than two thousand people and a livelihood for their families. There had never been a falling-out between management and any group of workers. The chinaware was the most respected in the industry, and the company was the largest of its kind. It was in sound shape financially.

My father had found his religion in service to others, particularly the people of Onondaga Pottery and Pass & Seymour. He would never have disowned them. As he resigned the presidency of the pottery to "younger men" in 1958, he took leave with this message: "As I have said, I am not leaving you. This is a thing I should never be able to do. . . . The people of the pottery and the welfare of the company have become so greatly a part of my life that I should be desolate without you. Blessings on you."

He died in 1964. I am glad the discontinuance of fine dinnerware, the unionization of the workers, the destruction of the Fayette Street plant, and the sale of the company did not happen on his watch.

I wish he could have flown a spaceship to the moon.

Identifying and Dating Syracuse China

Onondaga Pottery/Syracuse China has backstamped most of its ware with a trademark since 1871. It has dated its ware since August 1886 with a code for the year and month of manufacture. These backstamps and codes appear either separately or together, depending on the given system of marking ware at any one time. The dates that follow are based on the examination of the Syracuse China historical collection as well as company records, including price lists and factory orders.

This list attempts whenever possible to integrate backstamps, shapes, and decorations. In each of its sections it lists shapes offered to the public, in chronological order, with the range of years during which the shapes were manufactured or were at their peak of popularity. The shape's designer is given, if known. For dinnerware shapes made after 1910, generally available patterns are listed. After 1926, the specific decorations assigned to each shape are listed when they originated. As an aid to the identification of earthenware and fine china dinnerware shapes, the sugar bowl for each is illustrated below. The sugar bowl contains many stylistic characteristics that distinguish one shape from another (body contour, cover, handle, foot, moulded decorations, etc.) Only standard backstamps are included in this list. No attempt has been made to include the company's many hundreds of custom backstamps.

Earthenware: 1871–1896

The white earthenware manufactured by the company came in three distinctive types: CC (common or cream-colored), which went unmarked; white granite (ironstone); and improved white earthenware, which it called O.P.China. Each progressive change in the clay bodies of white granite and O.P.China was signaled by a new backstamp.

Earthenware Backstamps

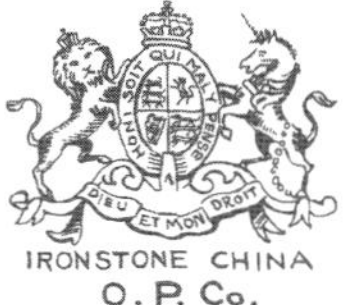

Lion and Unicorn Arms of England (1871–73)

Great Seal of State of New York (1873–97)

O.P./China broken line (1885–90)

O.P./China straight line (1890–95)

O.P./China Semi-Vitreous (1894–97)

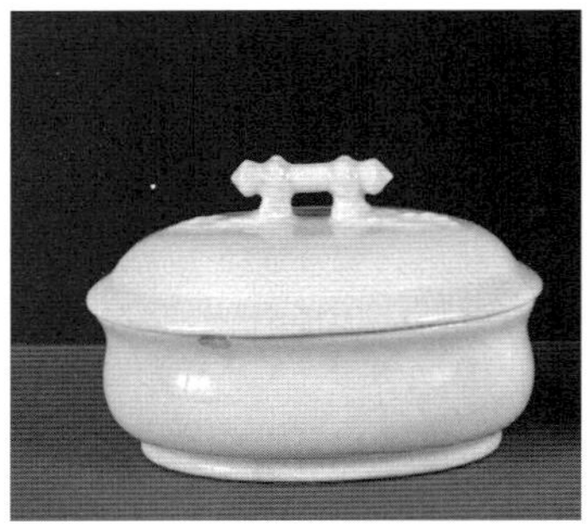

Cable

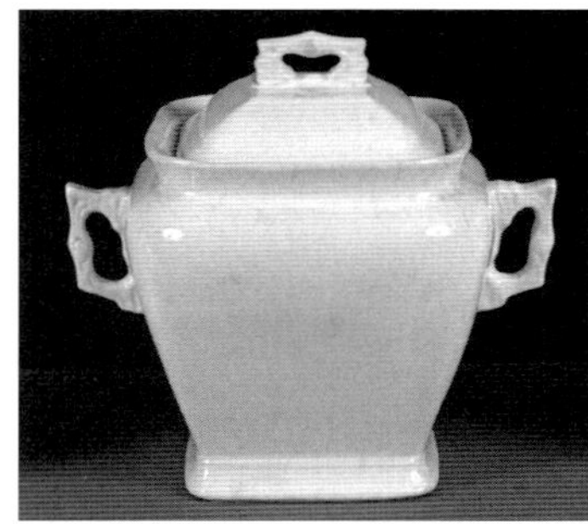

Square

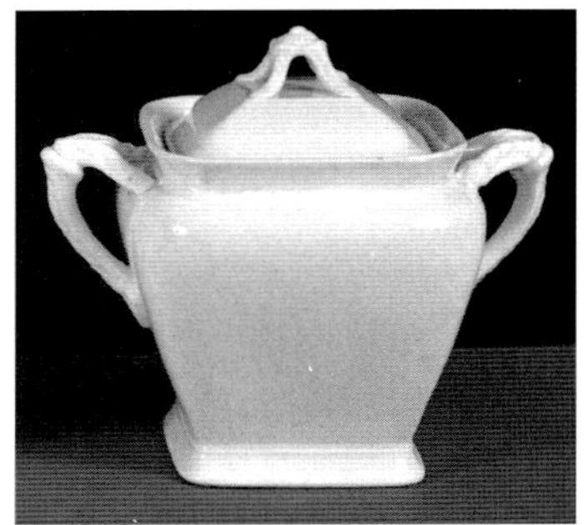

Fork Handle Square

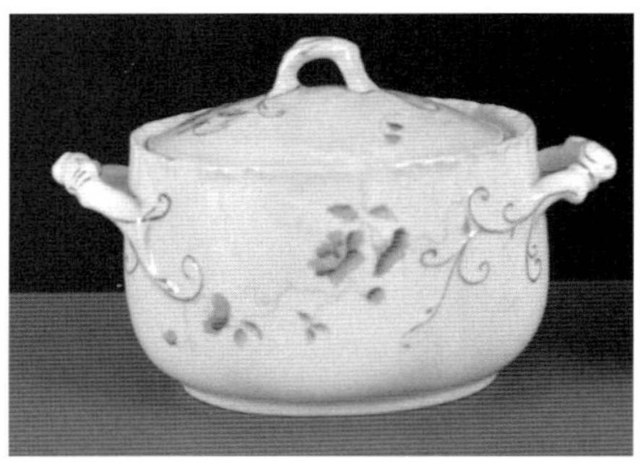

Doris

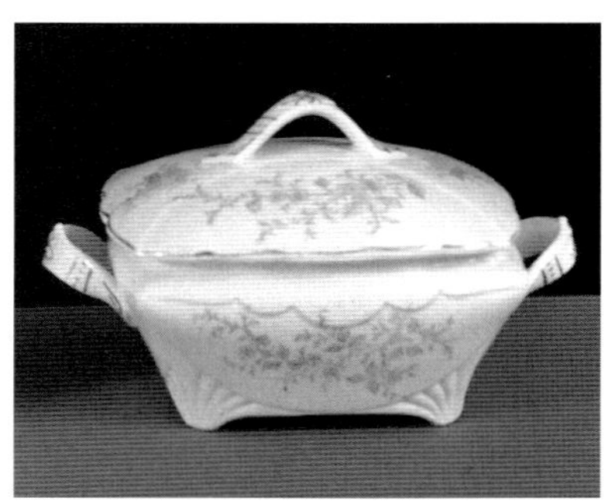

Juno

Earthenware Shapes

Cable (1873–93). Table and toilet ware made in CC and white granite.

Square (1880–93). Tableware made in white granite. Block–handled.

Syracuse (c.1880–96). Toilet ware made in white granite. (Fig. 3.3)

Fork Handle Square (1885–92). Tableware made in O.P.China.

Doris (1887–96). Tableware made in O.P.China. (Fig. 2.1)

"Grecian" Style (1887–89). Toilet ware made in O.P.China. (Fig. 2.2)

Juno (1887–96). Table and toilet ware made in O.P.China. (Figs. 2.3, 2.4, 3.3, plate 1)

Oneida (1891–93). Toilet ware made in white granite and O.P.China. Mark Haley. (Fig. 3.3)

Seville (1894–97). Tableware made in O.P.China. Mark Haley. (Fig. 3.5)

The company has marketed vitrified china products since 1891. It did not at first distinguish between dinnerware for household use and hotel ware. This division evolved after 1896 and was clearcut between 1920 and 1970. The backstamps are listed first, in chronological order with their inclusive dates of use. All shapes made of white flint body unless otherwise noted.

Hotel or Household Backstamps

Imperial Geddo (1891–94)

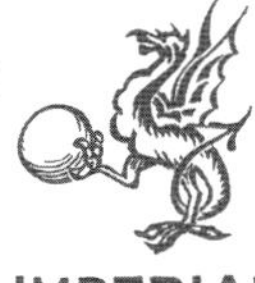

China Dragon (1892–95)

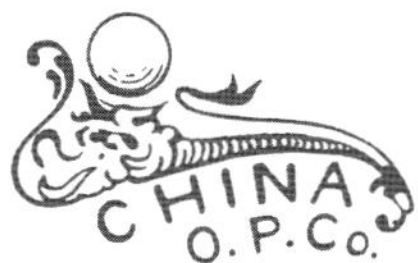

Western Hemisphere (1895–97)

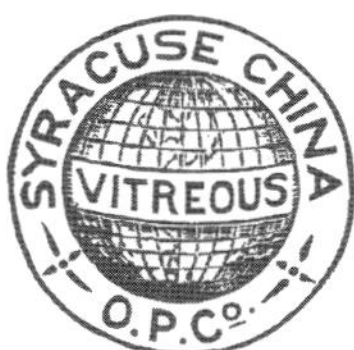

O.P.Co./Syracuse/China (Prototype March 1897 only)

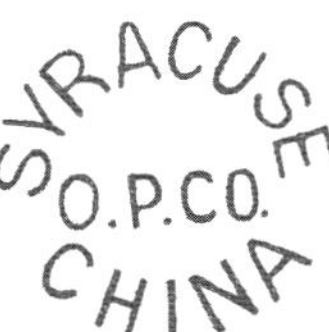

O.P.Co./Syracuse/China (1897–1920). On commercial ware with impressed date code; (1897–1926) on dinnerware.

O.P.Co./Syracuse/China (stamped date code) (1920–46). On commercial ware only with rubber-stamped date code.

Old Ivory (1927–60). On ivory colored body.

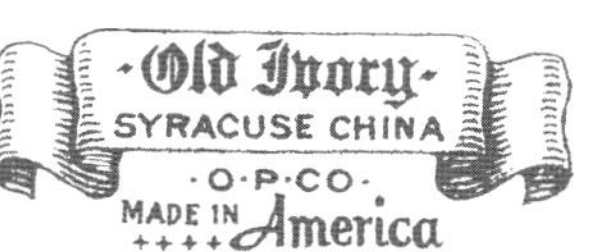

Adobe Ware (1932–72). On deep tan-colored body, for dinnerware and commercial ware.

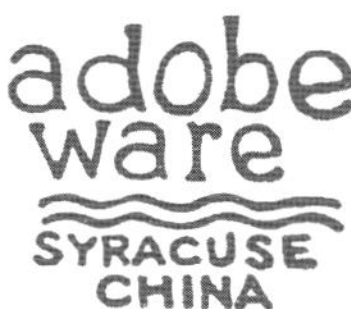

Econo-Rim (1933–67)

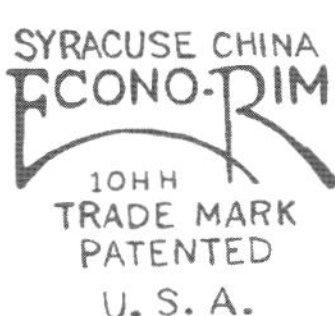

Morwel (1936–67)

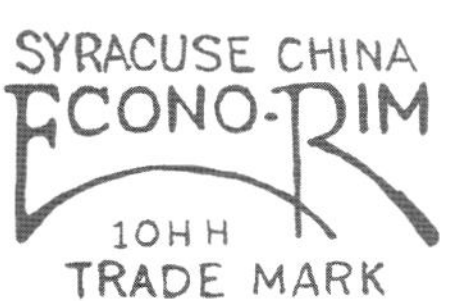

Syracuse China (1946–71). On commercial and dinnerware with many variations; replaced in 1962 by Peter Piening backstamp, then resumed.

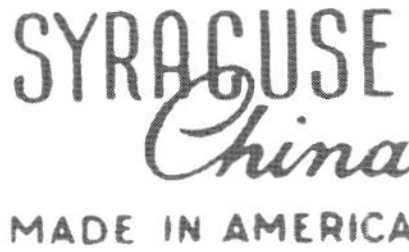

Peter Piening (April–August 1962). On commercial ware only.

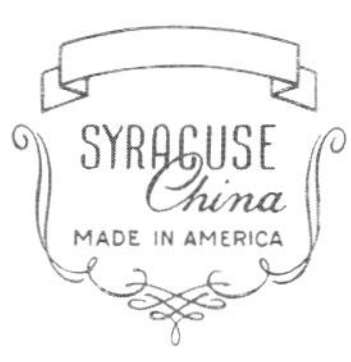

California, Carolina (1952–70)

Airlite (1946–49). R. Guy Cowan. (Fig. 8.3).

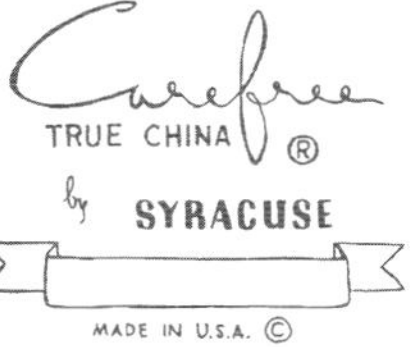

Carefree (1956–70)

Trend (1955–78)

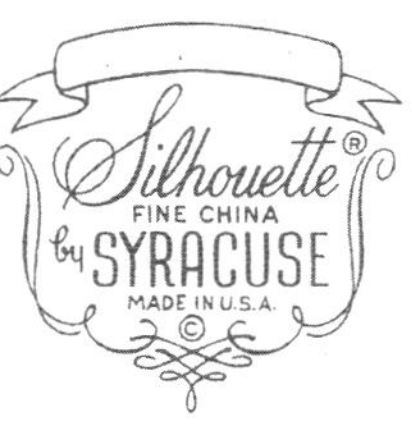

Silhouette (1960–70)

White Alumina Body (*Syralite*) (1964–97).

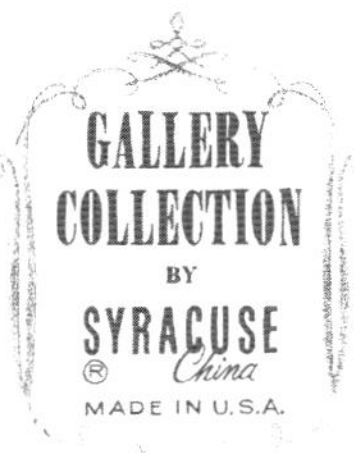

Gallery Collection (1962–70)

Household China Backstamps

Many of these backstamps included a pattern name but not a date code.

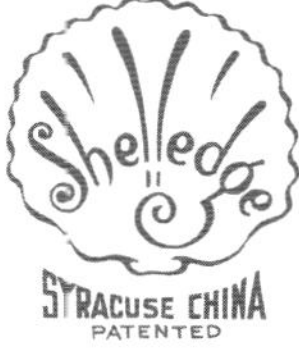

Shelledge (1936–49)

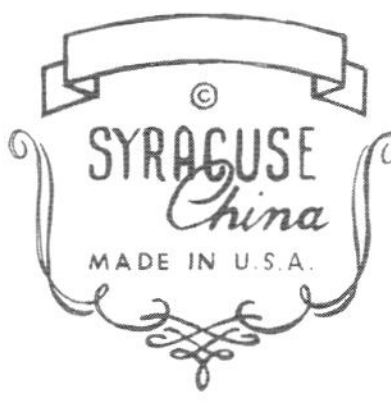

Empire, Wellington (1966–70)

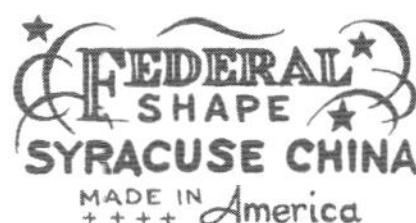

Federal (1938–69)

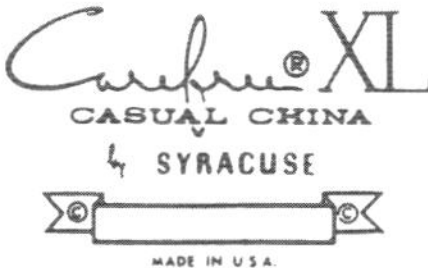

Carefree XL (1966–70)

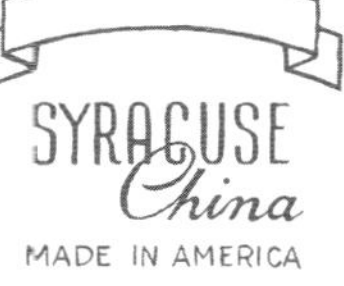

Paul Revere, Berkeley, California (1950–66)

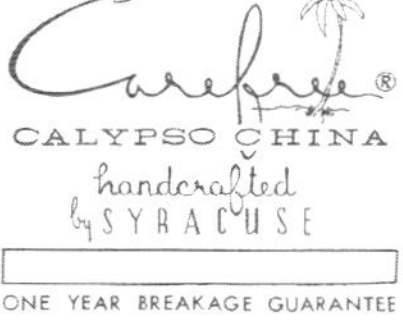

Calypso (1968–70)

Syracuse China Corporation USA new company logo (1971–97). With many variations.

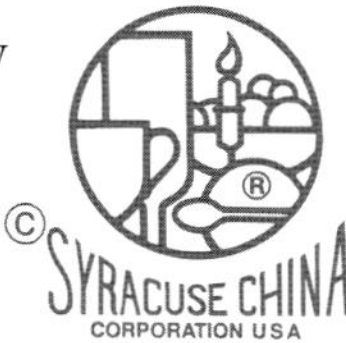

Country Ware (1973–78)

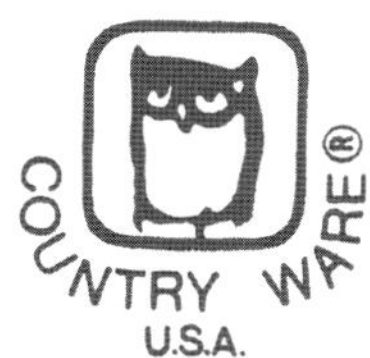

Gibraltar (1978–1997)

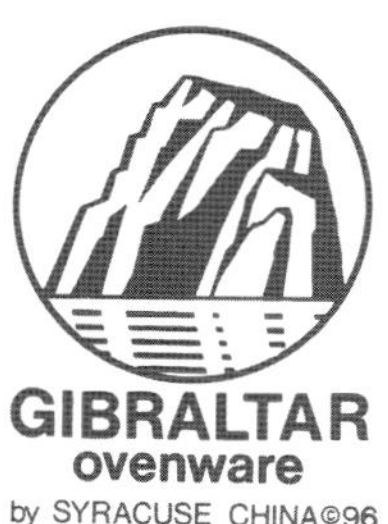

Royal Rideau Bright White Alumina Body (1993–97). Several variations, including *Luxor.*

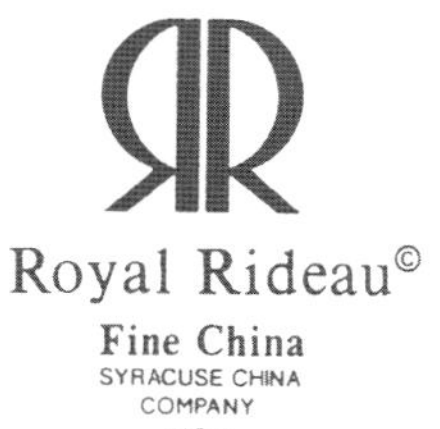

Aberdeen China (1927–69). Selected ware for factory store.

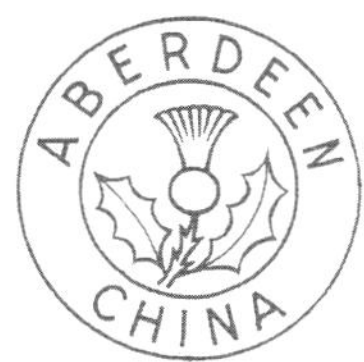

Vandesca (1959–94). Sometimes marked Syracuse China/Vandesca.

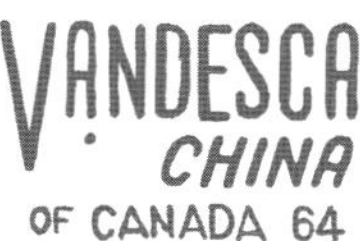

Celestial Fine China (1960s). Manufactured for a door-to-door sales company; "Fantasy," "Fernwood," "Granada," and "Serenity" patterns on special *Celestial* shape in *Carefree* style.

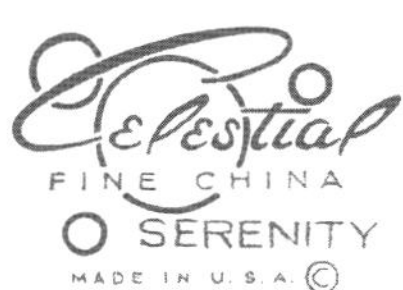

Hanover China (1950s and 1960s). Manufactured for a door-to-door sales company; "Primrose," "Pussy Willow," and "Cavalier" patterns on *Federal/Paul Revere* shapes, and "Anniversary," "Coronation," "Silver Rose," and "Springtime" on *Cambridge* shape.

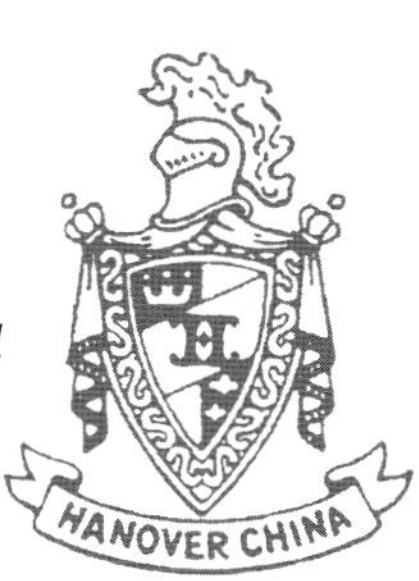

Lenore (1960s). For Sloan Company, probably for door-to-door sales; "Classic Rose" and "Spring Melody" patterns on *Carefree* shape.

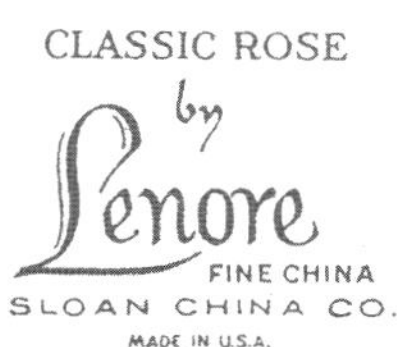

Imperial Geddo (1891–94).
Alexo, **Passo**, **Moro.** Hamlet
Bourne. (Fig. 3.1, plate 3). Usually imprinted on bottom with
coded cipher mark $\mathcal{J.P.}$ $>$L
or $\mathcal{J.P.}$ 10.

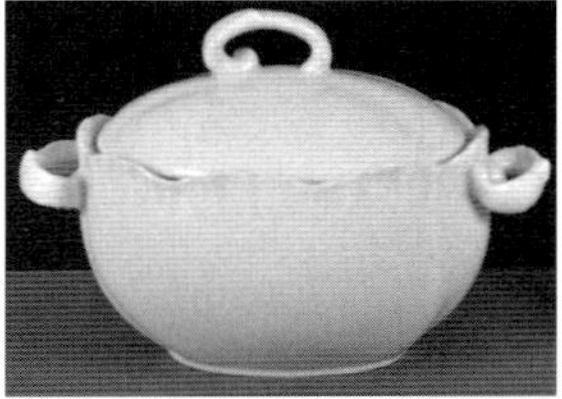
Alexo.

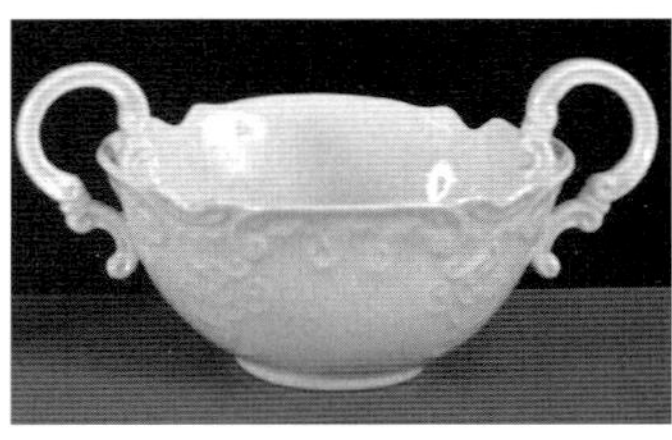
Passo.   Moro.

Imperial (1894–1906). Hamlet Bourne. Same shapes as
Imperial Geddo. (Plate 13)

Marmora (1891–96).
Mark Haley. (Fig. 3.2 ,
plate 1)

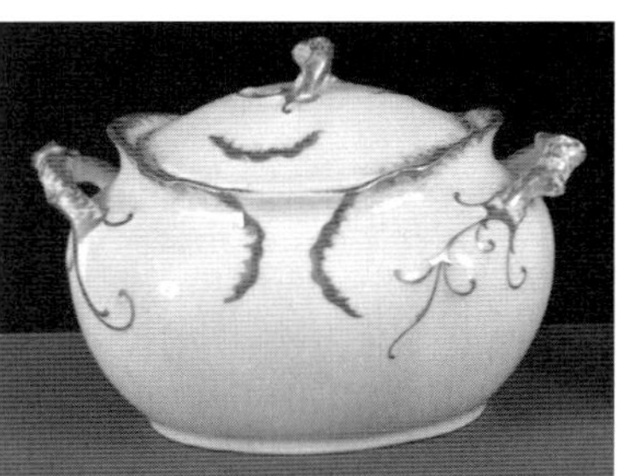

Plymouth (1895–1922).
Mark Haley. (Plate 4)

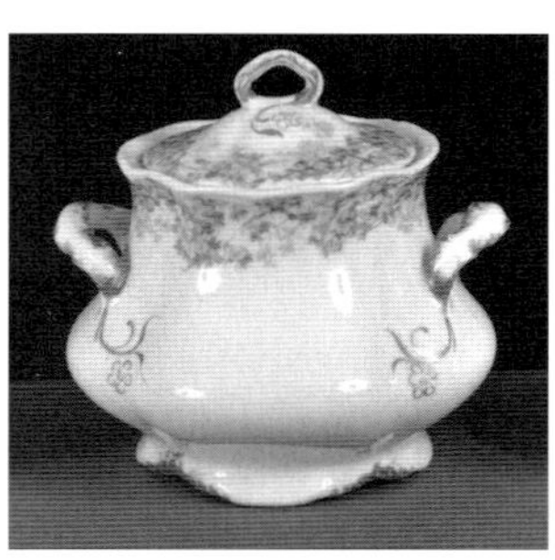

Puritan (1903–15). Mark
Haley. (Plate 16)

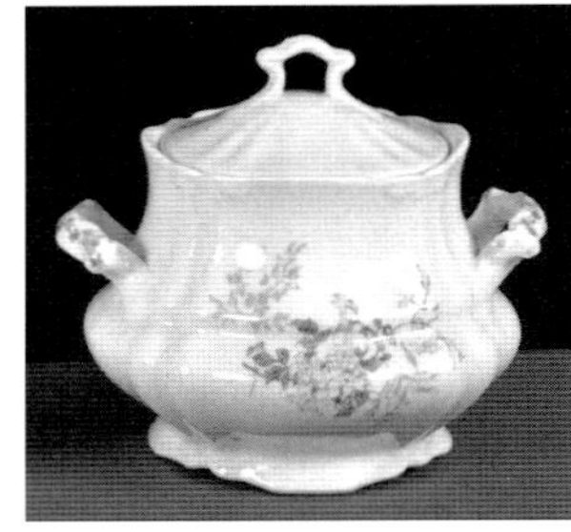

Mayflower (1906–40).
 Mark Haley.
(Fig. 5.2, plates 13, 19)

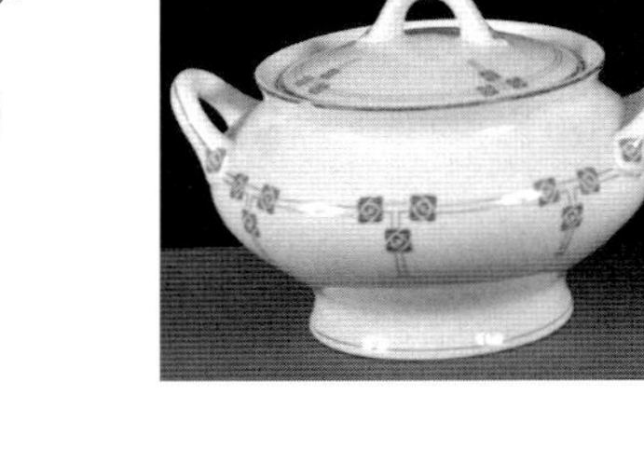

Arcadia
Art Pendant
Berkeley
Brentwood
Briar Rose
Burmese
Canterbury
Cloisonne
Corona
Empire
Fairfax
Fusan
Gardena
Gold treatments
Harvey print
The Hostess
Indian
Indian Tree
Lorraine
Melrose

Mistic
Old Colony
Old Haarlem
Onondaga
Oriental
Orleans
Palmyra
Priscilla
Regent
Rosemont
Roslyn
Somerset
Thebes
Tudor Rose

Mayflower (1927–40). Modified by Bertram Watkin.
On ivory body.

Berkshire
Chiquita
Meridale
Tapestry

Governor Winthrop
(1922–35). Mark Haley.
Also on adobe.
(Plate 20)

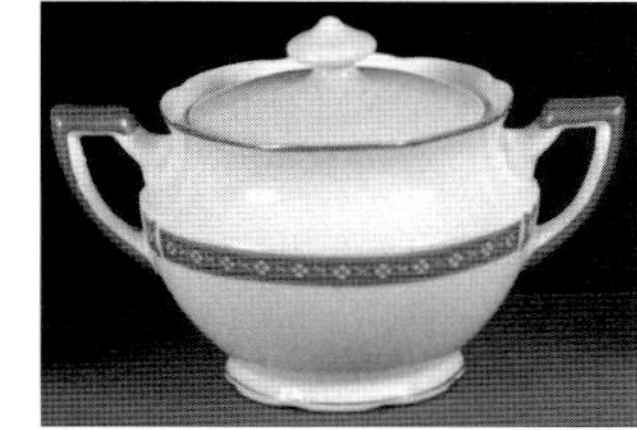

Alcora   Kenmoor
Bombay   Mandarin
Carmelita

Winchester (1928–52).
Bertram Watkin. On
ivory body. (Fig. 6.10)

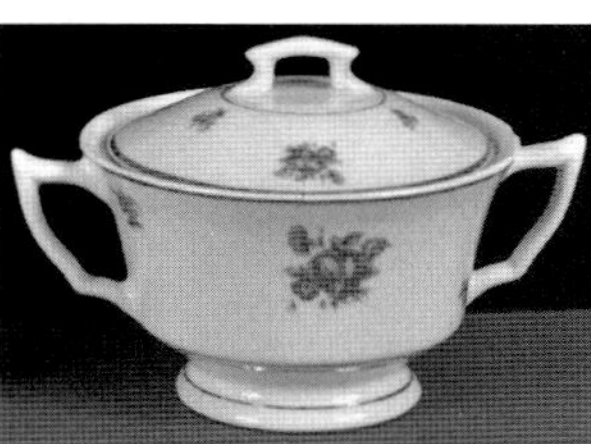

Arcadia
Avondale
California Poppy
Cathay
Cliftondale
Coralbel
E-100 Gold Acid-
    etched border
Elberta
Ensley
Fairburn
G-1405
Glenwood
Haddon
The Hostess
Lady Louise
Marion
Mayfair
Midlothian

Monticello
Nimbus Gold
Nimbus Platinum
Pembroke
Rose Marie
Selma
Shalimar
Sharon
Spring Garden
Springtime[2]
Wendover
Whitby Blue
Whitby Green
Windsor

Clinton (1932–33). Bertram
Watkin. Adobe body.
(Plate 20)

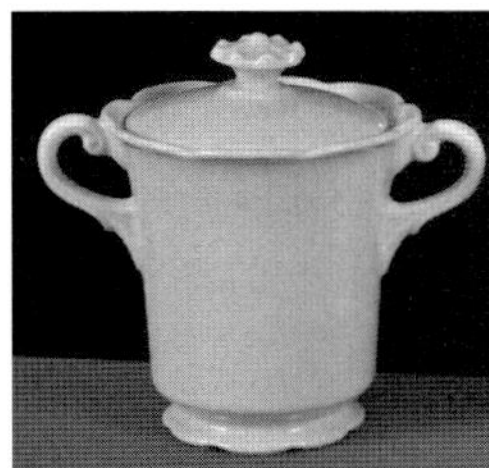

Birds (series of six)
Camelia
Chicory
Dogwood
Marguerite
Rainbow

Virginia (1937–70). R. Guy
Cowan. Ivory body. (Fig. 7.5)

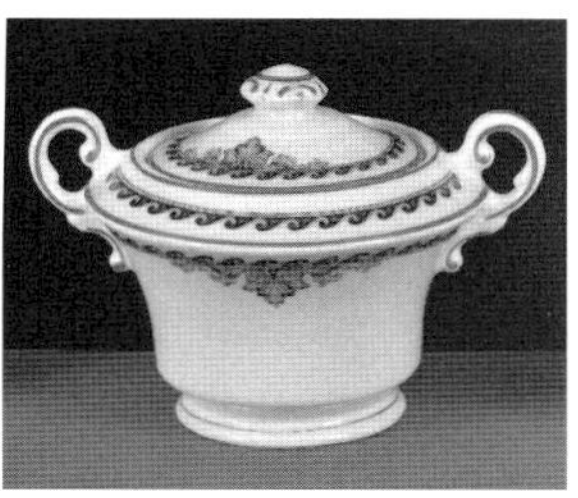

Arcadia
Athena
Avalon
Betsy Ross Blue
Betsy Ross Pink
Blossomtime
Bombay
Bracelet
Calhoun
Columbia
Coralbel
Coventry
Diane
Diane w/blue band
Dutchess
Edmonton
Festival
Governor Clinton
Greenwood
Hampton
Jefferson
Kent
King Arthur
Lady Mary
Madame Butterfly
Marchioness
Marlene

Monticello
Nimbus Gold
Nimbus Platinum
Norfolk Blue
Norfolk Maroon
Normandie
Queen Anne
Richelieu
Romance Green
Romance Maroon
Roseleaf
Royal Court
Salisbury
Selma
Sherwood
Sparta
Wayne
Wayne Blue
Wayne Green
Wayne Maroon
Webster

Shelledge (1937–41, original).
R. Guy Cowan. Moulded decorations: Fish intaglio, floral intaglio, fruit intaglio. (Fig. 7.4)

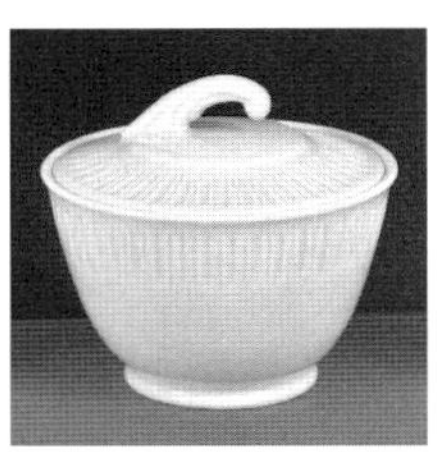

Classic White
  (undecorated)
Debutante
Dogwood
Only a Rose
Penelope
Vogue Blue
Vogue Coral

Federal (1938–69). R. Guy Cowan. Ivory body. (Fig. 7.5, plate 25)

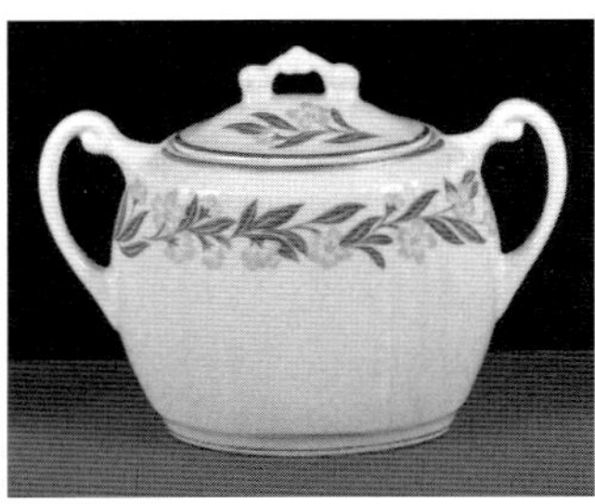

Allendale
Appleton
Beverly
Brantley
Briarcliff
Carvel
Cornwall
Cynthia
Dearborn
Forget-Me-Not
Honeysuckle
Lily-of-the-Valley
Marietta
Mayview
Milicent
Orchard
Pendleton
Portland
Radcliffe
Raleigh Green
Raleigh Pink
Riviera
Rosalie
Rosemoor
Santa Rosa
Sonja
Stansbury
Suzanne
Symphony
Victoria
Violet
Wardell
Wellesley
Winston

Shelledge (1949–69, remodeled). R. Guy Cowan.

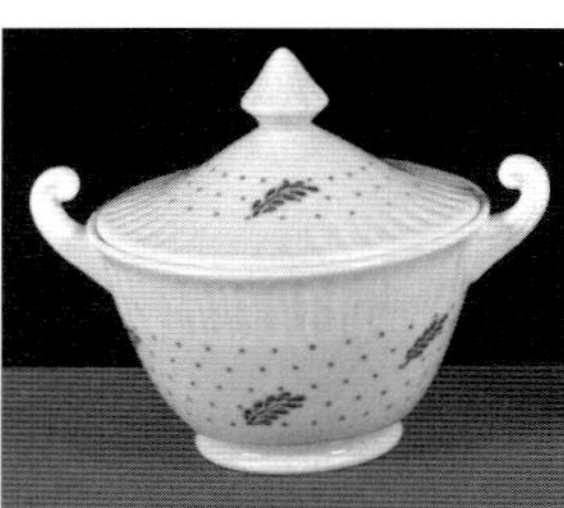

Bamboo
Classic
Dogwood
Flamingo Reeds
Floral Fantasy
Floral (intaglio)
Lancaster
Modern Plaid
Nantucket
Periwinkle
Platinum
Prelude
Raindrop
The Town
Vogue Blue
Vogue Pink
Wild Roses
Yellow Roses

Paul Revere (1950–66). R. Guy Cowan. Ivory body. (Fig. 8.10)

Arlington
Chester
Fleur-de-Lis Gray
Fleur-de-Lis
 Maroon
Graymont
June Rose
Magnolia
Westvale

Berkeley (1950–66). R. Guy Cowan. (Fig. 8.8)

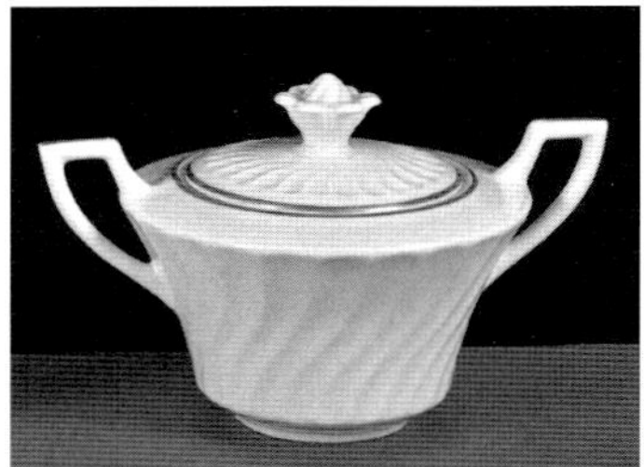

Apple Blossom
Baroque Gray
Baroque Maroon
Berkeley
 (undecorated)
Champlain
Charm
China Spring
Clover
Dawn
Elizabeth
Gardenia
Glory
Jewel Tree
Lilac Rose
Rhapsody
Standish
Temple Bells
Viking

California (1953–70).
Richard Garvin. (Fig. 8.12)

Alpine
Belaire
Bridal Rose
Caprice
Carousel
Champagne
Chevy Chase
Dorian
Elegance

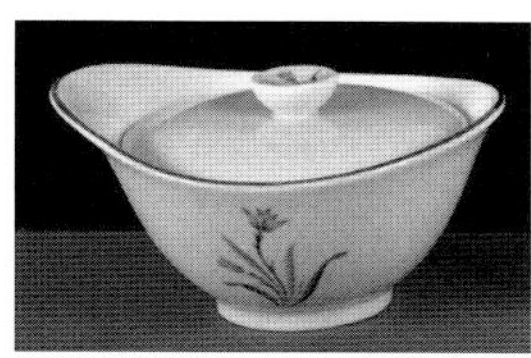

Evening Star
Golden Seeds
Monterey
   (undecorated)
Nocturne

Carolina (1952–70).
(Fig. 8.11)

Candlelight
Carolina (undecorated)
Celeste
Coronet
Countess
Debutante
Fascination
First Love
Grace
Grandeur
Harmony
Inspiration

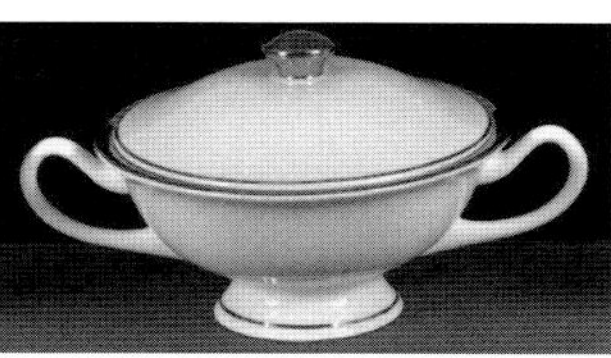

Jubilee
Lyric
Meadow Breeze
Melodie
Minuet
Patricia
Polaris
Regent
Splendor

Carefree (1956–70).
Richard Garvin.
(Fig. 8.13)

Blue Grass
Blue Mist
Concord Rose
Finesse
Flame Lily
Harvest Gold
Jackstraws
Lynnfield
Mayflower
Nordic
Old Cathay

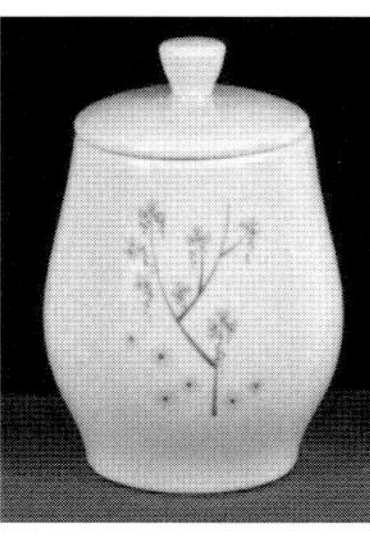

Serene (undecorated)
Vintage
Wayside
Windswept
Woodbine

Silhouette (1960–70).
(Fig. 9.1)

Beloved (undecorated)
Brae Loch
Courtship
Debonair
Engagement
Flirtation
Sonata
Stardust
Sweetheart
Wedding Ring
Westminster

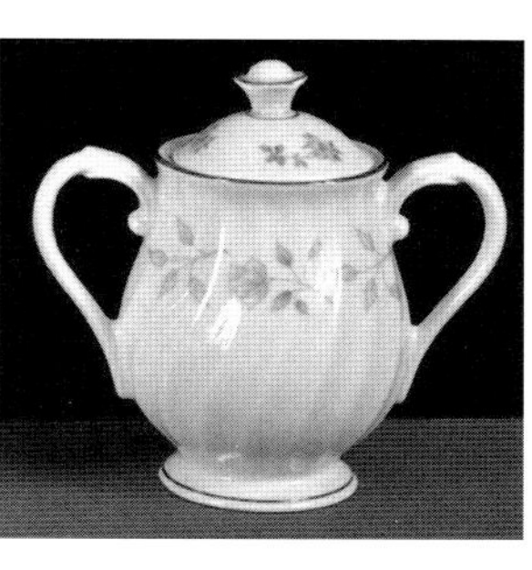

Gallery Collection (1962–70). (Fig. 9.2)

Baroque
Bedford
Concord
Deskey
Facet (Arthur Pulos)
Puritan
Shell

Empire (1966–70).
(Fig. 9.5)

Angelica
Belcanto
Belmont
Blue Riviera
Canterbury
Coronation
Empire (undecorated)
Jamestown
Marlborough blue
Medici
Yorktown

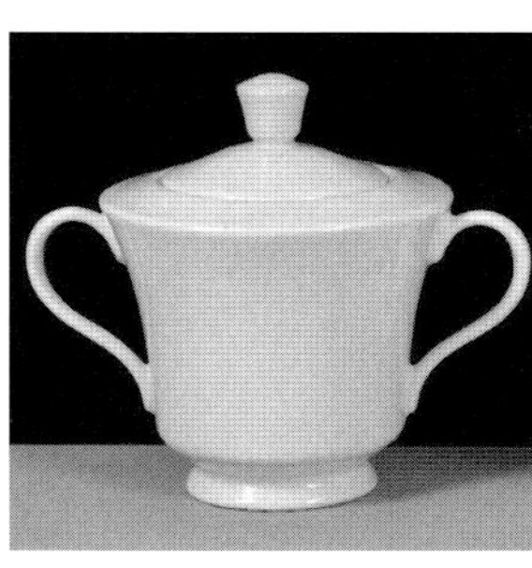

Carefree XL (1966–70).
Similar to **American** shape.

> Accent
> Puritan
> Seville

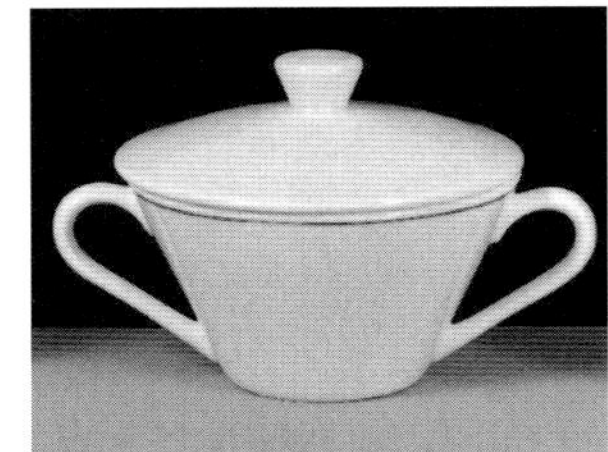

Calypso (1968–70). (Fig. 9.6)

> Aruba
> Barbados
> Largo
> Montego
> Surf White
> Trinidad

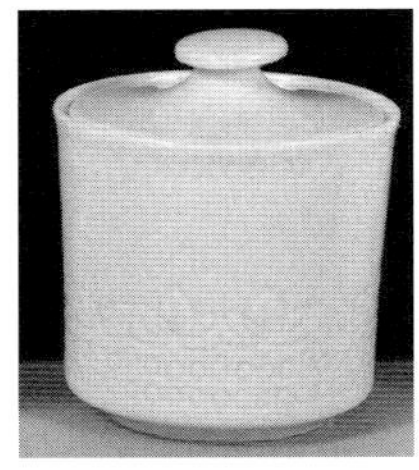

Wellington (1968–70). (Fig. 9.7)

> Black Marquesa
> Chatham
> Gold Riviera
> Kent
> Madison
> Marquesa
> Wellington(undeco-
>   rated)
> Wyndmoor

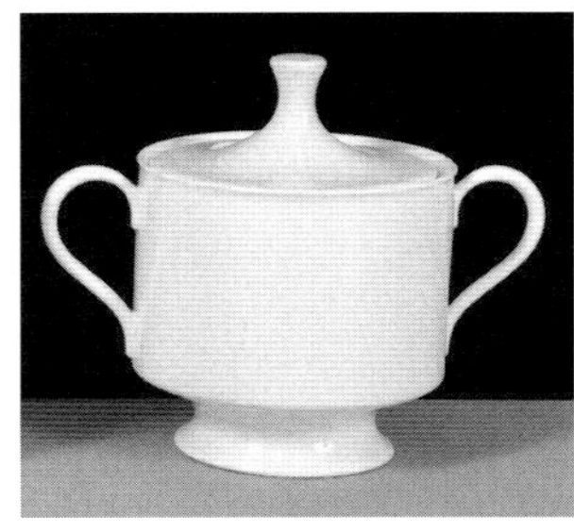

Vitrified China Hotel Shapes: 1896–1997

This list includes standard hotel ware shapes, their approximate inclusive dates of manufacture, and their designers, if known. Decorations on hotel ware (including custom ware) numbered in the thousands, too many to list. See figures and plates for examples of certain types. All carried standard backstamps except when otherwise noted.

Round Edge (1896–1908, original). Mark Haley. (Fig. 4.1)

Club Colonial (1901–21). Mark Haley. (Fig. 5.1)

Normandie (1901–29). Mark Haley(?). (Fig. 5.1)

Empire (1907–52). Mark Haley. (Fig. 6.2)

Round Edge, Rolled Edge (1908–97). Remodeled by Haley and others. (Fig. 4.1)

Pilgrim (1932–97). Bertram Watkin. (Fig. A.1)

Econorim (1933–97). R. Guy Cowan. (Figs. 7.1, A.2)

Morwel (1936–97) R. Guy Cowan. (Fig. A.3)

Doric (1936–69). R. Guy Cowan. (Fig. 7.2)

Airlite (1945–46). R. Guy Cowan. (Fig. 8.3).230

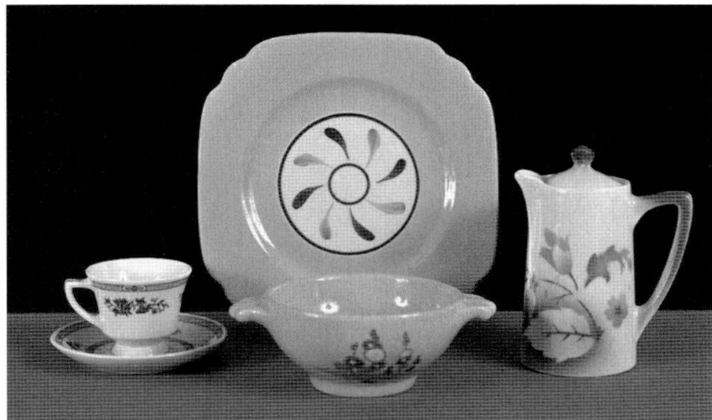

A.1. **Pilgrim** shape, cup and saucer, square salad, pureé bowl (handled), **Stanley** cocoa pot.

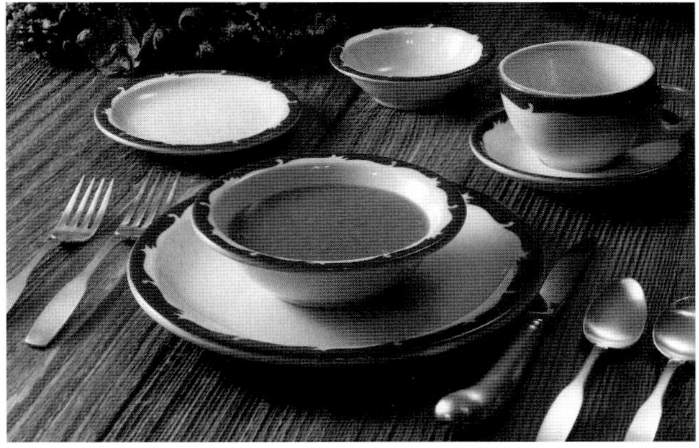

A.3. **Morwel** shape, "Wintergreen" *Shadowtone* pattern.

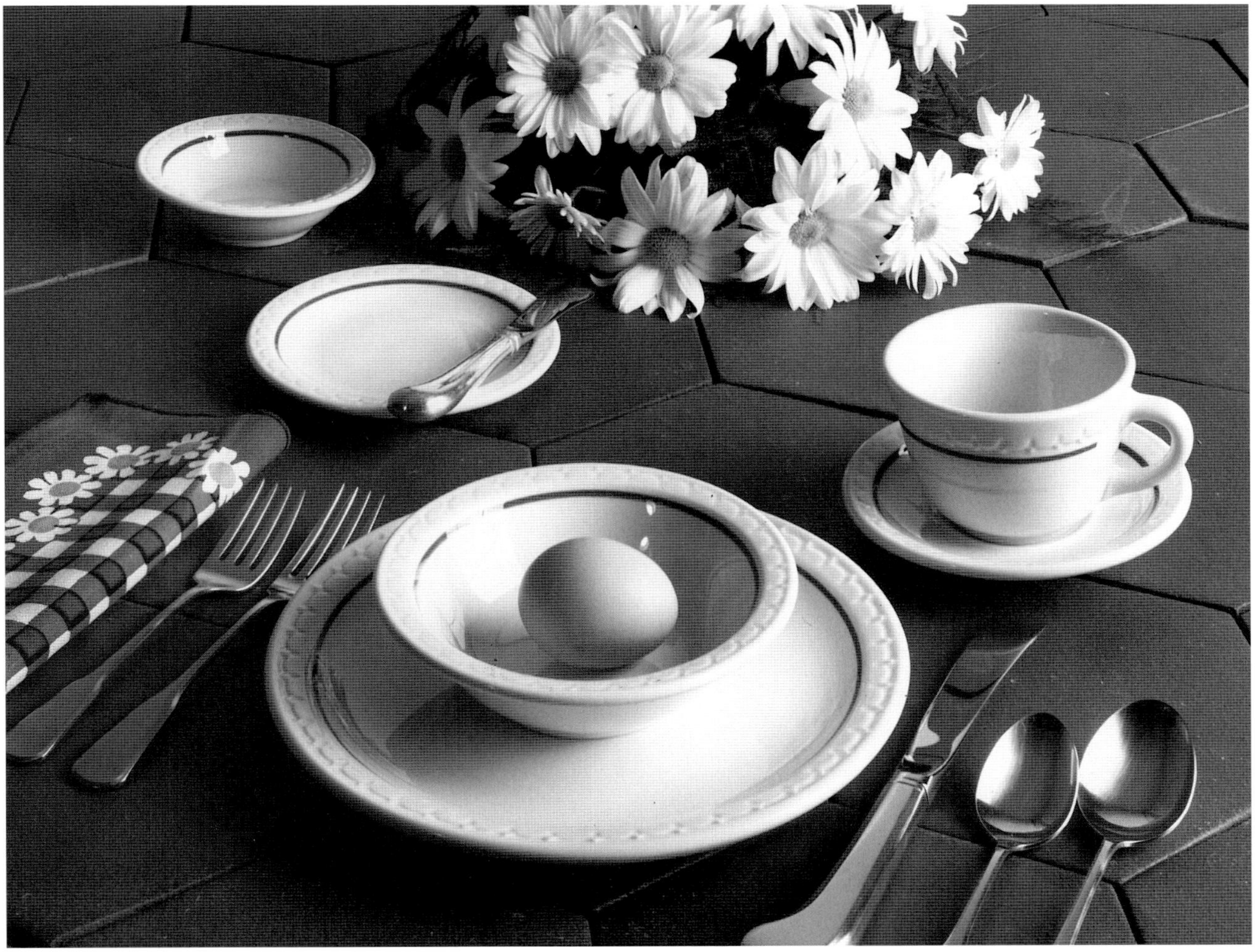

A.2. **Econo-Rim** shape, "Cardinal" (dark red) lines.

Winthrop (1949–97). R. Guy Cowan (Fig. A.4)

Essex (1953–97). (Figs. A.5, A.6)

Copa (1955–76). (Fig. A.6)

Trend (1955–78). (Fig. A.7)

Kent (1956–97).(Fig. A.6)

American (1964–97). (Fig. A.8)

Tudor (1966–72). Same as ***Essex*** but in *Syralite.* (Fig. A.9)

Oddshapes: *1871–1997*

The company made hundreds of "oddshapes." These were one-of-a-kind shapes, usually cups, saucers, jugs, creamers, teapots, and accessory pieces that were not part of the standard shapes listed above. They were made in both earthenware and china bodies and almost always as hotel ware. A few of the many popular oddshapes are pictured in fig. 1.8.

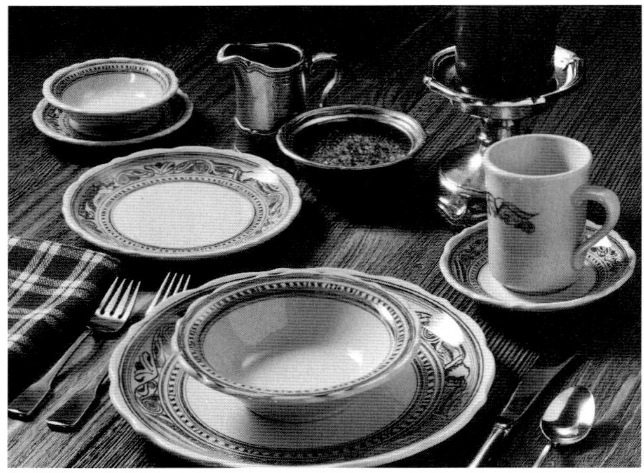

A.5. ***Essex*** shape (plates only), "Americana" pattern, ***Country Ware*** accessories, ***Newport*** cup, ***Winthrop*** saucer and grapefruit bowl.

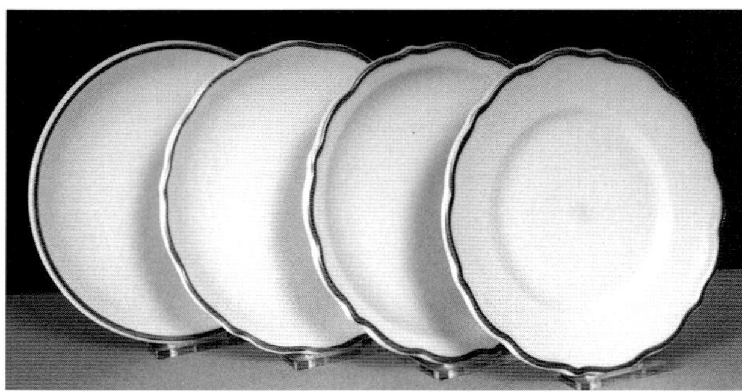

A.6. Rims of ***Copa***, ***Essex***, ***Kent***, ***Winthrop*** shapes compared. Shapes were distinguished specifically by their plate rims. Cups, saucers, and other holloware were interchangeable, but usually in ***Winthrop***. ***Copa*** was not a ***Winthrop***-related shape.

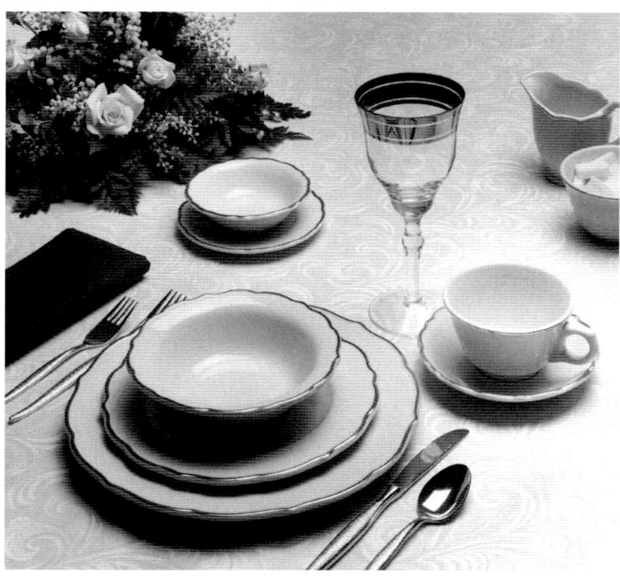

A.4. ***Winthrop*** shape, "Gourmet" gold-lined pattern.

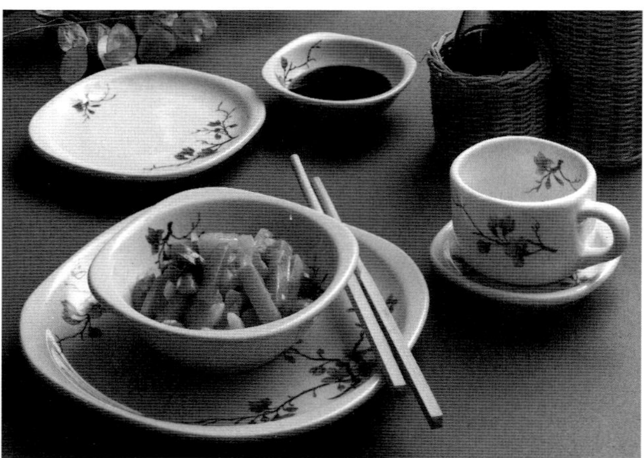

A.7. ***Trend*** shape, "Berkeley" pattern.

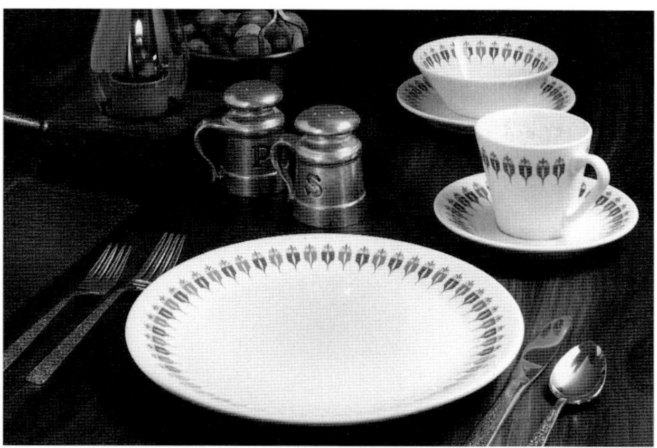

A.8. *American* shape, "Captains's Table" pattern with *Country Ware* accessories.

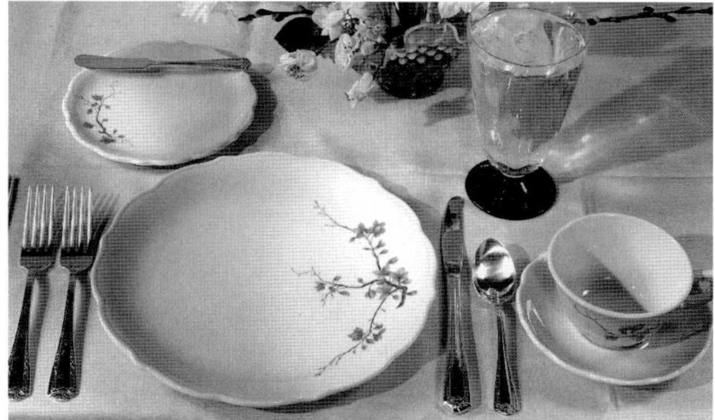

A.9. *Tudor* shape, "Beauval" pattern. *Tudor* was the same as the *Essex* shape, but made in *Syralite* instead of in the flint body.

New Company Shapes, Hotel Ware: 1971–1997

Country Ware (1973–78). (Figs. A.5, A.8, A.9, A.11)

Signet (1975–95). George Jensen. (Fig. A.10)

Olympus (1975–90). George Jensen. (Fig. A.11)

Gibraltar (1978–97). George Jensen. (Fig. A.12)

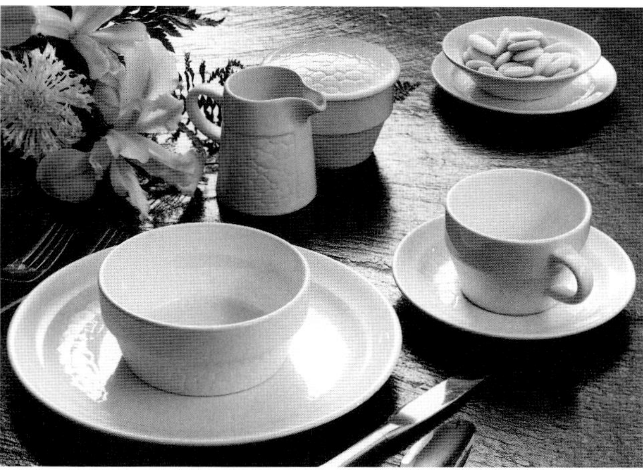

A.10. *Signet* shape, "White on White."

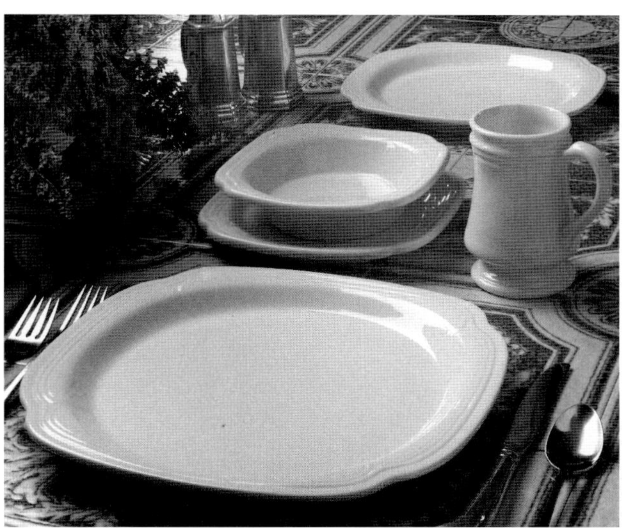

A.11. *Olympus* shape, *Country Ware* salt and pepper shakers.

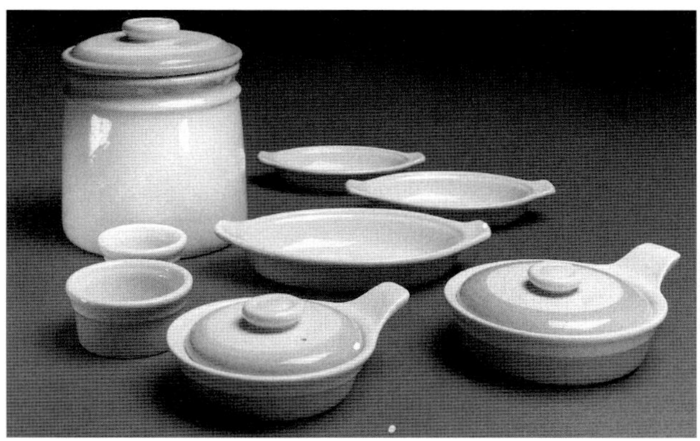

A.12. *Gibraltar* ware.

Syrene (1984–97). George Jensen. (Fig. A.13)

A La Carte (1985–97). Steve Unger. (Fig. A.14)

Savoy (1987–97). Steve Unger. (Fig. A.15)

Turina (1987–97). Steve Unger. (Fig. A.16)

Encore (1988–9?). Steve Unger. Narrow rim version of *Savoy.* (Not pictured)

Belmont (1989–97). Steve Unger. (Fig. A.17)

Tremont (1989–97). Steve Unger. (Fig. A.18)

Verona (1989–97). Steve Unger. (Fig. A.19)

Castleton (1993–97). Steve Unger. (Not pictured)

Justine (199?–96). Steve Unger. (Not pictured)

Barista Coffee (1995–97). Lucie Wellner. (Not pictured)

Cantina Spectrum (1996–97). Lucie Wellner. (Plate 30)

Mayer shapes carried over to New Company (not pictured):

Cord Edge
Chateau
Narco
Parliament

Shenango shapes carried over to New Company (not pictured):

Carlton
Fanfare
Staffordshire
Stylus

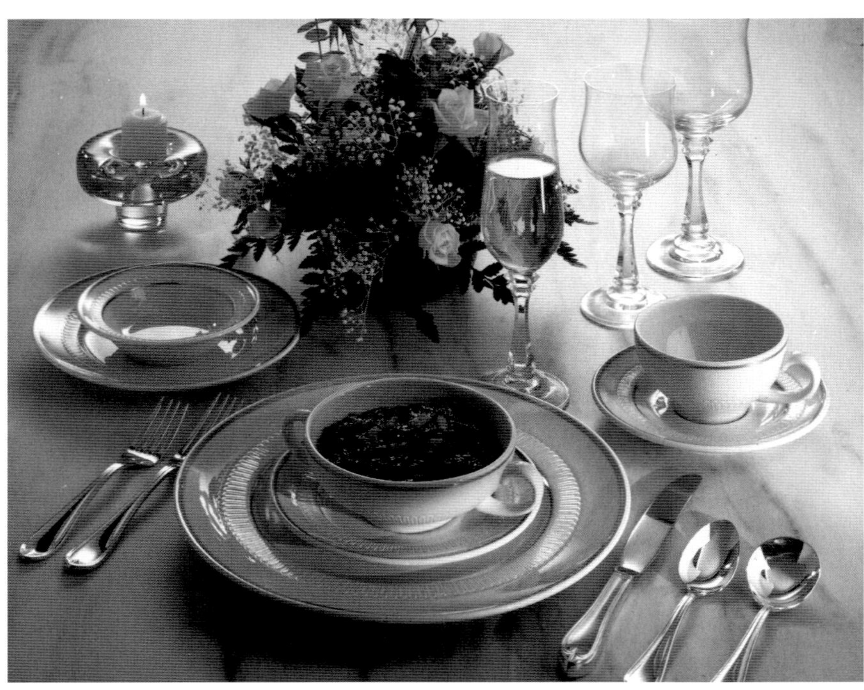

A.13. *Syrene* shape, "Arden" pattern.

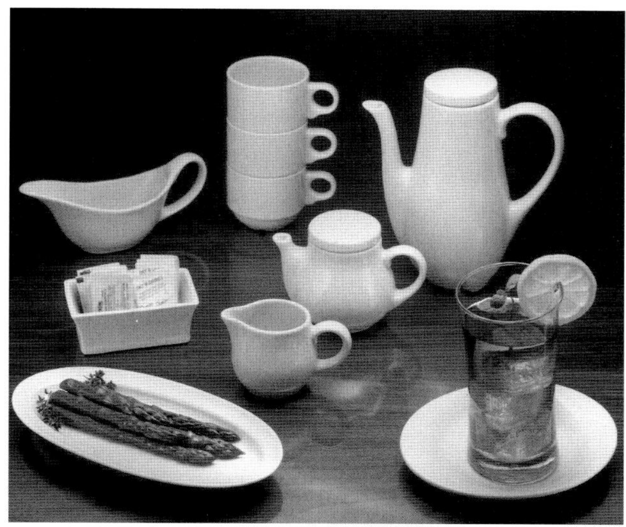

A.14. *A la Carte* ware, an accessory line.

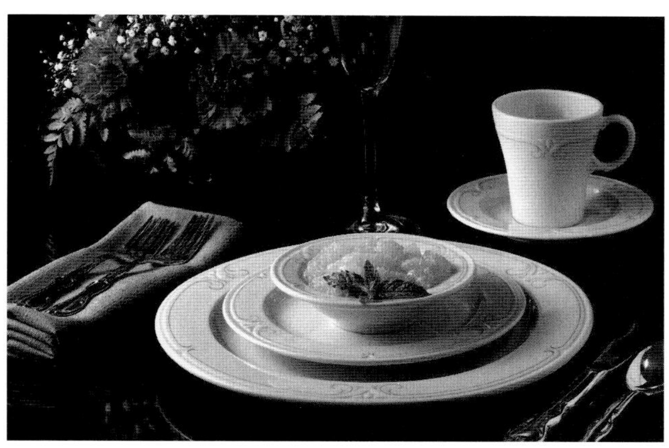

A.15. *Savoy* shape, "Melrose" pattern.

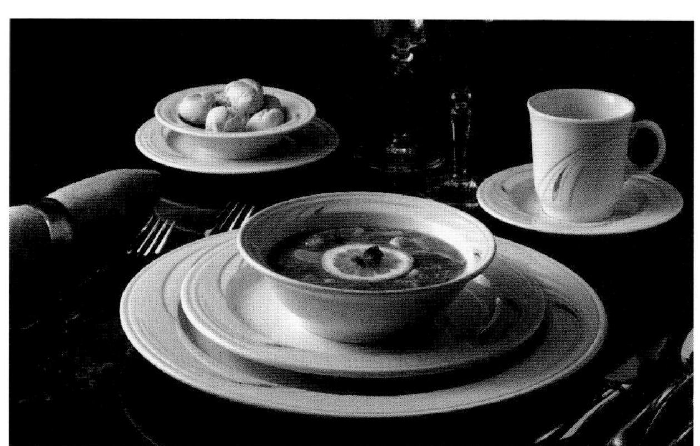

A.16. *Turina* shape, "Compton" pattern.

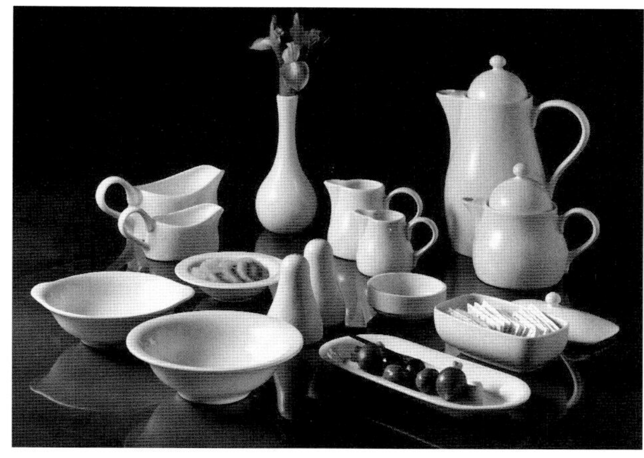

A.17. *Belmont* shape, "Clarion" pattern.

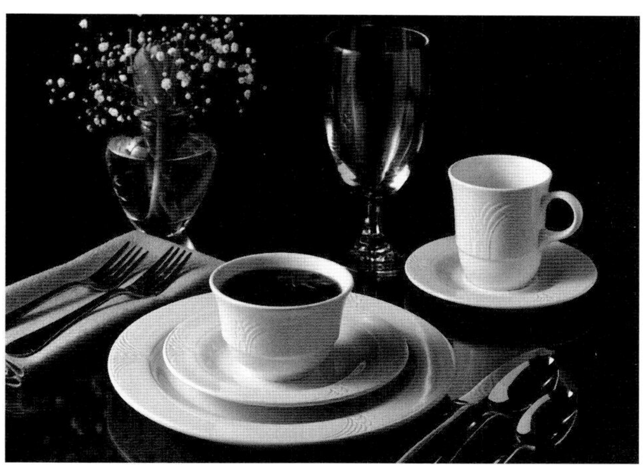

A.18. *Tremont* shape, "Cafe Royal" (undecorated white).

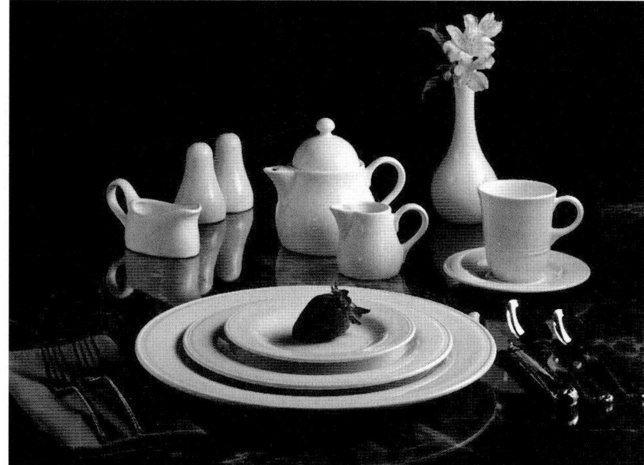

A.19. *Verona* shape, "Caroline" pattern.

The earliest earthenware was not dated. After August 1886, an imprinted date code in the form of a capital letter usually appeared on most ware, separate from the backstamp. In May 1895, numbers replaced the letters. Other marks, such as gold letters, ticks, or blue dots are the codes of unknown decorators. A decoration number applied by the decorator in gold sometimes appears.

Most ware made after August 1886 (until April 1895) bears an imprinted date specifying its month of manufacture.

August 1886 to September 1888: A–Z (alphabet, block capital letters with serif): 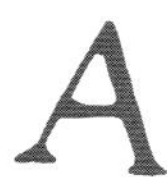

    August to December 1886: A–E
    January to December 1887: F–Q
    January to September 1888: R–Z

May 1895 to September 1903: Numbers 1–99 in sequence (99 was used in July, August, and September 1903). 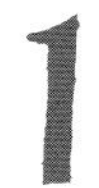

October 1903 to December 1911: Numbers 1–99 in sequence, enclosed in a circle. 

October 1888 to November 1890: A–Z (alphabet, italic capital letters with serifs): 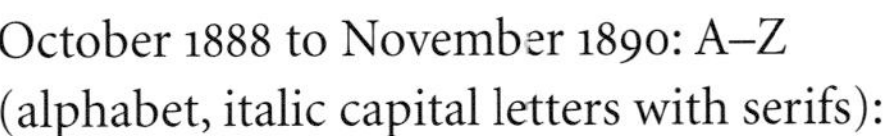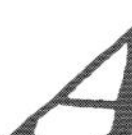

    October to December 1888: *A–C*
    January to December 1889: *D–O*
    January to November 1890: *P–Z*

January 1912 to June 1919: Numbers 1–90 in sequence, enclosed in a diamond. 

July to December 1919: Numbers 1–6 (new sequence of numbers repeated, then dropped to avoid confusion).

December 1890 to January 1893: A–Z (alphabet, fancy heavy asymetrical back-slanted block letters sanserif): 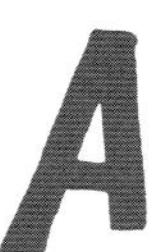

    December 1890 to December 1891: A–M
    January to December 1892: N–Y
    January 1893: Z

In January 1920 the imprinted date codes were replaced by rubber-stamped date codes that paired a capital letter (or double capital letters) for the year and a number for the month as an intregral part of the backstamp. These codes appeared on hotel ware, and very rarely on dinnerware. This practice continued through January 1960. Ware with letter first (K-1) originated at Fayette Street, number first (1-K) originated at Court Street.

    1930: K-1 through K-12
    1931: L-1 through L-12
    1933: N, etc.

February 1893 to March 1895: A–Z (Z used March and April) (alphabet, sanserif condensed): 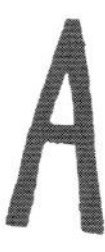

    February to December 1893: A–K
    January to December 1894: L–W
    January to March 1895: X–Z
    April 1895: Z

In 1946, letters were doubled.
    1946: AA-1 or 1-AA, etc., until
    January 1960: OO–1

Beginning in February 1960, the letter and number for the month was replaced by a line of twelve dots (points) under the company logo, with a missing dot indicating the month (e.g., a missing third dot indicates March). The year is indicated by a number corresponding to the years of the company's existence (e.g., the 89th year since 1871 = 1960).

    1960: February to December = 89
    1961: January to December = 90
    1962: January to June = 91

In July through December 1962, the designation of the months was changed from dots back to the letters A–L (A = January) placed under the company logo. This system continued through 1974. The new company developed its distinctive pictorial logo in 1971, continuing the pairing of numbers with letters. With the Fayette Street plant closed, there was no need to reverse the number/letter sequence. The number referred to the age of the company, counting from its origin in 1871.

From 1976 through 1985, the company numbered the year with only one digit (dropping the hundred-year reference).

    1974: 103A through 103L
    1975: 104A through 104L
    1976: 4A through 4L (instead of 104A–104L), etc.

From 1986 through 1997, the lettering was further simplified to A through D, indicating quarters of the year. 1986: 15A to 15D, etc.

Syracuse China Clay Bodies

White Flint Body (1891–97).

Old Ivory Body (1926–66). Many variations of backstamp.

Blue Body (1931 only).

Adobe Body (1932–72).

Silhouette Body (1961–97).

White Alumina Body (*Syralite*) (1964–97).

Royal Rideau, Bright White Alumina Body (1993–97).

China for Transportation

Compiled by Stan Skoczen

This list of transportation companies for which the Onondaga Pottery and Syracuse China companies made china has been compiled from company records, including order entry cards, print crest decoration records, and sample ware (unless otherwise noted).

Railroads

Atchison, Topeka & Santa Fe
Atlanta and West Point
Atlantic Coast Line
Auburn & Syracuse Electric
Baltimore & Ohio
Baltimore & Ohio Southwestern
Bangor & Aroostook
Boston & Albany
Boston & Maine
Buffalo, Rochester & Pittsburgh
Calles Private Car
Canadian National
Canadian Pacific
Central of Georgia
Chesapeake & Ohio
Chicago and Alton
Chicago & Eastern Illinois
Chicago & North Western
Chicago, Burlington & Quincy
Chicago, Indianapolis & Louisville (Monon)
Chicago, Milwaukee & Puget Sound
Chicago, Milwaukee & St. Paul
Chicago, Milwaukee Electric

Chicago, Milwaukee, St. Paul & Pacific
Chicago, North Shore & Milwaukee
Chicago, Rock Island & Pacific
Chicago, South Shore Electric
Colorado & Southern
Colorado Midland
Columbia River & Northern
D & C Steamer City of Detroit
Delaware & Hudson
Delaware, Lackawanna & Western
Denver & Rio Grande
Denver & Rio Grande Western
Detroit United
Duluth & Iron Range
Duluth, Missabe & Iron Range
Duluth, Missabe & Northern
Duluth, South Shore & Atlantic
El Paso & Southwestern
Empire United
Erie
Florida East Coast
Fonda, Johnstown & Gloversville
General Motors Train of Tomorrow
Grand Rapids & Indiana
Great Lakes Transit Company

Great Northern

Green Bay & Western

Gulf Mobile & Northern Railway

Gulf Mobile & Ohio

Harvey

Illinois Central

Illinois Terminal

International–Great Northern

Kansas City

Kansas City Southern

Lehigh & Hudson River

Lehigh Valley

Louisville & Nashville

Maine Central

Minneapolis, St. Paul & Sault Ste. Marie

Missouri-Kansas-Texas

Missouri Pacific

Munising, Marquette & Southeastern

Nashville, Chattanooga & St. Louis

New Central of Georgia[1]

New Hampshire Aerial Tramway

New York Central

New York Central & Hudson River

New York, Chicago & St. Louis

New York, New Haven & Hartford

New York, Ontario & Western

Norfolk and Western

Norfolk Southern Railway

Northern Alberta

Northern Pacific

Oahu Railway & Land Co.

Ontario Northland

Oregon Electric

Oregon Railway & Navigation Co.

Oregon Short Line

Pacific Electric

Panama Railroad & Steamship Line

Pennsylvania

Pere Marquette

Pittsburgh and Lake Erie

Pullman

Queen & Crescent

Reading

Richmond, Fredericksburg & Potomac

Richmond, Washington Line

Rutland

St. Louis, Iron Mt. Southern

St. Louis–San Francisco

St. Louis Southwestern

San Pedro, Los Angeles & Salt Lake

San Francisco–Oakland–San Jose R.R.

Seaboard Air Line

Southern

Southern Pacific

Spokane, Portland & Seattle

Texas & Pacific

Texas Midland

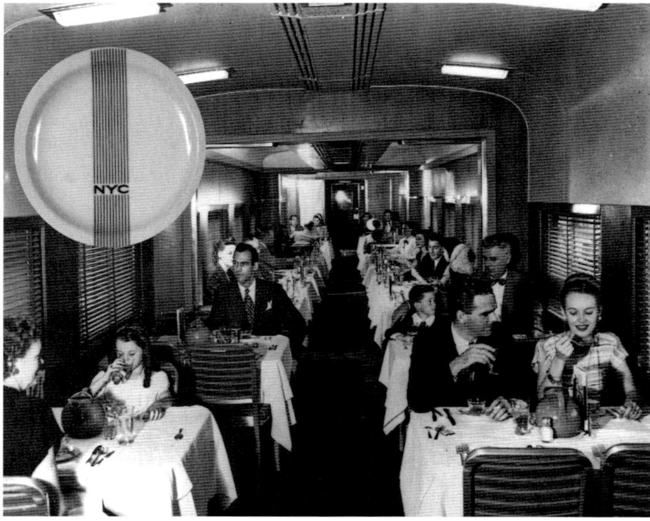

A.20. Interior of a New York Central dining car.

Toledo, St. Louis & Western

Uintah Railway

Union Pacific

Wabash

Washington Sunset Route

Washington Terminal

Western Pacific

Wisconsin Central Line

York Harbor

Steamship Lines[2]

Alaska Steamship Co.

Alcoa Steamship Co.

All America Cables

American-Banner Steamship Line

American-Hawaiian Steamship Co.

American President Lines

American Republic Lines

Atlantic & Pacific Steamship Co.

Atlantic Refining Co.

Baltimore Mail Steamship

Baltimore Steam Packet Co.

B-H Steamship Co.

Booth Line

Canada Steamship Lines

C & B Line

C.C. Packet Co.

Ceres Shipping Co.

Champlain Transportation Co.

Chesapeake & Ohio Train Ferry Service

Chesapeake Steamship Co.

Chicago, Duluth & Georgian Bay Transit

Colonial Line

Columbia & Okanogan

Columbia River & Northern

Continental Oil Co.

Cook Co.–Victoria Line

D & B

D & C

Dollar Steamship Co.

D S Steamship Order

Eastern Steamship Lines

Farrell Lines (originally South African Lines)

Flagship 29

Franklin Steamship Co.

Goodrich Steamship Lines

Grace Line

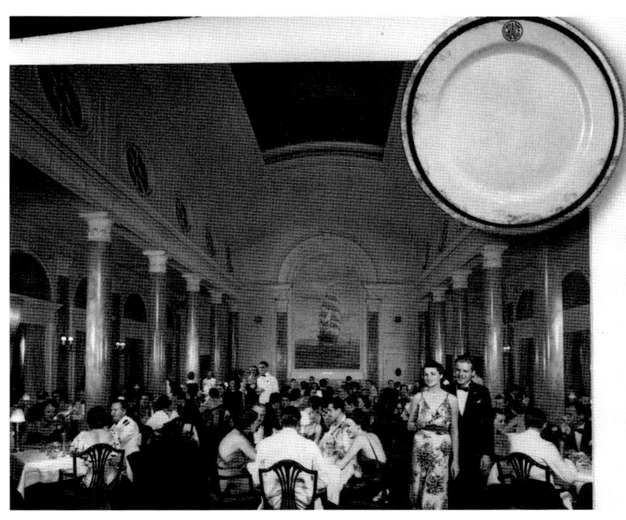

A.21. Advertisement for the Grace Line Steamship Company.

Great Lakes Engineering Works Steamer #9

Great Lakes Transit Co.

Great Northern Steamship Co.

Hawkeye Line

Hudson Navigation Co.

Humble Oil & Refining Co.

Interlake Steamship Co.

Lake Champlain Transportation Co.

Lake George Steamboat Co.

Leviathan

Maine Steamship Co.

Matson Navigation Co.

Merchants & Miners Transportation Co.

Michigan & Ohio Navigation Co.

Monticello Steamship Co.

Moore-McCormack Lines

Munson Steamship Line

New England Navigation Co.

New England Steamship Co.

New Shoreham Steamer

New York, New Haven and Hartford Marine District

Norfolk & Washington Steamboat Co.

Northern Steamship Co.

North Western Steamship Co. Ltd.

O.R. & H.

Oregon Railway & Navigation

Oriental Steamship Co. (Toyo Kisen Kaisha)

Pacific Coast Steamship Co.

Pacific Far East Line

Pacific Navigation Co.

Panama Mail Steamship Co.

Panama Railroad & Steamship Line

Pierce Petroleum Corp.

Plant Line

Prudential Steamship Lines

Puget Sound Navigation Co.

Red Star Lines

Robin Lines Steamship Co.

Sabine Towing Co.

Sinclair Pennant

Socony Vacuum Oil Co.

S.S. Co.

S.S. *Malolo*

S.S. *1926*

Standard Oil Co. U.S.

Steacy Service Steamship Co.

Steamer *Argyle*

Steamship *Frieda*

Texas Oil Co.

Thousand Islands Steamship Co.

Tidewater Oil Co.

United Fruit Co. Steamship Lines

Valley Camp Steamship Co.

White Star Line

Airlines

Air Canada (Vandesca)

American Airlines (see fig. 8.3)

Pan American Airways

Northwest Airlines

Trans Ocean Airlines

United Airlines

U.S. Air (Mayer)

Wardair (Vandesca)

Notes

Preface

1. William C. Gates, Jr., and Dana E. Ormerod, "The East Liverpool, Ohio, Pottery District: Identification of Manufacturers and Marks" *Journal of the Society for Historical Archaeology* 16, nos. 1–2 (1982).

2. The Homer Laughlin Pottery, still family-owned, has extensive historical records, but it has not been possible to compare the two archives.

One. Onondaga Pottery Comes into Being: 1871–1884

1. For early Syracuse potteries see William C. Ketchum, Jr., *Potters and Potteries of New York State, 1650–1900* (Syracuse, N.Y.: Syracuse Univ. Press, 1987), 303–9; and Richard G. Case, *Jugs and Jugtowns: 19th Century Onondaga County Stoneware,* exhibition catalogue (Syracuse, N.Y.: Onondaga Historical Association, 1987), 6–8.

2. George Dummer established the first successful American pottery in Jersey City in 1825 to produce porcelain. Reorganized as the American Pottery Company by David Henderson to make earthenware products using English models and immigrant potters and modelers, Jersey City acted as a springboard for the ensuing American pottery industry. Potters from Jersey City went on to establish the industry in Trenton; East Liverpool, Ohio; and other places. Ellen Paul Denker, "Jersey City: Shaping America's Pottery Industry," unpublished manuscript, Jersey City Museum, Supplementary Materials I, 1995. Much of the literature on the history of the nineteenth-century American ceramic tableware industry treats only Trenton and East Liverpool and leaves largely unidentified the several other smaller centers, such as Syracuse, that in combination produced a substantial amount of the nation's ware.

3. For Trenton, see Marc Jeffrey Stern, "The Potters of Trenton, New Jersey, 1850–1902: A Study in the Industrialization of Skilled Trades," Ph.D. diss., State Univ. of New York at Stony Brook, 1986; and Stern, *The Pottery Industry of Trenton: A Skilled Trade in Transition, 1850–1929* (New Brunswick, N.J.: Rutgers Univ. Press, 1994), 2, 27–28; for East Liverpool, see William C. Gates, Jr., *The City of Hills and Kilns: Life and Work in East Liverpool, Ohio* (East Liverpool, Ohio: East Liverpool Historical Society, 1984).

4. Peter's son, Charles, served as the nominal head. Other partners included Charles's brother William and Henry M. Case, a local investor associated with the salt industry.

5. *Gazetteer and Business Directory of Onondaga County, N.Y. for 1868–69,* compiled and published by Hamilton Child, 1869, 65.

6. Census of Geddes, 1870.

7. Onondaga Pottery, *Minutes,* 1871, Syracuse China archives.

8. They were N. Stanton Gere; his older brother Robert Nelson Gere; their partner in a contracting business, Edward Burgess Van Deusen; Cyrus D. Avery, a member of a pioneer Geddes family and an officer of the Onondaga County Milk Association; Adam C. Morey, partner of Morey and Manger Grocers; Mills P. Pharis, a salt manufacturer; Charles Mitchell, superintendent of the Geddes Blast Furnace; George W. Draper, a partner in Draper & Cole Iron works; and Charles E. Hubbell, who produced traditional stoneware in Geddes as a partner of Hubbell & Chesebro. Other investors included Henry Case, who had been a partner in the Coykendall Company; George A. Cool; and George Draper's brother, Horace, and brother-in-law, Dr. Wilfred W. Porter. Biographical files, Onondaga Historical Association; Syracuse China archives; Syracuse City Directories.

9. *Minutes,* 1871, Syracuse China archives.

10. *Syracuse China News,* 50th anniversary number, June–July 1921, 11.

11. *Crockery and Glass Journal,* 12 July 1877, estimated that the Arms of England were backstamped on "four-fifths of so-called American white granite."

12. Speech for American Ceramic Society, reprinted in *Syracuse Journal,* 31 Jan 1879. Oliver referred to the trademark of the Homer Laughlin Pottery in East Liverpool, Ohio.

13. The Erie Canal was replaced by the wider and deeper Barge Canal, which bypassed the city of Syracuse to the north.

14. *Syracuse Journal,* undated clipping in Syracuse China archives.

15. Edwin AtLee Barber, *The Pottery and Porcelain of the United States,* rev. ed. (New York: G. P. Putnam's Sons, 1901), 19.

16. Garrick Malleny, "Manners and Meals," *American Anthropologist* 1 (July 1888): 195.

17. See Regina Lee Blaszczyk, "The Aesthetic Moment: China Decorators, Consumer Demand, and Technological Change in the American Pottery Industry, 1865–1900," *Winterthur Portfolio* 29 (1994): 38.

18. For the history of the increase of available foods, see Harvey A. Levenstein, *Revolution at the Table* (New York: Oxford Univ. Press, 1988).

Two. The James Pass Era Begins: 1885–1890

1. "Onondaga Pottery's Seventy-fifth Anniversary Shows America Coming of Age in the Production of Fine China," *Ceramic Industry,* Mar. 1947, 91.

2. Syracuse University archives; Syracuse China archives; *Syracuse China News,* 50th anniversary number, June–July 1921, 6.

3. The diaries are part of the Syracuse China archives.

4. Although several attempts had been made previously, William Ellis Tucker was the first to produce American porcelain successfully, in 1825 in Philadelphia. See Barber, *Pottery and Porcelain,* for a discussion of early American ceramic production. Barber, who was America's first ceramic historian of note, interchanged the terms "china" and "porcelain." He wrote when the manufacture of American vitreous china on an industrial scale was in its infancy.

5. Quoted from Clarence Cook, *The House Beautiful* (Armstrong, N.Y.: Scribner, 1878), in Blaszczyk, "Aesthetic Moment," 128.

6. See Gates, *City of Hills,* and Stern, "Potters of Trenton."

7. American Crockery Company was decorating its ware by the early 1870s and Glasgow Pottery in Trenton by 1878. Most potteries still required the services of independent decorating shops. See Blaszczyk, "Aesthetic Moment," 137.

8. Carbon copies of Oliver's letters are in the Syracuse China archives.

9. Unpublished memoir, Syracuse China archives.

10. As late as the mid-twentieth century, employees in the lithographic department were given milk every day as a presumed antidote to lead.

11. Gates, *City of Hills,* 57; Stern, "Potters of Trenton," 375–79. In his diaries, Pass noted: "H . . . left without notice"; "G . . . left without notice after promising that he did not intend to leave"; "Setting-in day [in glost kiln] only five men to do the work—two of the five go off between lunch and dinner time, say nothing, leaving three men to do six days work setting in day"; "F . . . has been off drinking for several days. . . . Can't get him here and need his work." Richard Pass, "Early Days of the Potter's Club," *Syracuse China News,* 50th anniversary number," June–July 1921, 27–28.

Three. The Birth of American China: 1891–1896

1. R. L. Hobson, *Porcelain of All Countries* (Archibald Constable, 1908), 1–2.

2. Hobson, *Porcelain,* 117–18, 155–56.

3. Hobson, *Porcelain,* 178–80.

4. The English had tried this clay years earlier but gave it up because its impurities would not filter out of the clay mass.

5. The word "china" had already become a generic term for tableware, attached to sets of better earthenware to enhance their appeal. Pass had done this himself in 1885 when he named his best improved earthenware "O.P.China." But now he meant to make a "true" china.

6. Richard H. Pass and Charles F. Binns, "Turns of the Potter's Wheel in America," *Transactions of the [British] Ceramic Society* 29 (May 1930): 90.

7. Jean d'Albis, *Haviland* (Paris: Dessain et Tolra, 1988).

8. Homer Laughlin and Knowles, Taylor, Knowles were two of these.

9. Joe Weiss, quoted in *Syracuse China News,* 50th anniversary number, June–July 1921, 11.

10. Corry recorded Huber's story in a series of interviews in October 1954 and transcribed it as a serial "memoir" in eighteen parts for *Syracuse China News,* from March 1955 through April 1958. The original manuscript is in the Syracuse China archives.

11. W. L. Huber, "Thirty Years of Milestones," *Syracuse China News,* 50th anniversary number, June–July 1921, 2.

12. *Syracuse China News,* May–June 1955.

13. *Syracuse China News,* Sept.–Oct. 1955.

14. *Syracuse Journal,* 19 Sept. 1892.

15. From a Yates Hotel flyer in the collections of the Onondaga Historical Association, Syracuse, N.Y.

16. "Notes by Sidney Lodder on happenings around pottery since 1887," three typed pages, dated 1 October 1937, are in the Syracuse China archives.

17. Letter in Syracuse China archives.

18. *Syracuse China News,* Nov.–Dec. 1955.

19. *Syracuse China News,* Sept.–Oct. 1955.

20. Several potteries in Trenton also used a globe backstamp in the early 1890s, but not with a depiction of the earth sketched on them. See Barber, *Marks,* 1901.

21. The exposition's *Official Catalogue* lists the Onondaga Pottery Company as entered in Department H, Manufactures, Part VIII, Group 91, Class 576: semi-porcelain, white granite; and Class 577: translucent china, decorated ware.

22. *British Pottery Gazette* 1 Aug. 1893. Clipping in Syracuse China archives.

23. *Syracuse Journal* 23 Dec. 1893.

24. Onondaga Pottery minutes, 1894, vol 1. Syracuse China archives.

Four. A New Beginning: 1896–1900

1. Pass's trips have been documented from various accounts in the Syracuse China Archives. Pull-down jiggers had been in use in a few other potteries since the 1870s.

2. Stern, "Potters of Trenton," 36–37, 184–86.

3. Oliver letters in Syracuse China archives.

4. Pass and Binns, "Turns," 92. In America as well as Europe, the decalcomania process had been experimented with in ceramic decorating shops and some potteries, but with generally unsatisfactory results until the 1890s. When Pass imported French workers, he may also have imported

colors, papers, and other materials. All good lithographic stone came from Bavaria.

5. *Syracuse China News,* Nov.–Dec. 1956.

6. Vincent R. Bliss, "Why Many Hotels Are Obliged to Buy Chinaware by Guesswork," *Food Profits,* Oct. 1927, 10. Incomplete clipping in Syracuse China archives.

7. Syracuse China archives.

8. Greenwood had begun to make American china by this time (1897), but not in a round-edged shape. Its ware was a heavier product.

9. For Greenwood, see Stern, "Potters"; and Lois Lehner, *Lehner's Encyclopedia of U.S. Marks on Pottery, Porcelain and Clay* (Paducah, Ky.: Collector Books, 1988), 180.

10. For more about the Chittenango Pottery, see Michael Beardsley, "Chittenango Pottery Fires Interest among Collections," *Antique Week,* 6 Dec. 1993, 226.

Five. American China for the Twentieth Century: 1901–1913

1. For Syracuse as an Arts and Crafts community, see Cleota Reed, "'Near the Yates': Craft, Machine, and Ideology in Arts and Crafts Syracuse, 1900–1910," in *The Substance of Style: Perspectives on the American Arts and Crafts Movement,* edited by Bert Denker (Winterthur, Del.: Winterthur Museum, 1996), 359–74. For Robineau, see Peg Weiss, ed., *Adelaide Alsop Robineau: Glory in Porcelain* (Syracuse, N.Y.: Syracuse Univ. Press, 1981), though none of the contributors to this volume discusses her work for Onondaga Pottery. For Stickley, see Mary Ann Smith, *Gustav Stickley: The Craftsman* (Syracuse, N.Y.: Syracuse Univ. Press, 1983).

2. "New York Society of Keramic Arts," *Keramic Studio* 3 (Feb. 1902): 219–21.

3. Not to be confused with another design, "Robineau Moss Rose," named in the artist's honor. The names for many early designs are found in factory records and were used internally for identification purposes rather than as "official" names. The decorations sometimes carried the names of clients, but were often descriptive (e.g., "pink rose border").

4. Reed, "'Near the Yates,'" 361–68.

5. We are grateful to Mark Bassett and Victoria Naumann, authors of *Cowan Pottery and the Cleveland School* (Atglen, Pa.: Schiffer, 1997) for generously sharing their Cowan research. For Cowan, see also Christina Corsiglia,

"Ceramics Education and the Transformation of the Arts and Crafts Legacy in America: Charles Fergus Binns and R. Guy Cowan," *Journal of the Decorative Arts Society* (Great Britain) 17 (1995): 23–33. The Cowan archives and an extensive collection of his work are in the Rocky River (Ohio) Public Library (near Cleveland).

6. Binns had begun his career with the Royal Worcester pottery in England and then worked in Trenton for Walter Lenox at the Ceramic Art Studio, before moving in 1900 to head the new program at Alfred. He had an exceptional knowledge of the ceramic arts and science of all eras and cultures. His experience in industry in both England and America, as well as his ever-inquiring mind and his belief that American ceramic art was on the threshold of an era of greatness, had made him an ideal choice to organize a professional curriculum in the field.

7. Pass must have inquired further about Binns's brother. Three weeks later, Binns replied in a second letter to Pass that his brother was so well established as the artistic head of his company that he doubted that as much as $5000 would tempt him away. He added: "He has money invested in the concern . . . and moreover he has always been in the habit of having foremen under him who undertake all the routine work of administration such as pricing of work with the men settling wages and all such matters. The English methods are more exclusive than the American in these things. I do not use this as an argument on either side but simply to show that he would probably expect something similar to what he has always had." It is odd that Binns questioned Wigley's artistic ability. Wigley was one of the four artists responsible for creating the four Trenton vases exhibited by the Empire Pottery of Trenton at the St. Louis World's Fair in 1904. Wigley joined the Buffalo Pottery, which opened in 1903.

8. In fact, Onondaga Pottery decorators worked on piecework. For independent decorating shops in Trenton, see Stern, *Pottery Industry*, 35–36; for East Liverpool, see Gates, *City of Hills*, 84; and Blaszczyk, "Aesthetic Moment."

9. George Savage and Harold Newman, *An Illustrated Dictionary of Ceramics* (London: Thames and Hudson, 1974, reprint, 1992).

10. Both accounts are preserved in the Syracuse China archives.

Six. National Competition and Growth: 1913–1932

1. In 1921, Salisbury's contributions to the pottery industry and his qualities of leadership led to his election as president of the United States Potters Association. Through this organization he worked with E. L. Torbert, who spent much time in Washington lobbying in support of tariff legislation to protect the interests of American potteries from the lower labor costs of foreign manufacturers.

2. Correspondence files in Syracuse China archives show that the problem was still unresolved in 1920. Further discussion of the complicated issue is beyond the scope of this narrative.

3. The authors are grateful to Ruth Pass Hancock for contributions to her father's story.

4. Richard Pass's diaries are in the Syracuse China archives.

5. Pass & Seymour produced 500,000 aluminum components for gas masks.

6. Reproduced in Onondaga Pottery's 1918 price list, Syracuse China archives.

7. One of these saucers, decorated with Robineau's "Grape Border," is in the Syracuse China historical collection.

8. After James Pass had shown no interest in him in 1903 as a replacement for Lou Cowan, and Wigley had left Trenton to work at Buffalo China in Buffalo, N.Y., Pass, in 1911 brought him to Onondaga Pottery. He became assistant superintendent in 1944 and died in 1946. *Syracuse China News*, Dec. 1946.

9. For the early history of the air-conditioner, see the report in *Syracuse Herald-Journal*, 10 Oct. 1946. (The unit found a retirement home at the Smithsonian Museum in 1960.)

10. In 1921, 60,000 minors in the state between the ages of 14 and 15 and 222,000 between 16 and 17 had left school to work. Seventy-five percent of them had not completed one year of high school and very few had any direct preparation for any skilled job. The Continuation School Law extended to twenty-six states. The program was authorized by New York State's Compulsory Part-time or Continuation School Law. Statistics reported in *Syracuse China News*, Apr. 1925.

11. Two of Betty DeCerce's designs for the *Hospitality Line* in 1950 ("Chantilly" and "Clover") were among the most popular of the P-series and were still in use in 1996.

Lucy Dutton contributed the "Carnival" pattern and Rowena Thornhill designed "Clown" as part of the series. *Syracuse China News,* Mar.–Apr. 1952. A former employee, Chris Polifroni, told us that John Wigley adapted the pattern from her apron's printed fabric to create the "Roxbury" pattern.

12. Max Palmer, interview with Cleota Reed, 1988.

13. For Statler, see Floyd Miller, *Statler: America's Extraordinary Hotelman* (New York: Statler Foundation, 1968).

14. *Syracuse China News,* Jan.–May 1929.

15. There may have been one more, but it has not been located.

16. Richard W. Luckin, *Dining on Rails: An Encyclopedia of Railroad China* (Golden Colo.: RK Publishing, 1983), 193. For railroad china see also Douglas W. McIntyre, *The Official Guide to Railroad Dining Car China* (Marceline, Mo.: Walsworth Press, 1990). The Syracuse China archives provided the authors of both of these books with a significant amount of information.

Seven. Econo-Rim and the Great Depression: 1932–1941

1. For Cowan, see Bassett and Naumann, *Cowan Pottery;* see also Corsiglia, "Ceramics Education."

2. Bassett and Naumann, *Cowan Pottery,* chap. 1.

3. Charles Binns to James Pass, 23 Feb. 1903. Syracuse China archives.

4. Charles Binns to James Pass, 27 Feb. 1903. Syracuse China archives.

5. Basset and Naumann cite family sources indicating that Cowan worked for Solvay Process.

6. Bassett and Naumann. Lou Cowan remained in the Syracuse area until 1905. Then, with his son, he worked at the Central New York Pottery in Chittenango, where Guy was hired to help refurbish the old facility and develop a new china body.

7. He at first worked in Syracuse three days a week and commuted back to Cleveland, where he taught at the Cleveland School of Art and consulted for the Ferro Enamel Corporation.

8. He also brought some of his Cowan Pottery moulds, hoping that he might manufacture his products at Onondaga Pottery. Gorden Huggins, in an interview in 1993, stated that these moulds were destroyed when the Fayette Street plant was demolished in October 1971.

9. Renamed the Everson Museum of Art in the 1960s, the institution's still-growing collection of American ceramic art has since the 1980s been housed in specially built galleries funded by Canadian Pacific Limited (Syracuse China) and named the Syracuse China Center for the Study of American Ceramics. The museum had already begun to collect American ceramics in the 1930s with its purchase of Robineau's own collection of her unique porcelains.

10. Bevis Hillier, *Art Deco of the 1920s and 1930s* (London, 1968).

11. *Syracuse China News,* Mar.–Apr. 1957.

12. *Syracuse China News,* Mar.–Apr. 1969. A factory order dated 20 June 1939 announced the decision that "all items heretofore made in the narrow plain rim and designated as Statler will be known in the future as **Morwel.**"

13. The cups were called **Ideal, Chelsea, Copley,** and **Shelton.** The imprinted decoration was removed from the **Chelsea** cup to create the **Morwel** cup.

14. See epigraph, chap. 1.

15. In 1957, more colors were added.

16. The history of **Shelledge** items documents this unprecedented influence from an outsider on critical design decisions.

17. Unpublished manuscript in Syracuse China archives.

18. Decorating the Santa Fe Railroad Super-Chief for its maiden run on 12 May 1936 involved a collaborative effort among Coulter the Philadelphia architect Paul Cret, the Chicago designer S. B. McDonald, the Santa Fe advertising manager Roger Birdseye, the Chicago distributor E. A. Hinrichs, and Onondaga Pottery.

19. See Richard W. Luckin, *Mimbres to Mimbreño.* (Golden, Colo.: RK Publishing, 1992).

20. Laura Fry adapted a mouth atomizer to apply slips, a technique for which she was granted a patent in 1883. See Kenneth E. Smith, "Laura Ann Fry: Originator of Atomizing Process for Application of Underglaze Colour," *Bulletin of the American Ceramic Society* 17 (1938): 368–72.

21. The complete collection of *Shadowtone* masks has been carefully preserved by Syracuse China.

22. *Cowan Pottery Journal* 3, no. 2 (autumn 1995). This pride may have seemed sightly pompous to Cowan's assistants. Richard Smith, who worked at the pottery after graduating from Syracuse University in 1933, remembered that some of the designer's colleagues at the pottery referred to R. Guy Cowan privately, only half in jest, as "Our God Cowan."

23. Regina Lee Blaszczyk, "Imagining Consumers: Manufacturers and Markets in Ceramics and Glass, 1865–1965," unpublished Ph.D. diss., Univ. of Delaware, 1995.

24. This was the first mention of **Morwel** as a shape designation.

25. May 1939 and before.

26. Adler, unpublished manuscript, Syracuse China archives.

Eight. Richard Pass and Industrial Humanics: 1942–1958

1. He was inspired by the writings of Whiting Williams (1878–c.1955), the Cleveland writer and champion of workers who served as a consultant to several companies. Williams visited Pass in Syracuse, dining at his home on several occasions. One of his books, *What's on the Worker's Mind* (New York: C. Scribner's Sons, 1920) described his experiences working in coal mines and steel mills along with workers. He was a trustee of the Cleveland School of Art and may have known Guy Cowan when he taught there.

2. *Ceramic Industry,* Mar. 1947, 94.

3. The ordnance contract was terminated in February 1953, but the company continued its research on amplifiers and ceramic-based printed circuits for the peacetime commercial market. The company's Electronics Division began on 1 September, headed first by Richard Pass's brother James, who was also president of Pass & Seymour, and then, beginning in May 1954, by Conan A. Priest. Despite successes in developing special kilns, continuous rotary printers, and an automatic wax impregnator, the Electronics Division never paid its way and in 1959 was sold to Speer Carbon Company of St. Mary's, Pennsylvania.

4. Quoted in Frances Hannah, *Ceramics: Twentieth Century Design* (New York: Dutton, 1986), 77.

5. *Ceramic Industry,* Mar. 1947, 90.

6. Ibid.

7. *Syracuse China News,* May 1946, reported on the display of **Airlite** china in New York by Nathan Straus-Duparquet Inc., the distributor of the ware to American Airlines. *Syracuse China News,* again featured **Airlite** in its July–Aug. 1947 issue. See also Bassett and Naumann, *Cowan Pottery,* chap. 13.

8. *Syracuse China News,* Oct. 1947. This issue also announced that W. L. Huber's son, W. Bradley Huber, had been elected to the company's board of directors. He had come to the pottery in 1909, and worked his way up from the warehouse to become head of customer service. A third generation, William C. "Bill" Huber joined the sales department after the war.

9. Stanley S. Jacobs, quoted in *Syracuse China News,* Nov. 1947.

10. This coincided with Syracuse's centennial of its status as a city (it had been settled late in the eighteenth century). To mark the event, Ed Otis designed a centennial seal for service plates made especially for the Syracuse Chamber of Commerce.

11. *Syracuse China News,* June, July, Aug. 1950.

12. In 1963, the State Department awarded the pottery a contract to produce china services for the embassies and consulates of the United States government. For both services, the company had, in 1938, created elegant *Old Ivory* dinnerware embellished with a narrow border etched with a special 24-karat-gold design of stars and vertical bars, which bore in rich gold intaglio the Great Seal of the United States as a crest. The Embassy service plate had a cobalt blue band under the crest.

13. Since 1941, the pottery had issued small-format, illustrated price lists for each stock household dinnerware and hotel ware pattern for every available shape. These could be assembled into a pocket-sized notebook or distributed singly to customers on request. Magazine advertisements offered to send brochures of the latest patterns on request.

14. The total cost of the job was $186,000; the decals cost $35,000. The plates were boxed in two sets of four and sold for twenty-five dollars a set. The Syracuse China archives has a complete set of the correspondence for the job, a rare instance of comprehensive customer service records. The original edition carried a special backstamp. Later editions carried a standard Syracuse China backstamp.

15. *Syracuse China News,* Sept.–Oct. 1954. The magazines were *American Home, Better Homes & Gardens, Bride's Magazine, Good Housekeeping, Guide for the Bride, House Beautiful, House and Garden, Ladies' Home Journal, Living for Young Homemakers, McCall's, Modern Bride, Seventeen,* and *Woman's Home Companion.*

16. Ruth Pass Hancock, unpublished memoir.

Nine. The Final Years of the Old Company: 1959–1971

1. *Syracuse China News,* Jan.–Feb. 1962, mistakenly reported that these were revivals of the ware that won the medal at the Chicago World's Fair in 1893.

2. *Syracuse China News,* May–June 1964 devoted a commemorative issue to his life and work.

Ten. The New Company and New Challenges: 1971–1997 (and Beyond)

1. The sale was approved despite the dissent of Iroquois Industries, which was not associated with the Iroquois China Company. Iroquois Industries was Syracuse China's largest single shareholder. Five years earlier, it had made an unsuccessful bid to buy 50,000 shares of Syracuse China, and was owner of record of 23.2 percent of the firm's 115,892 outstanding shares of common stock. A two-thirds affirmative vote, or 77,262 shares, was necessary for approval of the proposal to sell the company to Ryacuss. At the stockholders' meeting, 81,954 shares voted in favor, 29,882 opposed, and 28,570 dissented with a demand for appraisal of their shares. This last figure included the Iroquois Industries shares. Approval came with the margin of 4,700 shares in favor. Ryacuss, Inc. was then dissolved.

2. He retired in 1973.

3. To illustrate this shift in emphasis, in 1890 Mark Haley had come to the position fresh from art school with training in sculpture. Guy Cowan had come to the job in the early 1930s from running his own art pottery business, with training in all phases of ceramic design and engineering. The first specialist in the broader field of industrial products design called on by the company was the late Arthur Pulos, a professor of industrial design at Syracuse University and a national leader in his field. In the 1960s, in an effort to upgrade the company's designs, he created **Facet** ware as part of the **Gallery Collection**. Pulos's *American Design Ethic: A History of Industrial Design to 1940* (Cambridge, Mass.: MIT Press, 1983) and *American Design Adventures, 1940–1975* (Cambridge, Mass.: MIT Press, 1988) remain authoritative histories of American design.

4. Shellenberger, who directed these convocations, had worked with Theis at Philco Corporation in Philadelphia in the 1950s and at Sylvania in Batavia, New York, in the 1960s. He then worked for Mitsubishi International Corporation before rejoining Theis at Syracuse China. He revitalized Syracuse China's marketing prowess by using the business strategies of major consumer durables product firms.

5. This was the company for whose predecessor James Pass had worked in 1882.

6. This was a new shape, not to be confused with the one of the same name by Lamberton China that the Shenango Pottery had taken over.

7. Syracuse China was the first to meet the strict standards of California's Proposition 65, Safe Drinking Water and Toxic Enforcement Act of 1986. California food service operators who could not prove that their ware met the state's standards were required to post this notice: "Warning: The particular pattern of dishes used here will expose you to lead, a chemical known to the State of California to cause birth defects or other reproductive harm."

Eleven. The Story of a Piece of China

1. The pottery always used this spelling for the word *mould.*

2. In the twentieth century, this process was replaced by continuous roll printing for large-volume orders; small jobs were still done on small plates.

3. The lithography press and over two thousand stones were given to Syracuse University's School of Art when the process was discontinued. The press and most of the stones are still in use, but now by students.

Afterword. My Pottery Heritage

1. See James Dugan, *The Great Iron Ship* (New York, Harper's: 1953).

2. *Syracuse China News,* 50th anniversary number, June–July 1921, 14.

3. See "The Story of Sadie McGown [*sic*]," *Local No. 33, Golden Anniversary 1897–1947,* 2 Dec. 1947, and Cheryl Weller Beck, ed., *The Twentieth Century History of Beaver County, Pennsylvania: 1900–1988,* (Beaver Falls, Pa.: Beaver County Historical Foundation, 1988), 110. One soap dish with the Economite Potteries backstamp survives in the Beaver Falls Historical Society Museum.

4. The Mayers were from Pass's native Burslem and were about his age. Their family went back more than a century in pottery making. See Annise Doring Heaivilin, *Grandma's Tea Leaf Ironstone: A History and Study of English and American Potteries,* n.p., n.d.; and *China from the Dawn of History* (Beaver Falls, Pa.: Mayer China Co., n.d.), 11.

5. Houck died in 1925. "Reminiscences of Louis Houck" appear in the newspaper *The Southeast Missourian,* 17 Apr., 14, 16 June 1969.

6. Richard Pass, "James Pass," 25 Dec. 1922, 2. Unpublished manuscript in the collection of Ruth Pass Hancock.

7. Ibid., 7.

8. Delivered to the Miami Air Station on 16 Mar. 1919, it was Curtis-designed, with a single Liberty engine, intended for coastal patrol.

9. In a letter of 26 Aug. 1945, my father wrote to me, "I found out that both the Americans and English use a method of navigation which I recommended in a written report to the Navy in March of 1919. They call it the multiple drift method."

10. All his life he remained a QB (Quiet Birdman) and for a while continued to fly with the Naval Reserve. He escorted the dirigible *Shenandoah* as it passed over Syracuse in the summer of 1924. When Lindberg touched down in Syracuse at the old Amboy airport after his triumphant Atlantic crossing, I was there perched on my father's shoulders singing "So take your hats off to Lucky Lindy," along with the rest of the crowd. I was about four years old.

11. For a wedding gift, the people of the pottery presented my parents with a sterling silver pitcher, treasured today by one of my children.

12. Mr. Dressler remained my father's life-long friend. He dined at our house and I remember his charming English manners.

13. He had been superintendent of production at Fayette Street since 1922 and, in addition, at Court Street since 1933.

14. When the company's president, Mills Pharis, died in 1891, it owed the Third National Bank $46,000 in notes personally signed by him, within $4,000 of the value of the capital stock.

15. The "troops" at the proving ground were always well-supplied with my mother's black bean soup. Her girdle had provided the elastic to bind the first mine.

16. From transcript by Dick Lewis of audio tapes made by William Salisbury recounting the history of Pass & Seymour.

Appendix A

1. Sometimes appears with word *Stone* in 1885. This marked experimental batches of special clay mixes.

2. Used with Aberdeen BS, also on **Federal** shape.

3. The examples cited are special cases and do not include the thousands of backstamps made for individual customers, some of which did not bear the company's name. It is beyond the scope of this book to list them. Atlas, Tatler, and other companies bought up thirds and redecorated some of them. As a result, some had a company backstamp plus the dealer's stamp on top of it or next to it. Before World War II, the company sold "dips" with slight imperfections (commercial ware only) and prices were lowered accordingly.

Appendix B

1. Taken over from Shenango in 1991.

2. In addition, Syracuse China supplied ware for numerous yacht clubs and personal yachts.

Bibliography

Archives

Alfred University Library, Alfred, N.Y.

American Ceramic Society, Ohio.

Chicago Historical Society, Chicago, Ill.

Corning Museum Library, Corning, N.Y.

Hagley Museum Library, Wilmington, Del.

Onondaga County Public Library, Syracuse, N.Y.

Onondaga Historical Association, Syracuse, N.Y.

R. Guy Cowan Archives. Rocky River Public Library, Rocky River, Ohio.

Syracuse China Archives. Syracuse China Company, Syracuse, N.Y.

Syracuse University Archives. Syracuse University, Syracuse, N.Y.

Winterthur Library, Winterthur, Del.

Books and Articles

Altman, Seymour, and Violet Altman. *The Book of Buffalo Pottery.* New York: Crown, 1969.

Barber, Edwin AtLee. *The Pottery and Porcelain of the United States.* New York: G.P. Putnam's Sons, 1893; revised and enlarged, 1901.

Bassett, Mark and Victoria Naumann. *Cowan Pottery and the Cleveland School.* Atglen, Pa.: Schiffer, 1997.

Beardsley, Michael. "Chittenango Pottery Fires Interest among Collectors." *Antique Week,* 6 Dec. 1993.

Beck, Cheryl Weller, ed. *The Twentieth Century History of Beaver County, Pennsylvania: 1900–1988.* Beaver Falls, Pa.: Beaver County Historical Foundation, 1988.

Binns, Charles F. "The Progress of Ceramic Art." *Transactions of the American Ceramic Society.*

———. *The Story of the Potter.* N.p., n.d.

Blaszczyk, Regina Lee. "The Aesthetic Moment: China Decorators, Consumer Demand, and Technological Change in the American Pottery Industry, 1865–1900." *Winterthur Portfolio* 29 (1994).

———. "Imagining Consumers: Manufacturers and Markets in Ceramics and Glass, 1865–1965." Unpublished Ph.D. diss., Univ. of Delaware, 1995.

Bliss, Vincent. "Why Many Hotels Are Obliged to Buy Chinaware by Guesswork." *Food Profits,* Oct. 1927. Incomplete clipping in Syracuse China Archives.

Case, Richard G. *Jugs and Jugtowns: 19th Century Onondaga County Stoneware.* Exhibition catalogue. Syracuse, N.Y.: Onondaga Historical Association, 1987.

———. *Onondaga Pottery.* Exhibition catalogue. Syracuse, N.Y.: Everson Museum of Art, 1973.

Child, Hamilton. *Gazetteer and Business Directory of Onondaga County, N.Y. for 1868–69.* Syracuse, N.Y.: Hamilton Child, 1869.

Cook, Clarence. *The House Beautiful.* Armstrong, N.Y.: Scribner, 1878.

Corsiglia, Christina. "Ceramics Education and the Transformation of the Arts and Crafts Legacy in America: Charles Fergus Binns and R. Guy Cowan." *Journal of the Decorative Arts Society* (Great Britain) 17 (1995): 23–33.

d'Albis, Jean. *Haviland.* Paris: Dessain et Tolra, 1988.

Denker, Ellen Paul. "Jersey City, N.J.: Shaping America's Pottery Industry." Unpublished manuscript. Jersey City Museum, supplementary materials, 1995.

———. *Lenox China: Celebrating a Century of Quality, 1889–1989.* Exhibition catalogue. Trenton, N.J.: New Jersey State Museum, 1990.

Dugan, James. *The Great Iron Ship.* New York: Harper's, 1953.

Elliot, Charles Wyllys. *Pottery and Porcelain.* New York, 1878.

Frackelton, Susan. *Tried by Fire.* N.p., 1886.

Gates, William C., Jr. *The City of Hills and Kilns: Life and Work in East Liverpool, Ohio.* East Liverpool, Ohio: East Liverpool Historical Society, 1984.

Gates, William C., Jr. and Dana E. Ormerod. "The East Liverpool, Ohio, Pottery District: Identification of Manufacturers and Marks." *Journal of the Society for Historical Archaeology* 16, nos. 1–2 (1982).

Grover, Kathryn, ed. *Dining in America 1850–1890.* Cambridge: Univ. of Massachusetts Press and Margaret Woodbury Strong Museum, 1987.

Hannah, Frances. *Ceramics: Twentieth Century Design.* New York: Dutton, 1986.

Heaivilin, Annise Doring. *Grandma's Tea Leaf Ironstone: A History and Study of English and American Potteries.* N.p., n.d.

Hillier, Bevis. *Art Deco of the 1920s and 1930s.* London, 1968.

Hobson, R. L. *Porcelain of All Countries.* London: Archibald Constable, 1908.

Ketchum, William C. *Potters and Potteries of New York State, 1650–1900.* Syracuse, N.Y.: Syracuse Univ. Press, 1987.

Lehner, Lois. *Lehner's Encyclopedia of U.S. Marks on Pottery, Porcelain and Clay.* Paducah, Ky.: Collector Books, 1988.

Lenox, Walter. *Proper Methods of Decorating and Firing China.* N.p., 1908.

Levenstein, Harvey A. *Revolution at the Table: The Transformation of the American Diet.* New York: Oxford Univ. Press, 1988.

Luckin, Richard W. *Dining on Rails: An Encyclopedia of Railroad China.* Golden, Colo.: RK Publishing, 1983.

———. *Mimbres to Mimbreño.* Golden, Colo.: R. K. Publishing, 1992.

Malleny, Garrick. "Manners and Meals." *American Anthropologist* 1 (July 1888): 195.

Mayer China Co. *China from the Dawn of History.* Mayer China Co.: Beaver Falls, Pa., N.d.

McIntyre, Douglas W. *The Official Guide to Railroad Dining Car China.* Marceline, Mo.: Walsworth Press, 1990.

Miller, Floyd. *Statler: America's Extraordinary Hotelman.* New York: Statler Foundation, 1968.

Myers, Susan. "A Survey of Traditional Pottery Manufacturing in Mid-Atlantic and North Eastern United States." *North Eastern Historical Archaeology* 6 (1977): 1–13.

"New York Society of Keramic Arts," *Keramic Studio* 3 (Feb. 1902).

Official Catalogue of the Centennial Exhibition. Philadelphia, 1876.

Pass, Richard H., and Charles F. Binns. "Turns of the Potter's Wheel in America." *Transactions of the [British] Ceramic Society* 29 (May 1930): 90.

Pulos, Arthur. *American Design Adventures, 1940–1975.* Cambridge, Mass.: MIT Press, 1988.

———. *American Design Ethic: A History of Industrial Design to 1940.* Cambridge, Mass.: MIT Press, 1983.

Reed, Cleota. "'Near the Yates': Craft, Machine, and Ideology in Arts and Crafts Syracuse, 1900–1910." In *The Substance of Style: Perspectives on the American Arts and Crafts Movement,* edited by Bert Denker, 359–74. Winterthur, Del.: Winterthur Museum, 1996.

Replacements. *China Identification Kit.* Greenboro, N.C., 1996.

Savage, George, and Harold Newman. *An Illustrated Dictionary of Ceramics.* London: Thames and Hudson, 1974. Reprint, 1992.

Smith, Kenneth E. "Laura Ann Fry: Originator of Atomizing Process for Application of Underglaze Colour." *Bulletin of the American Ceramic Society* 17 (1983): 368–72.

Smith, Mary Ann. *Gustav Stickley: The Craftsman.* Syracuse, N.Y.: Syracuse Univ. Press, 1983.

Spours, Judy. *Art Deco Tableware.* New York: Rizzoli, 1988.

Stern, Marc Jeffrey. "The Potters of Trenton, New Jersey, 1850–1902: A Study in the Industrialization of Skilled Trades." Ph.D. diss., State Univ. of New York at Stony Brook, 1986.

————. *The Pottery Industry of Trenton: A Skilled Trade in Transition, 1850–1929.* New Brunswick, N.J.: Rutgers Univ. Press, 1994.

Stiles, Helen E. *Pottery in the United States.* New York: E. P. Dutton, 1941.

Svec, J. J. *Pottery Production Processes.* Ceramic Industry Productions, 1946.

Tannahil, Reay. *Food in History.* New York: Crown, 1989.

Weiss, Peg, ed. *Adelaide Alsop Robineau: Glory in Porcelain.* Syracuse, N.Y.: Syracuse Univ. Press, 1981.

Williams, Whiting. *What's on the Worker's Mind: By One Who Put on Overalls to Find Out.* New York: C. Scribner's Sons, 1920.

Young, Jennie. *The Ceramic Art.* New York: Harpers Brothers, 1978.

Trade Journals, House Organs, and Newspapers

American Pottery and Glass Reporter

Boyd's City Directory, Syracuse

British Pottery Gazette

Ceramic Industry

China, Glass and Lamps

Cowan Pottery Journal

Crockery and Glass Journal

Crockery Journal

Journal of the American Ceramic Society

Southeast Missourian

Syracuse China News

Syracuse Daily Journal

Syracuse Herald Journal

Syracuse Journal

Syracuse Post Standard

Syracuse Standard

Transactions of the American Ceramic Society

Index